Biocatastrophe

The Legacy of Human Ecology

Toxins, Health Effects, Links, Appendices, and Bibliographies

Ephraim Tinkham

Engine Company No. 9

Radscan-Chemfall

Est. 1970

Phenomenology of Biocatastrophe
Publication Series Volume 3

ISBN 10: 0-9846046-1-8
ISBN 13: 978-0-9846046-1-6
Davistown Museum © 2010, 2013

Third edition, second printing

Cover photo by Dallas Austin, OH-58D Kiowa Warrior Platoon leader.

Engine Company No. 9
Radscan-Chemfall
Est. 1970

Disclaimer

Engine Company No. 9 relocated to Maine in 1970. The staff members of Engine Company No. 9 are not members of, affiliated with, or in contact with, any municipal or community fire department in the State of Maine.

This publication is sponsored by
Davistown Museum
Department of Environmental History
Special Publication 69
www.davistownmuseum.org

Pennywheel Press
P.O. Box 144
Hulls Cove, ME 04644

Other publications by Engine Company No. 9

Radscan: Information Sampler on Long-Lived Radionuclides

A Review of Radiological Surveillance Reports of Waste Effluents in Marine Pathways at the Maine Yankee Atomic Power Company at Wiscasset, Maine--- 1970-1984: An Annotated Bibliography

Legacy for Our Children: The Unfunded Costs of Decommissioning the Maine Yankee Atomic Power Station: The Failure to Fund Nuclear Waste Storage and Disposal at the Maine Yankee Atomic Power Station: A Commentary on Violations of the 1982 Nuclear Waste Policy Act and the General Requirements of the Nuclear Regulatory Commission for Decommissioning Nuclear Facilities

Patterns of Noncompliance: The Nuclear Regulatory Commission and the Maine Yankee Atomic Power Company: Generic and Site-Specific Deficiencies in Radiological Surveillance Programs

RADNET: Nuclear Information on the Internet: General Introduction; Definitions and Conversion Factors; Biologically Significant Radionuclides; Radiation Protection Guidelines

RADNET: Anthropogenic Radioactivity: Plume Pulse Pathways, Baseline Data and Dietary Intake

RADNET: Anthropogenic Radioactivity: Chernobyl Fallout Data: 1986 – 2001

RADNET: Anthropogenic Radioactivity: Major Plume Source Points

Integrated Data Base for 1992: U.S. Spent Fuel and Radioactive Waste Inventories, Projections, and Characteristics: Reprinted from October 1992 Oak Ridge National Laboratory Report DOE/RW-0006, Rev 8

"Be fruitful and increase, fill the earth and subdue it, rule over the fish in the sea, the birds of heaven, and every living thing that moves upon the earth." (Genesis 1, 28).

Cover photographer Dallas Austin is a pilot from Norcross, GA. At the time the photo was taken, he was assigned to C Troop, 4/6 Air Cavalry Squadron based out of Fort Lewis Washington. The picture was taken between Bayji and Tikrit, Iraq along the Tigris River. The oil leak originated from the Bayji Oil Refineries dumping millions of gallons of oil into the Tigris. An oil worker tried to get rid of the oil by burning it off using a flare, resulting in a huge cloud of smoke and fire that set farmer's fields on fire up to 75 miles south of Bayji.

Biocatastrophe

The Legacy of Human Ecology
and the
Collapse of Global Consumer Society

Volume 3 Table of Contents

Preface

This is the third volume in the three volume *Phenomenology of Biocatastrophe* publication series. This publication series explores the biohistory of the imposition of human ecosystems (industrial, social, political, and economic) on or within natural ecosystems. This series is, in essence, the narration of selected stories about what humans are doing to the biosphere. The essays, definitions, databases, editorial opinions, etc. in these texts explore the impact of human activities on the viability of natural ecosystems in our vulnerable, finite, World Commons. No environmental issue is more important than the unfolding tragedy of the depletion of potable water supplies and the contamination of the atmospheric water cycle. The concomitant spectacle of developed, and now aging, western market economies in crisis is intimately connected with the evolving tragedy of the mad rush to oblivion of an out-of-control global consumer society in a biosphere with finite natural resources.

This is the hard copy edition of our frequently updated online series. Suggestions, corrections, and additional information are welcomed. Please send your feedback to the editor at:

CaptainTinkham@gmail.com

The changing text of the *Phenomenology of Biocatastrophe* three volume series may be accessed at:

www.biocalert.org

Printed and eBook editions of this three volume series may be purchased from:

amazon.com

Printed editions may also be purchased from:

www.davistownmuseum.org/publications.html

I. Biocatastrophe as an Ongoing Historical Event

The Dynamics of Biocatastrophe

We are now living in the early stages of biocatastrophe. We feel its growing impact in the form of local, regional, and transcontinental panoramas of unexpected natural disasters (e.g. hurricane Katrina), accidents (Chernobyl and the Gulf oil spill), or warfare and genocide (Vietnam, Iraq, Darfur, Zimbabwe). The early stages of biocatastrophe are further experienced in diminishing fossil fuel reserves, worldwide food stress for almost two billion people with incomes less than two dollars per day, a rapidly accelerating world clean water shortage, reduced ecosystem diversity and productivity, increasing commodity prices, and rapidly growing oceanic dead zones. The potential worldwide spread of a complex variety of newly emerging or remerging antibiotic resistant bacteria (ABRB) and viral infections provide a growing threat to world health. More invisible is the rapid invasion of genetically modified foods into our diets, not just those of the well fed middle and upper classes of world consumer society, but for all world citizens except those living in the most isolated rural communities. All of these phenomena occur in the context of a growing global consumer society characterized by rapidly changing political, social, and economic institutions. The Western model of finance capitalism fueled military-industrial market economies has reached its natural peak of productivity in the context of a biosphere with finite resources, a large world population, and a bankrupt world economy. We have in fact overdrawn our collective checking account in a mad rush for global economic growth. The Reaganomics drag-race of a speeding consumer society is crashing and there are many casualties. Chemical fallout may seem like an unimportant footnote to the implosion of the viability of the Western model of finance capitalism, but a closer inquiry will reveal its essential role in our rapidly changing world landscape.

Legacy of Global Warfare

The dislocations of the unexpected events of the dawn of the Age of Biocatastrophe (the smoking ruins of 911 were just a prelude to the smoking ruins of Iraq and Afghanistan) form the backdrop of the necessity for individuals, families, and communities to confront and understand the threat posed by the contamination of food and water supplies by ecotoxins of every description. The spectacle of the Gulf oil spill disaster is a highly visible chapter in the ongoing narration of the environmental history of petrochemical-industrial society. A large but often unacknowledged source of ecotoxin contaminant signals has military (US, Russia, China) origins. The public health threat of the global contaminant pulse of perchlorates (rocket fuel), one of many ecotoxins derived from military activities, may soon join Gulf War Syndrome, other emerging diseases of military origin, and the ongoing spread of the global contaminant pulses of biologically significant radioisotopes produced during the arms race as one of the major legacies global warfare. The petrochemistry and petropolitics of global warfare (1916-?) are the prime movers of the Age of Biocatastrophe. The historical driving force behind the brilliant discoveries and innovations of petrochemists was and remains the quest to manufacture improved weaponry. The highly profitable production of industrial and consumer product ecotoxins is a natural outgrowth of our finesse at petrochemisty. The legacy of global warfare is, in turn, only one slice of the biocatastrophe pie chart. The majority of slices of the bioC-pie chart derive from the evolution of a robust consumer culture and its industrial agricultural systems. The ecotoxic contaminant pulses generated by modern industrial consumer society are a burden to future generations, which may only be matched by the accelerating indebtedness of a non-sustainable global consumer culture. The delayed external costs of the legacy of global warfare are an important hidden component of this accelerating indebtedness.

Rainfall Events

The hemispheric transport and deposition of the invisible clouds of many ecotoxin contaminant pulses as chemical fallout plumes occur in association with snow and rainfall events. Ecotoxin contaminant signals in the global atmospheric water cycle (rainfall, public water supplies, terrestrial, lacustrine, and estuarine waters, and eventually in maternal cord blood and breast milk) are measured in a range of concentrations from parts per trillion to parts per billion. Many persistent organic pollutants (POPs) are lipophilic (fat soluble) and biomagnify in upper tropic (food chain) levels. The increasing presence and mobility of these POPs and other ecotoxins in the food chain, and in the public water supplies, often derived from household consumer products and electronic equipment, constitute a public health threat to pregnant women, children, and adults of all ages, lifestyles, and socioeconomic status. The rapid increase in autism and other developmental problems in children is a direct consequence of the synergistic effect of multiple ecotoxins that have bioaccumulated in pathways to human consumption. An important constituent of our ongoing ritual of denial about the dynamics of biocatastrophe is our failure to acknowledge that maternal cord blood, breast milk, and human tissues are components of an atmosmopheric water cycle that we, as an industrial civilization, have efficiently contaminated in a few centuries. The consequences of this desecration have vast health physics and public safety manifestations.

Community Awareness

An awareness of the public safety issues implicit in the phenomena of biocatastrophe on the community level is the very foundation for the moral necessity of governmental entities to attempt to confront and document the rapid and accelerating environmental changes that are now part of global daily experience. What can individuals do to learn to evaluate chemical fallout events that may threaten their families? The proliferation of ecotoxin contaminant pulses in pathways to human consumption raises the question about the specific steps individuals and families can take to mitigate the impact of ecotoxins, especially in the home environment. Community first responders and public educators will have the future responsibility of advising and alerting community members to the presence of or emerging threats of chemical fallout toxins in local water supplies and consumer products and of answering the question: what are the ecotoxins and pathogens in the community environment and how do they affect children and families? Awareness of the public safety impact of the many specific ecotoxins that characterize the Age of Biocatastrophe begins with individual consideration of ecotoxins in the home, yard, garden, and local community and their discussion by families, in public meetings, and in public school science and history courses. Financial dislocations in our changing economic, social, and political landscapes should not be allowed to distract attention from evolving public health threats that affect all cultural and socioeconomic classes. The highly visible impact of the Gulf oil spill disaster on the Gulf of Mexico fishing communities and wildlife is the tip of an iceberg of chemical fallout that urgently needs more documentation, acknowledgement, and public debate.

The Illusion of Regulatory Mitigation

The mitigation of the impact of biocatastrophe by governmental and non-governmental organizations with respect to either global warming or the release and bioaccumulation of ecotoxins is a hoped for event of the future. The Montreal protocol of 1987, which limited the release of chlorofluorocarbons to avoid ozone layer depletion, is an example of concerted governmental efforts that had a significant impact in mitigating this threat. The elimination of tetraethyl lead and the consequent reduction of lead as an environmental threat, and the halt in the production and use of many pesticides (herbicides and insecticides) that are the most dangerous POPs, are important examples of how concerned public citizens and scientists can join with government to mitigate important environmental threats to public health and safety. An intense international effort is now underway to reduce greenhouse gas emissions and limit the increases in global warming, sea level rise, and climate change. Attempts to reach international agreement on actions intended to prevent cataclysmic climate change (the global climate crisis) have not proven successful, as exemplified by the Kyoto Accord and the recent Copenhagen climate change summit. It is now becoming increasingly evident, given the long chemical life and biogeochemical mobility of many persistent synthetic chemicals and the large quantities of new forms of chemicals being introduced into the biosphere, that governmental action cannot remediate industrial and military releases of chemicals that have already occurred or will soon occur. Future control of cataclysmic climate change by governmental actions, policy changes, new regulations, including mandated changes in lifestyle or energy consumption habits seem unlikely to mitigate global warming or to control the public safety threats posed by chemical fallout, antibiotic resistant bacteria, viruses, and genetically modified organisms. Awareness of the growing threat of cataclysmic climate change (CCC) and global ecotoxin contaminant pulses (GECPs) begins on the community level and is the basis for any governmental efforts to document and mitigate the radical changes now impacting the Earth's biosphere.

Public Safety Information Databases

The first step in the attempt to mitigate the public safety impact of the phenomenon of biocatastrophe is the identification of ecotoxins and their sources, pathways, and contaminant signals in abiotic and biotic environments, including their concentration ratios and health effects. A vast network of governmental, university, and private research organizations has produced a huge database documenting the chemical composition and biological significance of many ecotoxins. A small selection of this information is reprinted in the *Appendices* in this *Volume*. The production of some of these ecotoxins was discontinued in the 1980s (e.g. PCB, chlordane). As persistent organic pollutants, they still constitute an important component of ecotoxin contaminant pulses. The first stage in confronting the threat of biocatastrophe cannot be taken without the not-always-obvious necessity of using these existing databases as sources of critical information. Despite the robust databases already available pertaining to the bioaccumulation of these ecotoxins in biotic media, including humans, it is unlikely that most governmental or NGO efforts will be successful at mitigating the likelihood of full-blown biocatastrophe. The huge information databases already compiled, some of which are cited in the information sources listed in this publication, are the foundation of any public safety effort to mitigate the impact on children and families of the growing global contaminant pulses of ecotoxins in our food, water, and human tissues, blood, and breast milk. The current media focus on CO_2 emissions and global warming as the primary threat to the world's environment diverts attention from the complexity of the synergistic components constituting biocatastrophe. Global military/industrial/consumer society is in the process of self-annihilation. As a Titanic human ecosystem based on the ecology of money, modern industrial society has hit its iceberg but has not yet sunk – genobiocide in progress. The evolving world financial crisis is one symptom of a disease that has deep roots in our cultural lifestyle and values. The study of human ecology is in fact a narration of the biohistory of a Western civilization that spread its market economy philosophy and ecotoxins to all corners of the globe in the late 20th century. Not to follow the dots of this biohistory to its natural culmination in biocatastrophe is to evade the ecological and moral consequences of the evolution of industrial society in the late Holocene.

Questions

Is there any hope that the vast majority of participants in our global consumer culture, especially resource-devouring Americans, will recognize and accept the fact that a growth-based, ecotoxin-emitting, global consumer society is no longer viable in a world of finite resources? Will American political leaders as well as its business and financial community acknowledge the fact, given our huge current account deficits and the unresolved problems of how to fund the future costs of Medicaid, Medicare, and Social Security, that any further growth in the gross national productivity can only be achieved by further growth of our national indebtedness? Will a global consensus be achieved such that governments, corporations, communities, and individuals can work together to create and implement sustainable economies that will halt the further destruction of the World Commons? How can sustainable economic activity as well as the equitable distribution of health care, education, and employment opportunities be achieved in a world challenged by a global financial crisis caused by the evolution of a narcissistic consumer culture whose debts far outstrip its assets? How can we counter the genesis of biocatastrophe in the context of a global military/industrial/consumer society whose wealth is based on the systematic exploitation of nonrenewable natural resources, the accelerating contamination of the World Commons with anthropogenic ecotoxins, and the relegation of most world citizens to the role of consumers of the products of a predatory corporate industrial agricultural complex?

II. The Chemistry of Ecotoxins

This synopsis of the major chemical fallout contaminant groups includes a brief review of some of the most significant ecotoxins produced and released by anthropogenic activity. While some industrial activity remobilizes naturally occurring ecotoxins, such as mercury, lead, and arsenic, the key constituents of biologically significant chemical fallout (BSCF) are the thousands of ecotoxins produced by the petrochemical industry. Included in each abstract are the sources of the chemicals, the means by which each is transported through the environment and human populace, and a sampling of levels of the chemical detected in various media. A more detailed listing of other ecotoxins follows our summary of the major contaminant groups. Our contaminant group sketch should not be construed as a complete portrait of ecotoxins in abiotic and biotic media, including human body burdens; there are over 80,000 synthetically produced chemicals in use today with roughly a thousand more being developed annually, few of which are regulated or scrutinized. The *Appendices* in this *Volume* provide a more detailed sampling of the many hazardous and extremely hazardous chemicals monitored by the CDC (*Appendix A*) and the EPA (*Appendix C*); later *Appendices* provide additional biomonitoring data collected by state and NGO organizations.

Classification of Petrochemicals

Six major groups of chemicals characterize all petrochemical production efforts (*Encyclopedia of Science and Technology*, Vol. 13, 2002).

- Aliphatic Compounds: methane, ethane, propane, and butane.

- Olefins: ethylene, propylene, butylene, and butadiene.

- Aromatic hydrocarbons: benzene, toluene, xylene, ethylbenzene, and naphthalene.

- Chemicals from natural gas: also a source of ethane and propane as intermediaries in petroleum products.

- Synthesis gas production: a mixture of carbon monoxide and hydrogen is used to produce a wide range of ecotoxic petrochemicals.

- Inorganic petrochemicals: sulfur, sulfites, and nitrates. As inorganic chemicals, these products contain no carbon and are therefore less biologically significant as ecotoxins due to their incompatibility with elemental nutrient cycling.

The Problem

TABLE 1. Hydrocarbon intermediates used as sources of petrochemical products

Carbon number	Hydrocarbon type		
	Saturated	Unsaturated	Aromatic
1	Methane		
2	Ethane	Ethylene	
		Acetylene	
3	Propane	Propylene	
4	Butanes	n-Butenes	
		Isobutene	
		Butadiene	
5	Pentanes	Isopentenes	
		(isoamylenes)	
		Isoprene	
6	Hexanes	Methylpentenes	Benzene
	Cyclohexane		
7		Mixed heptenes	Toluene
8		Di-isobutylene	Xylenes
			Ethylbenzene
			Styrene
9			Cumene
12		Propylene	
		tetramer	
		Tri-isobutylene	
18			Dodecylbenzene
6–18		n-Olefins	
11–18	n-Paraffins		

Figure 1 McGraw-Hill. (2002). *McGraw-Hill encyclopedia of science & technology: 13.* pg. 20.

The systematic documentation of the worldwide distribution of ecotoxic petrochemicals and other anthropogenic (man-made) ecotoxins in abiotic and biotic media has yet to be initiated. Human body burden surveys by the Centers for Disease Control (CDC) and research organizations such as the Environmental Working Group (EWG) represent preliminary compilations of some of the most important contaminant groups in small population samples. A 1987 study of human adipose tissue by Onstot, one of the first surveys of body burden ecotoxins, revealed 700 detectable contaminants. If contemporary minimum ecotoxin detection levels in biotic media could be extended from the current level of nano units (parts per billion) to pico units (parts per trillion), the number of anthropogenic ecotoxins present in the highest trophic levels of the food chain (predatory birds, humans) as a component of the molecular structure of living organisms would be approaching 10,000 contaminants. Most of these ecotoxins are derived from the post-1940 production of petrochemicals, which facilitated the rapid growth of global military/industrial/consumer society.

Sources of Petrochemicals

In the second half of the 20th century, petrochemical use became a pervasive aspect of industrial, commercial, agricultural, and personal activities in every developed nation in the world. Most of these chemicals are derived from crude oil, or petroleum, a naturally-occurring, non-renewable slurry of hydrocarbons that has become a crucial resource in the global economy. [1] A brief survey of how petrochemicals are produced from petroleum and a summary of the complex diversity of their forms begins with the fact that petrochemicals are derived from the entire range of petroleum fractions and from natural gas (methane). Petrochemicals may be divided into groups by their molecular makeup, specifically the types of bonds in the atoms. The largest portions of petrochemicals are derived from hydrocarbon gasses with the lowest carbon numbers: methane (1 carbon atom), ethane (2 carbon atoms), propane (3 carbon atoms), and the butanes (4 carbon atoms). The fundamental process for producing the complex range of petrochemicals used by modern industrial society is cracking, the thermal decomposition of petroleum into its myriad byproducts by heating at various temperatures in the presence of a wide variety of catalysts, hence the complex variety of ecotoxic products. Generally, the larger and more complex a hydrocarbon is, the more greenhouse gasses and hazardous byproducts are produced as a result of its combustion or degradation. When combusted, hydrocarbons follow a fairly consistent formula, which, in its simplest form, yields carbon dioxide and water. For example, the simplest hydrocarbon, methane, undergoes the following reaction: $CH_4 + 2O_2 => CO_2 + 2H_2O$. [3] This is the least polluting combustion reaction involving a hydrocarbon and results in a single molecule of carbon dioxide for each molecule of methane consumed. The most environmentally harmful hydrocarbons are extremely stable due to the halogens (non metallic elements such as chlorine, bromine, fluorine, etc.) that have been incorporated in their structures. This process of halogenation produces the highly ecotoxic persistent organic pollutants (POPs) described below and frequently mentioned throughout this publication series. Figure 2 summarizes the hydrocarbon intermediates used to produce petrochemicals.

TABLE 2. Major sources of petrochemical intermediates	
Hydrocarbon	Source
Methane	Natural gas
Ethane	Natural gas
Ethylene	Cracking processes
Propane	Natural gas, catalytic reforming, cracking processes
Propylene	Cracking processes
Butane	Natural gas, reforming and cracking processes
Butene	Cracking processes
Cyclohexane	Distillation
Benzene	
Toluene	Catalytic reforming
Xylene(s)	
Ethylbenzene	
Alkylbenzenes	Alkylation
>C_9	Polymerization

TABLE 3. Chemicals produced from methane (natural gas)	
Basic derivatives and sources	Uses
Carbon black	Rubber compounding Printing ink, paint
Methanol	Formaldehyde (mainly for resins) Methyl esters (polyester fibers), amines, and other chemicals Solvents
Chloromethanes	Chlorofluorocarbons for refrigerants, aerosols, solvents, cleaners, grain fumigant

Figure 2 McGraw-Hill. (2002). *McGraw-Hill encyclopedia of science & technology: 13.* pg. 21.

Aromatic Hydrocarbons

Of particular interest as a widely produced petroleum derivative are the highly ecotoxic aromatic hydrocarbons derived from the higher carbon number saturated gasses hexane and cyclohexane, which are produced through the conversion of non-aromatic hydrocarbons by the direct dehydrogenation of naphthenes (see below.) While lower carbon ethylene is used to produce the largest quantities of petrochemicals, aromatic hydrocarbon petrochemicals are among the most ecotoxic products of human activity (Figure 3).

Olefins

Refinery gasses are particularly significant because they contain large quantities of olefins, which due to their double bonded chemical structure, are a more useful raw material for producing many petrochemicals than the more immutable paraffin hydrocarbons. The thermal cracking process is used for the production of gasoline and other liquid fuels, but it also produces the olefins, which include the key hydrocarbons ethylene and butylene. Not present in large quantities in natural petroleum, the olefin groups are produced as intermediates by refining processes such as distillation, solvent extraction, and synthesis gas production (see Figure 2 and Figure 3).

The Special Case of Gasoline

Gasoline is only one of the many products of the thermal cracking and distillation of petroleum. Used almost entirely for transportation and industrial agricultural purposes, the United States alone averaged a consumption rate of 376,636,400 gallons a day in 2007. [2] Isooctane, the "octane" listed on consumer gasoline, undergoes the following reaction when combusting perfectly: $C_8H_{18} + O_2 = H_2O + CO_2$. Automobile gasoline, however, consists of over 500 types of hydrocarbons, many of which do not readily or cleanly combust. As a result of this imperfect combustion, byproducts such as carbon monoxide are released into the environment. Other hydrocarbons present in gasoline, such as benzene and aromatic hydrocarbons, are emitted as particulates in the auto exhaust along with the carbon dioxide and water. [5]

Persistent Organic Pollutants (POPs)

Among the most significant of the petrochemical production strategies that result in the production of the ubiquitous persistent organic pollutants are the conversion of non-aromatic hydrocarbons to aromatic hydrocarbons by dehydrogenation and the addition of chlorine and bromine to olefins by halogenation, which results in the production of halogenated hydrocarbons. *Appendix A* contains what might be considered a preliminary listing of the persistent organic pollutants (POPs) of most concern to governmental biomonitoring agencies. This listing is misleading in the sense that hundreds and, possibly, thousands of POPs can now be identified as a threat to world health. As volatile organic pollutants, all are transported by the global atmospheric water cycle. All the environmental chemicals listed in *Appendix A*, except heavy metals, are POPs. Of particular interest are the many chemicals in each chemical group. The EPA covers the same grouping with its list of the "dirty dozen" (Agency for Toxic Substances and Disease Registry 2000). Many, but not all, of its extra hazardous chemicals are POPs. Another take on POPs is the ATSDR 2008 definition of "critical pollutants," all of which are POPs. Their definition of critical pollutants is synonymous with the EPA definition of POPs: that they persist in the environment for long periods of time, bioaccumulate in fish and wildlife, and are toxic to humans and animals (Agency for Toxic Substances and Disease Registry 2008, ix).

Figure 2 and Figure 3 provide a vivid example of the complexity of the many labyrinths of modern petrochemical production, all of which produce a wide variety of potent ecotoxins. Most POPs are produced as components of the processes listed in these graphic illustrations. Unfortunately, these illustrations are sketches of processes infinitely more complicated than anything that can be illustrated on a printed page. Chemicals such as dioxin-furans, as products of incineration, are entirely omitted from these sketches, as are the thousands of cogeners and metabolites derived from each chemical group. Our inventive industrial capacity for creating new ecotoxins entirely outstrips our ability to evaluate their toxicity and health physics impact.

Dehydrogenation

Dehydrogenation is an endothermic (heat dependent) process that detaches hydrogen from a molecule to produce other chemicals. Vapor phase conversion of primary and secondary alcohols to aldehydes and ketones, the catalytic reforming of naphthas and naphthenes, and side chain dehydrogenation are typical dehydrogenation strategies used to produce billions of pounds of organic chemicals annually. Related aromatization processes include high temperature condensation and dehydroisomerization of naphthenes. The Age of Chemical Fallout is in part marked by the widespread production of benzene, toluene, xylenes, and other ecotoxic compounds after 1945 by the dehydrogenation process (see Figure 2 and Figure 3).

Halogenation

Halogenation is one of the most controversial strategies used for producing petrochemicals. The naturally occurring halogens fluorine (FL), chlorine (CL), bromine (BR), iodine (I) and astatine (AT) are combined with the olefins listed in Figure 2 and Figure 3. The extreme stability of

TABLE 4. Chemicals from ethylene	
Basic derivatives and sources	Uses
Ethylene oxide	Ethylene gycol (polyester fiber and resins, antifreeze)
	Di- and triethylene glycols
	Ethanolamines
	Nonionic detergents
	Glycol esters
Ethyl alcohol	Acetaldehyde
	Solvents
	Ethyl acetate
Polyethylene	
Low-density	Film, injection molding
High-density	Blow molding, injection molding
Styrene (from ethylene and benzene)	Polystyrene and copolymer resins
	Styrene-butadiene rubber and latex
	Polyesters
Ethylene dichloride	Vinyl chloride
	Scavenger in antiknock fluid
	Ethyleneamines
Ethyl chloride	Tetraethyllead
	Minor amounts for ethylations (such as ethyl cellulose)
Ethylene dibromide	Scavenger in antiknock fluid
Acetyls	Plastics and chemical intermediates
Linear alcohols and n-olefins	Detergents, plasticizers

Figure 3. McGraw-Hill. (2002). *McGraw-Hill encyclopedia of science & technology: 13.* **pg. 21.**

the resultant chemicals is the basis for their industrial usefulness as fire retardants, insulating agents, pesticides, and insecticides, especially since 1945. These petrochemicals are now more well known as the POPs (persistent organic pollutants). All of the chemicals targeted by the Stockholm Convention of 2001 as POPs fall into the category of halogenated hydrocarbons and have chlorine as their halogen: Aldrin, Chlordane, DDT, Dieldrin, Endrin, Heptachlor, Hexachlorobenzene, Mirex, Toxaphene,

Polychlorinated Biphenyls (PCB), Polychlorinated dibenzo-p-dioxins (Dioxin), and polychlorinated dibenzo-p-furans (Furans).[6] The thermal chlorination of methane is now a worldwide industrial activity that results in the production of a wide variety of chlorinated ecotoxins, hence their ubiquitous presence as chemical fallout contaminant pulses in most biotic media.

Gas synthesis

A wide variety of petrochemicals are produced by a process related to thermal cracking and petroleum decomposition, the synthesis of carbon monoxide and hydrogen into a wide variety of petrochemicals. Figure 4 illustrates the complicated series of interrelationships between gas synthesis and ethylene production. Gas synthesis is one of several strategies for producing ethylene and other olefins, which are then reconstituted into the wide variety of aromatic hydrocarbons, such as the benzene family of ecotoxins whose ease of formulation, diminished saturation, and molecular stability make them such industrially useful ecotoxins.

Ethylene

Ethylene is the most produced organic chemical in the world due to its importance as a chemical building block for polymers and polymer-related products. Exposure to the most prevalent form, ethylene oxide, a carcinogen, is potentially fatal in large doses and can cause respiratory, muscular, skeletal, and nervous system damage even at lower levels of exposure. [8]

In terms of gross quantity, ethylene is the most produced petrochemical (and organic compound) in the world, with world production at the end of 2006 at 121 million metric tons. [4] This chemical is also an important naturally-occurring plant hormone, small quantities of which are emitted by refrigerated or decaying vegetables and fruits. The environmental impact of a sudden, rapid influx of large quantities of ethylene into the atmosphere has yet to be documented. Ethylene is frequently used in the synthesis of halogen containing petrochemicals and is used to create products such as methyl chloroform (chlorothene), an industrial solvent banned by the Montreal Protocol, tetrachloroethylene (perchloroethylene), a dry cleaning fluid, and polyvinyl chloride, used in PVC pipes and clothing. The CFCs (chlorofluorocarbons), banned by the Montreal Protocol due to their ozone depleting and greenhouse gas simulating properties are another form of ethylene-derived ecotoxin. [7]

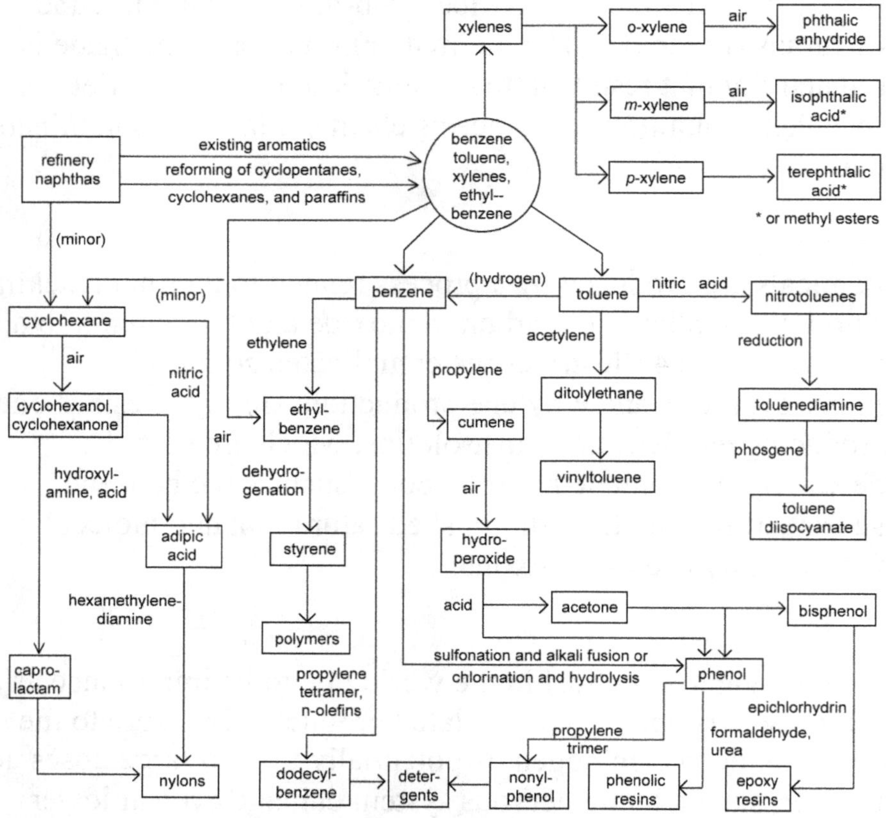

Fig. 3. Chemicals from cycloaliphatic compounds and from aromatic compounds.

Figure 4 McGraw-Hill. (2002). *McGraw-Hill encyclopedia of science & technology: 13.* **pg. 217.**

Hydrogenation

Hydrogenation is the chemical reaction of hydrogen with unsaturated organic compounds under the influence of temperature and pressure in the presence of catalysts. Products are hydrogenated to alter the molecular structure of a fat so that the fat becomes hardened. Hydrogenation is often accompanied by the removal of another element such as nitrogen, sulfur, or a halogen. Hydrogenation is used for the production of liquid fuels, hydrogenated vegetable oils, fatty alcohols, and soap. The process of hydrogenation converts fat into a type of cellulose that is not unlike plastic; its resistance to oxidation (rancidity) for long durations, sometimes measured in years, is a principal commercial benefit of the hydrogenation process.

EPA *List of Lists*

In 1986, Congress passed the Emergency Planning and Community Right-To-Know Act (EPCRA). This congressional legislation, along with Section 122(r) of the Clean Air Act, mandated that industries producing hazardous wastes above the reportable quantities (RQ) listed in the legislation notify the EPA of both their production quantities and any documented spillage. The passage of this legislation signaled heightened awareness by the general public of the dangers of thousands of chemicals now being produced for industrial agriculture, as petrochemicals for a multiplicity of applications, and for consumer products of every description. As of 2008, the EPA *List of Lists* (2006) includes over 3,000 chemicals required to be reported for emergency planning and community right-to-know purposes. Of particular note is the category of extremely hazardous substances (EHS), which are listed in a separate column in the EPA *List of Lists* (see *Appendix C*). EPCRA section 302 states specifically "for section 302 EHSS Local Emergency Planning Committees (LEPCs) must develop emergency response plans and facilities must notify the State Emergency Response Commission (SERC) and LEPC if they receive or produce the substance onsite at or above the EHS's TPQ" [threshold planning quantity] (United States Environmental Protection Agency. 2006). The EPA regulations go on to mandate a variety of other reporting requirements, some of which are reprinted in *Appendix C* as an introduction to our sample of EPA *List of Lists* of hazardous chemicals. Of particular note is that the EPA database is focused on the industrial production of hazardous chemicals above a certain manufacturing quantity (threshold planning quantity TPQ). There are thousands of other chemicals that the EPA does not include in its *List of Lists*. The EPA database does not deal with the ultimate destination of the hazardous chemicals it tracks in the manufacturing process after they are distributed through the wide variety of commercial networks ranging from pesticide use to toxic chemicals in consumer products. The CDC does selected biomonitoring in human subjects for some of the chemicals listed in the EPA *List of Lists* (see *Appendix A*). There remains, however, a huge vacuum in our knowledge of the inventories and pathways of chemical ecotoxins in the environment.

The Problem of Ecotoxicity

Our sketch of the presence of 19 selected ecotoxins in the biosphere in the following section of this text includes two ubiquitous forms of the dioxin family of environmental chemicals. The toxicity of the dioxins and their many chemical metabolites highlights the grim reality of ecotoxicology. It is impossible to link a specific health effect, whether a particular type of cancer or more generalized immune system dysfunction, to the presence of one specific microcontaminant, such as a dioxin-furan metabolite, which is inevitably present in most abiotic and biotic media, including human blood and serum, in conjunction with numerous other human ecotoxins (see *Appendix A*). Medical science does not have, nor is it likely ever to have, the ability to trace the etiology of specific diseases and syndromes to one metabolite or chemical form of an ecotoxin, such as dioxin, when it is only one of several hundred contaminants in a living organism. This contrasts with our ability to test a specific chemical for its adverse health effects by subjecting living organisms, often mice, to exposure to controlled levels of a particular environmental chemical. Determining the specific cause of a breast tumor or autism in a child is an entirely different as well as a probably impossible task. This impossibility helps us rationalize the systematic contamination of the biosphere with the microcontaminants of industrial society. In any short term economic analysis, the benefits of the use of most chemicals outweigh the risks; our ability to limit the impact of industrial activity on the environment is constrained by the synergistic impact of thousands of ecotoxins as microcontaminants generated by human activity. Our systematic contamination of the global atmospheric water cycle is a self-limiting human activity and is now becoming a scientifically verifiable reality, and thus, part of our growing awareness of the phenomenon of biocatastrophe. Control of the production and proliferation of most microcontaminants is as yet a social, political, and scientific impossibility.

Citations:

1. Speight, James G. (1999). *The chemistry and technology of petroleum*. Marcel Dekker Inc., New York, NY.

2. http://tonto.eia.doe.gov/dnav/pet/pet_cons_prim_dcu_nus_a.htm

3. http://www.chm.davidson.edu/ChemistryApplets/Calorimetry/HeatOfCombustionofMethane.html

4. http://www.purchasing.com/article/CA6436655.html

5. http://www.elmhurst.edu/~chm/vchembook/514gasoline.html

6. http://www.epa.gov/oppfead1/international/pops.htm

7. http://www.epa.gov/fedrgstr/EPA-AIR/1995/October/Day-11/pr-1117.html

8. http://www.atsdr.cdc.gov/tfacts137.html

III. Selected Major Anthropogenic Ecotoxins of Interest

Hotel California, Today's Menu Part 1 – Entrées

Thousands of anthropogenic ecotoxins are playing, or will play, a major role in biocatastrophe. Their health physics impact is heightened due to the tens of thousands of cogeners and metabolites that derive from their initial production. Comprehensive cataloging and biomonitoring of many of these environment chemicals may never be possible because they remain unidentified. The CDC reports on human exposure to environmental chemicals illustrate this quandary. Many PCB and dioxin-furans cogeners have been identified but remain below the limit of detection of CDC biomonitoring. Thousands of other biologically significant chemicals have yet to be identified. The synergistic health physics impact of these unidentified chemicals, including those having a presence below the limit of detection (LOD), is, and may remain, beyond our ability to evaluate. The possibility of the unavailability of additional public funding may limit the future ability of the CDC to monitor and evaluate the numerous emerging nanotoxins, the production of which characterizes modern global consumer society. As individual ecotoxins, their health physics impact is invisible; the synergism of their interrelationships and impact on human health is also beyond the capability of our health systems to evaluate or mitigate.

Some important biologically significant, persistent organic pollutants have been banned from production in western countries and their impact has been partially mitigated. This summary of selected ecotoxins of major interest includes several of these ecotoxins because of their continuing presence in the environment. This listing is a sampling of major ecotoxins and in no way constitutes an adequate description of the many anthropogenic ecotoxins that now threaten human health as contaminant signals in most ecosystems. These sketches, the entrees on our Hotel California menu, as it were, are followed by a brief listing of some other ecotoxins of interest, i.e. today's menu of appetizers. Both our list of selected major ecotoxins of interest, and our sketch of other ecotoxins and pathogens are but a brief sampling of the thousands of ecotoxins now being transported in the global atmospheric water cycle. Any comprehensive study of anthropogenic chemical fallout would require hundreds of experts, years of research, and tens of thousands of pages of data. The space limitations of our preliminary survey on a very complex topic – biocatastrophe – restrict commentary on ecotoxins to the brief summaries that follow.

Category: Aromatic Hydrocarbons

Chemical: Benzene, C_6H_6

Sources: Benzene is a precursor, byproduct, and constituent of numerous consumer products, including plastics. It is also a byproduct of the combustion of materials containing carbon, including cigarettes and gasoline. The US produced 3.109 billion gallons of benzene in 2004 with a projected annual increase of 4%. In 2004, China alone produced 2.5563 million tons of benzene.

Transport Vectors: Benzene is highly volatile, transports easily in the air, is very lipid soluble and thus present in the lipids of aquatic organisms, but does not appear to bioaccumulate.[1][3] Cigarette consumption vastly increases exposure to benzene. Despite biodegrading rapidly in air, water, and soil, benzene's constant release and high volatility allow it travel over long distances, causing minute exposure to all biotic media, including humans, no matter how remote their location. While cigarettes account for 0.1% of benzene emissions, they account for 40% of exposure, followed by car exhaust at 18%. [1][7][8]

Airborne:	Source Points:	Food Products, ppb:	Human Fat tissue:
Smoker homes: 3.3 ppb or 10.5 µg/m³ **Nonsmoker homes:** 2.2 ppb or 7 µg/m³ **Smoking Bar:** 8.08 – 11.3 ppb[1]	**Los Angeles traffic tunnels, 1998 High:** 118,420 ppm [7] **Cigarette smoke, Mainstream:** 5.9 to 73 ug/cigarette **Sidestream:** 345 – 653 ug/cigarette [6]	**Food from 1990 without benzoates:** <2 ng/g **" with benzoates:** <1 - 38 ng/g [5]	**2001:** 0.020 - 11.70 µg g^{-1} [4]

Health Effects: Benzene has been widely recognized as carcinogenic by the US Department of Health and Human Services and the EPA (Environmental Protection Agency).

Benzene reaches highly toxic levels at many of its source points; in one instance, over 10% of the air in an LA traffic tunnel was found to be benzene, over five times the predicted lethal threshold of 20,000 ppm. The average smoker inhales 0.657 grams of benzene a year; comparatively, the average nonsmoker inhales 1/10th that amount.

Citations:

1. http://www.atsdr.cdc.gov/toxprofiles/tp3-c5.pdf.

2. http://ntp.niehs.nih.gov/ntp/roc/eleventh/profiles/s019benz.pdf.

3. http://www.npi.gov.au/database/substance-info/profiles/12.html.

4. Russo MV, Campanella L. (2001). Static Headspace Analysis by GC-MS (in SIM mode) to determine benzene in human tissues. *Anaytical Letters*. 34. pg. 883-91.

5. Grob K, Frauenfelder C, Artho A. (1990). Uptake by foods of tetrachloroethylene, trichloroethylene, toluene, and benzene from air. Z Lebensm Unters Forsch 191. pg. 435-441.

6. Brunnemann KD, Kagan MR, Cox JE, et al. 1990. Analysis of 1,3-butadiene and other selected gasphase components in cigarette mainstream and sidestream smoke by gas chromatography-mass selective detection. *Carcinogenesis*. 11. pg. 1863-1868.

7. Fraser MP, Cass GR, Simoneit BRT. 1998. Gas-phase and particle-phase organic compounds emitted from motor vehicle traffic in a Los Angeles roadway tunnel. *Environmental Science and Technology*. 32. pg. 2051-2060.

8. http://www.hpa.org.uk/webc/HPAwebFile/HPAweb_C/1194947380489.

Category: Aromatic Hydrocarbons

Chemical: Bisphenol-A, $C_{15}H_{16}O_2$

Sources: Bisphenol-A is used as a plasticizer in numerous consumer products, including medical equipment, canned food, baby bottles, beverage bottles, and electronics equipment. It has also been used as a fungicide. [1] In 2003, nearly 3 billion kilograms were produced worldwide. [3]

Transport Vectors: Despite being identified as an endocrine disruptor in the 1930s, Bisphenol-A has largely escaped regulation due to its relatively low direct toxicity. It is pervasive in consumer products, making its way directly into humans and back into the ecosystem via landfills. It is nonetheless possible through careful choice of consumer products to avoid ingestion of bisphenol-A. Its erratic uptake is evidenced by its presence in only 8 of 44 subjects in an EWG study, yet bisphenol-A was found in 100% of German drinking water samples in a 2001 study.[1][4][5] Bisphenol-A and a number of its cogeners were also found in the body burden survey of Maine residents, the summary of which is reprinted in *Appendix L*.

Canned Foods, ug/kg:	Products:	Intake, ug/kg/day:	Human blood:
UK Pasta: 67.3-129.5[6] **US Solid Foods:** ND – 192 [7]	**Taiwan Infant Formula:** 44-113 ug/kg [8]	**Infant, formula fed:** 1-11 **Breast-fed:** 0.2-1[1]	1.08 ng/mL (wet weight) [5]

Health Effects: Bisphenol-A has been shown to be a potent hormone and endocrine disrupting chemical when administered to lab animals at very low doses. It is suspected of causing reproductive damage and birth defects that may lead to prostate and breast cancer in humans. Studies have found that BPA can have adverse health effects at levels thousands of times lower than what the EPA considers safe and, according to some studies, small and repeated exposures to bisphenol-A can have an amplified effect on the human body by mimicking human sex hormones, or promoting cell growth. Bisphenol-A has been found to cause estrogen changes in animal cells at the same concentrations that are being found in pregnant women and their fetuses.

Controversy over toxicity exists between public health advocates and the plastics industry, which describes Bisphenol-A as a weak estrogenic and expresses little concern about human exposure levels. Of the 11 published studies funded by industry between 1998 and 2005, none reported adverse effects at low level exposures, while 94 of 104 (90%) government-funded studies found statistically significant adverse effects at levels to which many US residents are currently exposed, levels much lower than the EPA's current acceptable level. [11]

Pregnant women who consume BPA could be putting their unborn children at risk of developing cancer and other diseases when they reach adulthood. Exposure within the womb to bisphenol A (BPA) caused changes linked with diseases such as obesity, cancer, and diabetes, according to studies on pregnant mice by a team from Duke University Medical Centre, North Carolina. When mouse moms were fed BPA, slightly over half were born with yellow coats, an indicator for cancer. "The fact that the mice fed BPA had a yellow coat and likely would grow to be obese as adults demonstrates that this single substance had a system-wide effect," said Dr Dana Dolinoy, one of the team. [9]

"On October 31 [2008], the Science Board of the US Food and Drug Administration… embraced EWG arguments that bisphenol-A (BPA), a synthetic estrogen used to make polycarbonate plastics and epoxy resin may be a threat to human health. The panel forced FDA to retreat from its stance that trace levels of BPA are safe in food packaging, including infant formula cans and baby bottles. EWG scientists testified, wrote comments and served on the expert panel for the Science Board. In September [2008], the National Institutes of Health's National Toxicology Program (NTP) declared that BPA, shown in laboratory tests to disrupt the endocrine system, may alter brain development, cause behavioral problems and damage the prostate glands in fetuses, infants and young children." [10]

Citations:

1. Okada, H, et al. (2008). Direct evidence revealing structural elements essential for the high binding ability of bisphenol A to human estrogen-related receptor-gamma. *Environmental Health Perspectives*. 116 (1). pg. 32-38. PMID: 18197296.

2. *Draft Screening Assessment for the Challenge Phenol, 4,4'-(1-methylethylidene)bis-(Bisphenol A) Chemical Abstracts Service Registry Number 80-05-7.* Environment Canada; www.ec.gc.ca.

3. Kuch HM, Ballschmiter K. (2001). Determination of endocrine-disrupting phenolic compounds and estrogens in surface and drinking water by HRGC-(NCI)-MS in the picogram per liter range. *Environmental Science and Technology.* 35. pg. 3201-3206.

4. http://www.bodyburden.org/chemicals/chemical.php?chemid=100357.

5. Goodson A, Robin H, Summerfield W, Cooper I. (2004). Migration of bisphenol A from can coatings--effects of damage, storage conditions and heating. *Food Additive Contamination.* 21. pg. 1015-1026.

6. Wilson NK, Chuang JC, Morgan MK, Lordo RA, Sheldon LS. (2006). An observational study of the potential exposures of preschool children to pentachlorophenol, bisphenol-A, and nonylphenol at home and daycare. *Environmental Research.*

7. Kuo H-W, Ding W-H. Trace determination of bisphenol A and phytoestrogens in infant formula powders by gas chromatography-mass spectrometry. *Journal of Chromatography A.* 1027. pg. 67-74.

8. National Toxicology Program, U.S. Department of Health and Human Services. (November 26, 2007). *CERHR Expert Panel Report for Bisphenol A* Retrieved on [[2008-04-18]].

9. www.treehugger.com/files/2007/08/bisphenol_a_and.php.

10. www.enviroblog.org.

11. Alliance for a Clean and Healthy Maine; http://www.cleanandhealthyme.org/BodyofEvidenceReport/TheChemicals/BisphenolAHormoneDisrupter/tabid/99/Default.aspx.

Category: Organochloride

Chemical: Chloroform, CHCl₃

Sources: Chloroform is a halogenated hydrocarbon that is used as a solvent, as well as a constituent precursor for refrigerants, dyes and pesticides. [1] It has also recently been discovered that triclosan, a common ingredient in toothpaste and soap, reacts with chlorinated drinking water to form chloroform. [2] Prior to the discovery of its toxicity, it was also used as an anesthetic and still has pharmaceutical applications as a sedative in cough medicines and other uses. While chloroform occurs in trace amounts naturally, most of its environmental presence derives from human activity. Biodegradation of other organochlorides can also result in the production of chloroform. The EPA recently raised the legally safe level of chloroform in drinking water from 0 to 300 ppb, though many states have adopted more stringent standards.

Transport Vectors: Chloroform is highly volatile, lipid soluble, and commonly found in the environment in water, air, and soil. All but two continental states have at least one NPL site with chloroform contamination.[1] Recent data has shown a drastic increase in chloroform contamination in individuals who frequently swim in chlorinated pools.

Water (ppb)	Food	Air	Humans
Public Pools: up to 163 [3] **Hot tub:** 15 - 674 [4]	**Milled Grain:** 1.4 – 3,000 ug/kg **Soft drinks:** 2.7 – 178 ug/kg [5][6]	**Above hot tub:** $8 \times 10^{-4} - 1.5 \times 10^{-1}$ ppm [4]	**Swimmers, blood plasma:** 0.8 – 25.1 nmol/L (.9549 – 2.996 mcg/L)

Health Effects: Breathing about 900 parts of chloroform per million parts air (900 ppm) for a short time can cause dizziness, fatigue, and headache. Breathing air, eating food, or drinking water containing chloroform may result in metabolic disturbances and in damage to the heart, liver, and kidneys. Chloroform is an important biologically significant ecotoxin and a primary component, along with chlorine, of the production of trihalomethane and its many cogeners. Large amounts of chloroform can cause sores after skin contact. According to the ATSDR, birth defects and other reproductive health effects have not yet been documented in people [1], but animal studies have shown that miscarriages occurred in rats and mice that breathed air containing 30 to 300 ppm chloroform during pregnancy and also in rats that ate chloroform during pregnancy. Offspring of rats and mice that breathed chloroform during pregnancy had birth defects. Abnormal sperm were found in mice that breathed air containing 400 ppm chloroform for a few days. Chloroform is a probable constituent of the synergistic interaction of chlorine containing ecotoxins in abiotic and biotic media, which results in the formation of new cogeners and ecotoxic metabolites, of which the trihalomethane family of contaminants in drinking water is only one example.

The U.S. Department of Health and Human Services (DHHS) has determined that chloroform may reasonably be anticipated to be a carcinogen. Rats and mice that ate food or drank water with chloroform developed cancer of the liver and kidneys. [9]

Citations:

1. Agency for Toxic Substances and Disease Registry. (1997). *Toxicological profile for chloroform*. Atlanta: US Department of Health and Human Services.
2. Rule KL, Ebbett VR, Vikesland PJ. (May 1, 2005). *Environmental Science and Technology*. 39(9). pg. 3176-85.
3. Barnes D, Fitzgerald PA, Swan HB. (1989). Catalyzed formation of chlorinated organic materials in waters. *Water Science and Technology*. 21(2). pg. 59-63.

4. Benoit FM, Jackson R. (1987). Trihalomethane formation in whirlpool spas. *Water Research*. 2(13). pg. 353-357.

5. Abdel-Rahman MA. (1982). The presence of trihalomethanes in soft drinks. *Journal of Applied Toxicology*. 2. pg. 165-166.

6. Lovegren NV, Fisher GS, Lehendre MG, et al. (1979). Volatile constituents of dried legumes. *Journal of Agricultural and Food Chemistry*. 27. pg. 851-853.

7. Heather J Whitaker, Mark J Nieuwenhuijsen, and Nicola G Best. (May 2003). The relationship between water concentrations and individual uptake of chloroform: a simulation study. *Environmental Health Perspectives*. 111(5). pg. 688–694.

8. Aggazzotti G, Fantuzzi G, Tartoni PL, et al. (1990). Plasma chloroform concentrations in swimmers using indoor swimming pools. *Archives of Environmental Health*. 45(3). pg. 175-179.

9. CDC, ATSDR; http://www.atsdr.cdc.gov/tfacts6.html#bookmark05.

Category: Organochloride (Chlorinated Hydrocarbon)

Chemical: DDT, $C_{14}H_9Cl_5$

Sources: DDT is a halogenated hydrocarbon, the production of which utilizes hydrogen chlorine in its manufacture. First widely used for lice control in WWII, DDT became well known as a potent ecotoxin after publication of Rachel Carson's *Silent Spring* in 1962. Its use as a pesticide was banned in the U.S. in 1972. Technical grade DDT was 65-80% DDT combined with 14 other chemicals. The estimated total world production of DDT is 2 million tons to date. [2] As one of the POPs (persistent organic pollutants) banned by the Stockholm Convention of 2001, due to the stability and chemical longevity, significant quantities of DDT will continue to circulate in the biosphere for centuries. It remains in production and use in 3rd world disease control operations.[1]

Transport Vectors: DDT is highly lipid-soluble and as a result bioconcentrates and biomagnifies in biotic media, especially at higher trophic levels. It is not water soluble, but as a result of attaching to particulates and its high volatility, it is easily transported by both wet and dry deposition, "distilling" from warmer to cooler regions as it breaks down into (DDE) and (DDD), which are also toxic. Most human exposure is through food consumption. DDT infrequently decays at its theoretical chemical half-life and in many contemporary samples the levels of DDT still exceed those of DDE. More than 30 years after its ban, residual DDT is still a widespread contaminant in many ecosystems. [2] It has a bioconcentration factor (BCF) of up to 690,000 in some species of mussel. [3] DDT is one of many POPs subject to hemispheric transport as a component of the global atmospheric water cycle. DDT, along with PCBs and many other chemicals, was discovered by Swedish and other researchers in Antarctic seabirds in the 1960s. The residue of DDT found in many Arctic species had been declining in late 20th century, but recently, scientists in Antarctica reported that Adélie penguins now have a constant, and in sometimes increasing, level of DDT in their body fat. It appears that the birds are being re-exposed to DDT that was previously deposited in ice and snow that is now melting, as well as being subject to ongoing tropospheric fallout of DDT from continuing world production and use. [8]

Prior to Ban:	Marine Media:	Fish:	Human blood:
1967-68 US Surface Water (27 of 224 samples): .005 - .316 ug/L [4]	St. Lawrence River, 1997 (mean): 500 pg/L [5]	Pelagic: 200 ug/kg lipid Abyssal: 1,465 ug/kg lipid [7]	Sao Paulo workers, DDT in DDT Appliers, ug/L:13.5 Unexposed: 1.5 [6]

Health Effects: DDT affects the nervous system, resulting in excitability, tremors, and seizures. A study in humans showed that women who had high levels of a form of DDE in their breast milk were unable to breast feed their babies for as long as women who had low levels of DDE in their breast milk. Another study in humans showed that women who had high amounts of DDE in breast milk were more likely to deliver premature babies. In animals, short-term exposure to large amounts of DDT in food affected the nervous system and exposure to small amounts of DDT or its breakdown products may also have harmful effects on reproduction. Long-term exposure to smaller amounts may affect the liver. Studies in DDT-exposed workers did not show increases in cancer, but studies in animals given DDT with their food have shown that DDT can cause liver cancer. The Department of Health and Human Services (DHHS) has determined that DDT may reasonably be expected to be a human carcinogen and the International Agency for Research on Cancer (IARC) has determined that DDT may possibly cause cancer in humans. The EPA determined that DDT, DDE, and DDD are probable human carcinogens. There are no studies on the specific health effects for children exposed to DDT, DDE, or DDD. It could reasonably be assumed that children exposed to large amounts of DDT will have adverse health effects similar to those seen in adults. It is not yet known whether children are more susceptible to the effects of these substances. Studies in rats have shown that DDT and DDE can affect the endocrine system by mimicking natural hormones, affecting the development of the reproductive and nervous systems. Puberty was delayed in male rats given high amounts of DDE as juveniles and a study in mice showed that exposure to DDT during the first weeks of life may cause neurobehavioral problems later. [9]

Citations:

1. John Paull, Toxic Colonialism. (November 3, 2007). *New Scientist*. 25(2628).
2. Agency for Toxic Substances and Disease Registry. (2002). *Toxicological profile for DDT, DDE and DDD*. Atlanta: US Department of Health and Human Services.

3. Geyer H, Politzki G, Freitag D. (1984). Prediction of ecotoxicological behavior of chemicals: Relationship between n-octanol/water coefficient and bioaccumulation of organic chemicals by alg*Chlorella. Chemosphere*. 13(2). pg. 269-284.

4. Lichtenberg J, Eichelberger J, Dreeman R, et al. (1970). Pesticides in surface waters of the United States: A 5-year summary, 1964-68. *Pesticides Monitoring Journal*. 4 (2). pg. 71-86.

5. Poissant L, Koprivnjak J-F, Matthieu R. (1997). Some persistent organic pollutants and heavy metals in the atmosphere over a St. Lawrence River valley site (Villeroy) in 1992. *Chemosphere*. 34 (3). pg. 567-585.

6. Minelli E, Ribeiro M. (1996). DDT and HCH residues in the blood serum of Malaria control sprayers. *Bulletin of Environmental Contamination and Toxicology*. 57. pg. 691-696.

7. Hargrave BT, Harding GC, Vass WP, et al. 1992. Organochlorine pesticides and polychlorinated biphenyls in the Arctic Ocean food web. *Archives of Environmental Contamination and Toxicology*. 22. pg. 41-54.

8. http://www.nytimes.com/2008/06/27

9. http://www.atsdr.cdc.gov/tfacts35.html

Category: Dioxins (Chlorinated Hydrocarbons)

Chemical: Polychlorinated Dibenzo-p-dioxins (CDD), $C_4H_4O_2$ (1,4-dioxin or p-dioxin) variants

Sources: Polychlorinated Dibenzo-p-dioxins (CDDs) are a type of chlorinated hydrocarbon centered around $C_4H_4O_2$ with 75 cogeners. They are often referred to as "dioxins" along with chlorodibenzofurans. The deliberate production of dioxin as a component of chlorinated pesticides and other petrochemicals has been outlawed in the United States and Europe but not in other countries. Of most significance is their ubiquitous accidental worldwide production as a result of combustion (e.g. municipal and industrial waste incinerators, internal combustion engines, power plants, cigarette smoke, etc.) or from millwork due to paper bleaching. Levels persist in the environment from past pesticide use and as a result of continuous worldwide proliferation from the incineration of consumer product wastes of every description, including most plastics. [1][2]

Transport Vectors: CDDs are highly volatile, lipid soluble, persistent organic pollutants subject to rapid uptake and hemispheric transport in the global atmospheric water cycle as adsorbed contaminants on particulate aerosols, as well as an adsorbed (but not absorbed) component of rainfall. CDDs do not dissolve easily in water, but as highly lipophilic ecotoxins readily biomagnify in both marine and terrestrial ecosystems. As adsorbed components of nutrient elements, CDDs are ingested by microorganisms and are then consumed by progressively larger organisms. CDDs can also be ingested as adsorbed microcontaminants at any point in the global atmospheric water cycle where biotic media consume fresh water. Such contamination is well below most current LODs (limits of detection) and is often initially present in concentrations of parts per quadrillion. CDDs readily biomagnify within food webs; dairy products and fish consumption account for the majority of human exposures, as evidenced by the differences in contamination in at-risk groups. [2]

Source Points:	Time-sensitive Data:	Food Products	Humans
BC Ditch Sediment Utility Pole Adjacent: 2,576 ppt 4 Meters Down: 14 ppt 4 Meters Upstream: None[7]	Green Lake Core Sample 1860-1930: < 10 pg/g 1960s: 1,300 pg/g 1986-1990: 750 pg/g (total CDD/CDF)[4]	NJ Crab meat: 100 – 120 ppb [5] Cow Milk: 20.9 ppb, lipid Chicken: 69.21 ppb, lipid [6]	Breast milk, ppt: Inuit Women: 39.6 Southern Quebec Women: 14.6 [18] Blood Serum, pg/g: Non-fish eaters: 17.5 Heavy fish eaters: 63.5 [7]

Health Effects: Dioxin has been the subject of an intensive health reassessment by the EPA for more than 10 years; dioxins and dioxin-like compounds such as PCBs are classified as "known carcinogens." The EPA concludes that the <u>average</u> American may face a 1 in 1,000 risk of cancer due to these compounds, with the risk possibly as high as 1 in 100, because the collective simultaneous exposure to the multiple forms of dioxin, dioxin-furans, and related BCB/PBB cogeners and metabolites has become so widespread in our general food supply, particularly in meat, dairy, poultry, fish, eggs, and other fatty animal products. It is possible that high-temperature pan frying or grilling of fish or other PCB-contaminated foods could convert some of the PCBs to more-toxic dioxins and furans. (The PCBs would also volatilize into the kitchen air at high rates.) The EPA reports that some of the toxic effects of rice-oil PCB poisoning incidents in Taiwan and Japan involved heat conversion of PCBs to furans (EPA 1997).

Dioxins have biological significance as ecotoxins at extremely small concentrations levels and are included in the EPA (Environmental Protection Agency) *List of Lists* (2006) as EHSs (extremely hazardous substances). Consumption of even tiny amounts of dioxin or inhalation of dioxin fumes can result in an elevated risk of cancer. The accumulated exposure to the dioxin family of ecotoxins may play a role in the increasing incidence of cancer in highly technometabolic communities, which consume significant quantities of high protein food.

Other health effects of exposure to dioxin and related POPs include chloracne, changes in blood and urine that may indicate liver damage, alterations in the ability of the liver to metabolize hemoglobin, lipids, sugar, and protein, slight increases in the risk of diabetes and abnormal glucose tolerance, and possible reproductive or developmental effects in people.

The U.S. Department of Health and Human Services (DHHS) has determined that it is reasonable to expect that 2,3,7,8-TCDD may cause cancer. The International Agency for Research on Cancer (IARC) has determined that 2,3,7,8-TCDD can cause cancer in people, but that it is not possible to classify other CDDs as to their carcinogenicity to humans. The EPA has determined that 2,3,7,8-TCDD is a possible human carcinogen when considered alone and a probable human carcinogen when considered in association with phenoxy herbicides and/or chlorophenols. The EPA has also determined that a mixture of CDDs with six chlorine atoms (4 of the 6 chlorine atoms at the 2, 3, 7, and 8 positions) is a probable human carcinogen.

Citations:

1. Environmental Protection Agency. (1999). *Fact sheet on polychlorinated dibenzo-p-dioxins and related compounds update: Impact on fish advisories.* Washington, DC: Office of Water. http://www.epa.gov/waterscience/fish/files/dioxin.pdf
2. Agency for Toxic Substances and Disease Registry. (1998). *Toxicological profile for chlorinated dibenzo-p-dioxins.* Atlanta: US Department of Health and Human Services. http://www.atsdr.cdc.gov/toxprofiles/tp104.html
3. Wan MT, Van Oostdam J. (1995). Utility and railroad rights-of-way contaminants: Dioxins and furans. *Journal of Environmental Quality.* 24(2). pg. 61-69.
4. Smith, R.M.; O'Keefe, P.W.; Hilker, D.R.; Bush, B.; Connor, S.; Donnelly, R.; Storm, R.; Liddle, M. (1993). The historical record of PCDDs, PCDFs, PAHs, PCBs, and lead in Green Lake, New York – 1860 to 1990. *Organohalogen Compounds.* 24. pg. 141-145.
5. Bopp RF, Gross ML, Tong H, et al. (1991). A major incident of dioxin contamination: Sediments of New Jersey Estuaries. *Environmental Science and Technology.* 25. pg. 951-956.
6. Beck H, Eckart K, Mathar W, et al. (1989a). PCDD and PCDF body burden from food intake in the Federal Republic of Germany. *Chemosphere.* 18. pg. 417-424.
7. Svensson B-G, Nilsson A, Hansson M, et al. 1991. Exposure to dioxins and dibenzofurans through the consumption of fish. *New England Journal of Medicine.* 324(1). pg. 8-12.

Category: Organochlorides

Chemical: Chlorodibenzofurans (CDF), $C_{12}H_{8-x}OCl_x$

Sources: Polychlorinated Dibenzo-p-dioxins (CDFs), one of many chemical groups in the dioxin family of ecotoxins, are a type of chlorinated hydrocarbon centered around $C_4H_4O_2$ with 135 isomers. They are often referred to as "dioxins" along with CDDs.

They are undesired byproducts of industrial chemical production, rubbish incineration, paper mill bleaching, and photolysis of PCBs, PCDE, and polychlorinated benzenes. [1]

Transport Vectors: CDFs are highly volatile, lipid soluble, persistent organic pollutants. CDFs are transported hemispherically as components of the global atmospheric water cycle due to their adsorption by airborne particulate aerosols. Larger isomers all possess an atmospheric lifetime of >10 days, which helps explain their worldwide distribution. Levels in unpolluted rural air and soil are generally too low to measure; concentrations at dump sites are thousands to millions of times higher. CDFs were found in all of the subjects of a recent EWG study[6] and some of their many variants (cogeners) are the subject of a CDC biomonitoring with LODs (limits of detection) as low as pg/g, or parts per trillion. Many forms of CDF metabolites are present in the environment in fg/g, parts per quadrillion, and are, thus, below the limit of detection.

Source Points (ppt):	Marine Media (ppt):	Food Products (ppt):	Human Fat tissue (ppt):
Chloralkali bleaching mill: <52,000 2,3,7,8-tetraCDF 81,000 (octaCDF) [2]	Six American Rivers, Sediment: 5 – 97[1] Lake Michigan sediment: 24 [3]	Blue Crab Pancreas: 12,122.7[5] NY Supermarket Meat: .14 - 7.0 Fish: .07 - 1.14 Dairy: .3 – 5 [4]	Sweden: 85.9 Germany: 84.9 Canada: 127.9 USA: 146.6 [1]

Health Effects: As extremely hazardous chemicals that biomagnify at higher tropic levels, CDFs are biologically significant at *any level of contamination in any biotic media*, and are included in the EPA's *List of Lists* (2006) as extremely hazardous substances (EHS). Consumption of even tiny amounts of dioxins or dioxin-furans can result in a variety of health effects, many of which cannot be differentiated from the health effects of other ecotoxins due to the small quantities causing the effects.

Citations:

1. Agency for Toxic Substances and Disease Registry. (1994). *Toxicological profile for chlorinated chlorodibenzofurans*. Atlanta: US Department of Health and Human Services.
2. Eitzer BD, Hites RA. 1989. Polychlorinated dibenzo-p-dioxins and dibenzofurans in the ambient atmosphere of Bloomington, Indiana. *Environmental Science and Technology*. 23. pg. 1389-1395.
3. Smith LM, Schwartz TR, Feltz K, et al. (1990). Determination and occurrence of AHH-active polychlorinated-biphenyls, 2,3,7,8-tetrachloro-para-dioxin and 2,3,7,8-

tetrachlorodibenzofuran in Lake Michigan sediment and biota. The question of their relative toxicological significance. *Chemosphere*. 21. pg. 1063-1085.

4. Schecter A, Startin J, Wright C, et al. 1993. Dioxin levels in food from the United States with estimated daily intake. In: Fiedler H, Frank H, Otto H, et al., eds. *Organohalogen compounds*. Vienna: Federal Environmental Agency. 13. pg. 93-96.

5. McLaughlin DL, Pearson RG, Clement RE. 1989. Concentrations of chlorinated dibenzo-p-dioxins (CDD) and dibenzofurans (CDF) in soil from the vicinity of a large refuse incinerator in Hamilton Ontario Canada. *Chemosphere*. 18. pg. 851-854.

6. http://www.bodyburden.org/participants/participant-group.php?group=bb2

Category: Organochlorides

Chemical: Polychlorinated Biphenyls (PCB) $C_{12}H_{10-x}Cl_x$

Sources: "PCB" refers to 209 cogeners, roughly 130 of which were used for a variety of products including lighting fixtures, electrical insulation, various oils, adhesives, paints, pesticides, sealants, paper, and in transformers as an insulating barrier. Despite termination of the production of PCBs by the Monsanto Company in 1977, PCBs continue to enter the environment due to improper disposal of products containing them as well as possible undocumented production in Far Eastern locations. PCBs are in the EPA's "Dirty Dozen," [1] as reflected in the toxicological profiles issued by the Agency for Toxic Substance and Disease Registry (ATSDR), and are among the most ubiquitous ecotoxins produced by industrial activities.

Transport Vectors: PCBs cycle easily between water, soil and air; its many cogeners have a half-life ranging from a few years to as long as twenty years.[3] They are highly volatile as well as lipid soluble and are subject to rapid uptake and hemispheric transport within the global atmospheric water cycle as documented by Miller[17] in the landmark publication *Chemical Fallout*. As first documented in the Miller treatise, PCBs, as a lipophilic persistent organic pollutant, are rapidly transported and biomagnified in food webs, as demonstrated by their ubiquitous appearance in Antarctic seabirds. Food consumption contributes the most to human PCB exposure; fortunately, the average daily exposure for humans has dropped between 1978 and 1991 from .027 ug/kg to <0.001 kg [1], but PCBs are still a ubiquitous component of human urine, blood, and serum (*Appendix A*) and are a predicator for other newly emerging POPs, such as their chemical cousins, polybrominated diphenyl ethers (PBDEs).

Sample Concentration Levels

Abiotic Media:

Water (ng/L)	Soil	Air (ng/m^3)	Sediment
US Tap Average: <0.1 **Lake Superior 1978:** .63 – 3.3 **Lake Superior 1993:** .070 - .10 [1] **Rural Rain 1970s–1980s:** 1-50 [9] **1980s–1990s:** .5-20 [8] **Urban Rain 1970s–1980s:** 10 -250 [9] **1980s–1990s:** 10 [8]	**US PCB disposal facilities:** **Low:** 100 ug/m^2 (74% of facilities) **High:** 180,000 ug/m^2 **NY Hazardous Waste site:** 740 mg/kg[6]	**Cape Cod, MA homes low:** >1 **High:** 5 **Median:** < 3 [2] **Baltimore Urban:** .38 – 3.36 **Baltimore Rural:** .02 - .34 [4] **Arctic:** .074 ng/m^3 [5]	**NY Hazardous Waste Site:** 41,500 mg/kg[6] **Upper Hudson 1960s:** 350 ppm **1991:** 34 ppm [10] **Lake Ontario 1982/1983:** 1300 – 1900 ng/g **1985/1986:** 80-290 ng/g [11] **St. Lawrence River, 1995 max:** 5,700,000 ng/g

Biotic Media:

Fish (ug/g wet, mean)	Marine Mammals (ug/g lipid weight)	Birds	Food
US North Coasts mean: 1.64 [7] **Sierra Nevada:** .018 - .430 [1] **Buffalo River Carp Young:** 2.40 **Middle age:** 4.30 **Old:** 5.00 [12] **Superfund Site Bullhead:** 20.55 [1]	**Irish Sea Porpoise:** 6.19 **Gulf Coast bottlenose dolphins** **Adult male:** 93 **Juvenile:** 49 **Suckling:** 41 **Fetus:** 4 [1]	**NY Bufflehead Duck:** 0.15±0.19 **Canada Geese:** 0.05±0.01 **Canadian Mallard:** 0.161 [1]	**Soy Baby Food:** 10.25 ng/g mean [13] **Fish:** .892 ppm **Shellfish:** .056 ppm **Eggs:** .072 ppm **Milk:** .067 ppm [1]

Human:

Serum, non-occupationally exposed, non-fish consumers (ppb)	Serum, non-occupationally exposed, fish consumers (ppb)	Breast milk (ng/g lipid)
1996 Great Lakes Mean: 1.2 **High:** 9.7 **Low:** .5 [14] **1982 – 1984 Long Beach, CA Work Force Mean:** 4 **High:** 37 **Low:** <1 [15] **1983 Fairmount, WV Mean:** 5 **High:** 23 **Low:** 1 [16]	**1995 Great Lakes Mean:** 4.8 **High:** 58.2 **Low:** 0.7 [14] **1982 New Bedford, MA Mean:** 13.34 **High:** 87.97 **Low:** 1.40 [17] **1973 Triana, AL Mean:** 22.2 **High:** 158 **Low:** 3	**1972 Sweden:** 1,090 **1985 Sweden:** 600 **1992 Sweden:** 380 **Akwesasne Indian Reservation 1986-89:** 602 **Reservation 1992:** 254 **1992 Canada:** 238 **Massachusetts near superfund site:** 1,107 – 2,379 [1]

Health effects: **(WIP)** Previous studies in workers exposed to PCBs have shown changes in blood and urine that may indicate liver damage. In 1968 in Japan, rice bran oil contaminated with PCBs was used as chicken feed, resulting in a mass poisoning known as Yusho Disease in more than 14,000 people. Symptoms included skin and eye lesions, irregular menstrual cycles, lowered immune system responses, and less commonly, fatigue, headache, cough, and unusual skin sores. In children, there were reports of poor cognitive development. [19, 20, 21]

Citations:

1. Agency for Toxic Substances and Disease Registry (ATSDR). (2000). *Toxicological profile for Polychlorinated Biphenyls (PCBs)*. Atlanta, GA: U.S. Department of Health and Human Services, Public Health Service.
2. Rudel, R A, Seryak, L M, and Brody, J G (2008). PCB-containing wood floor finish is a likely source of elevated PCBs in resident's blood, household air and dust: a case study of exposure. *Environmental Health*.
3. http://dioxin2004.abstract-management.de/pdf/p311.pdf.
4. Offenberg JH, Baker JE. (1999). Influence of Baltimore's urban atmosphere on organic contaminants over the northern Chesapeake Bay. *Journal of Air Waste Management Association*. 49. pg. 959-965.
5. Harner T, Kylin H, Bidleman TF, et al. (1998). Polychlorinated naphthalenes and coplanar polychlorinated biphenyls in Arctic air. *Environmental Science and Technology*. 32. pg. 3257-3265.
6. ATSDR. 1995. *Exposure to PCBs from hazardous waste among Mohawk women and infants at Akwesasne*. Atlanta, GA.: U.S. Department of Health and Human Services, Public Health Service, Agency for Toxic Substances and Disease Registry.
7. Kennish MJ, Ruppel BE. 1996. Polychlorinated biphenyl contamination in selected estuarine and coastal marine finfish and shellfish of New Jersey. *Estuaries*. 19(2A). pg. 288-295.
8. Eisenreich SJ, Baker JE, Franz T, et al. 1992. *Atmospheric deposition of hydrophobic organic contaminants to the Laurentian Great Lakes*. In: Schnoor JL, ed. *Fate of pesticides and chemicals in the environment*. New York, NY: John Wiley & Sons, Inc. pg. 51-78.
9. Eisenreich SJ, Looney BB, Thornton JD. 1981. Airborne organic contaminants in the Great Lakes ecosystem. *Environmental Science and Technology*. 15. pg. 30-38.
10. Bopp RF, Chillrud SN, Shuster EL, et al. 1998. Trends in chlorinated hydrocarbon levels in Hudson River Basin sediments. *Environmental Health Perspectives Supplement*. 106(4). pg. 1075-1081.
11. Oliver BG, Charlton MN, Durham RW. 1989. Distribution, redistribution, and geochronology of polychlorinated biphenyl congeners and other chlorinated hydrocarbons in Lake Ontario sediments. *Environmental Science and Technology*. 23. pg. 200-208.
12. Loganathan BG, Kannan K, Watanabe I, et al. 1995. Isomer-specific determination and toxic evaluation of polychlorinated biphenyls, polychlorinated/brominated dibenzo-p-dioxins and dibenzofurans, polybrominated biphenyl ethers, and extractable organic

halogen in carp from the Buffalo River, New York. *Environmental Science and Technology*. 29(7). pg. 1832-1838.

13. Ramos L, Torre M, Laborda F, et al. 1998. Determination of polychlorinated biphenyls in soybean infant formulas by gas chromatography. *Journal of Chromatography*. 823:365-372.

14. Anderson HA, Falk C, Hanrahan L, et al. 1998. Profiles of Great Lakes critical pollutants: A sentinel analysis of human blood and urine. *Environmental Health Perspectives*. 106(5). pg. 279-289.

15. Sahl JD, Crocker TT, Gordon RJ, et al. 1985a. Polychlorinated biphenyl concentrations in the blood plasma of a selected sample of non-occupationally exposed southern California working adults. *Science of the Total Environment*. 46. pg. 9-18.

16. Welty ER. (August 8, 1983). *Personal communication*. (As cited in Kreiss 1985).

17. Condon SK. (August 25 and 28, 1983). *Personal communications*. Commonwealth of Massachusetts Department of Public Health.

18. Miller, Morton W. and Berg, George G., Eds. 1970. *Chemical fallout: Current research on persistent pesticides.* Charles C. Thomas, Publisher, Springfield, IL.

19. http://shippai.jst.go.jp/en/Detail?fn=2&id=CB1056031; www.ncbi.nlm.nih.gov/sites/entrez?cmd=Retrieve&db=PubMed&list_uids=11386736&dopt=Abstract

20. http://rarediseases.info.nih.gov/GARD/Disease.aspx?PageID=4&diseaseID=8326

21. www.healthgoods.com/Education/Health_Information/General_Health/environmental_diseases.htm

---

Category: Phthalates

Chemical: DEHP, $C_{24}H_{38}O_4$

Sources: di(2-ethylhexyl) phthalate (DEHP) is a common plasticizer used in textiles, toys, medical tubing and supplies, packaging, wires, and cables. It is a biologically significant ecotoxin released from polyvinyl chloride plastic piping, the incineration of which also produces dioxin-furan-like cogeners. 291,000 pounds of DEHP were released in 1997 from industrial sources and much more substantial quantities, possibly in the tens or even hundreds of thousands of tons, are now utilized in the production of a wide variety of plastic consumer products and PVC piping.

Transport Vectors: DEHP binds strongly to soil and particulates, not dissolving in air or water readily. As a result, much like PBDEs, it is found bound to house dust and particulate matter. Due to its proliferation in consumer products, it reaches humans through a number of routes including inhalation, ingestion, dermal absorption from clothing, from medical equipment, and by leaching into soil and ground water. It can also be transported in the global atmospheric water cycle once adsorbed onto particulate matter. It was found in 77% of subjects in an EWG study. [1][7] DEHP is now an ubiquitous worldwide contaminant in abiotic and biotic media.

Source Points:	Contamination points:	Sediments:	Human blood:
Plastic Toys, 2000: 11-19% [6] **Breast milk:** 0-110 ug/kg [8]	**Plasma from blood bag:** 4.3 – 1,230 ppm [3] **House Dust:**640 ug/mg[5]	**Passaic River, NJ:** 960-27,000 ug/kg[4]	**Serum, geometric mean:** 79.6 ug/g lipid weight [7]

Health Effects: Most of the studies on the health effects of DEHP have been on rats and mice given high amounts of DEHP and/or prolonged exposures. However, absorption and breakdown of DEHP in humans is different than in rats or mice, so the effects seen in rats and mice may not occur in humans. Brief oral exposure to very high levels of DEHP damaged sperm in mice, which was reversed when exposure ceased, but sexual maturity was delayed. High amounts of DEHP damaged the liver of rats and mice, but whether or not DEHP contributes to human kidney damage is still unclear. Small children can be exposed to DEHP by sucking on or skin contact with plastic toys and pacifiers that contain it. In pregnant rats and mice exposed to high amounts of DEHP, researchers observed birth defects and fetal deaths. [9] The US DHHS has determined that DEHP may reasonably be anticipated to be a human carcinogen and the EPA has determined that DEHP is a probable human carcinogen, based on liver cancer in the studies on rats and mice. The International Agency for Research on Cancer (IARC) has stated that DEHP cannot be classified as to its carcinogenicity to humans.

Citations:

1. Agency for Toxic Substances and Disease Registry. (2002). *Toxicological profile for DI(2-ETHYLHEXYL)PHTHALATE*. Atlanta: US Department of Health and Human Services.
2. Kruopienė, Jolita and Šemetienė, Dorina. (2004). *Use and Substitution of DEHP in Lithuanian Furniture Industry*. Kaunus: University of Technology, Institute of Environmental Engineering. http://www1.apini.lt/includes/getfile.php?id=191
3. Vessman J, Rietz G. (1974). Determination of di(ethylhexyl)phthalate in human plasma and plasma proteins by electron capture gas chromatography. *Journal of Chromatography*. 100. pg. 153-163.
4. Iannuzzi TJ, Huntley SL, Schmidt CW, et al. (1997). Combined sewer overflows (CSOs) as sources of sediment contamination in the lower Passaic River, New Jersey. I. Priority pollutants and inorganic chemicals. *Chemosphere*. 34(2). pg. 213-231.
5. Øie L, Hersoug L-G, Madsen JO. (1997). Residential exposure to plasticizers and its possible role in the pathogenesis of asthma. *Environmental Health Perspectives*. 105(9). pg. 972-978.
6. Stringer R, Labunska I, Santillo D, et al. (2000). *Concentrations of phthalate esters and identification of other additives in PVC children's toys. Environmental Science and Pollution Research International*. 7(1). pg. 27-38.

7. http://www.ewg.org/sites/humantoxome/chemicals/chemical.php?chemid=70005

8. FDA. (2001). *Safety assessment of di(2-ethylhexyl)phthalate (DEHP) released from PVC medical devices*. Rockville, MD: Center for Devices and Radiological Health, U.S. Food and Drug Administration. http://www.fda.gov/cdrh/ost/dehp-pvc.pdf.

9. CDC, ATSDR; www.atsdr.cdc.gov/tfacts9.html#bookmark05

Category: Organochlorides (Chlorinated Hydrocarbons)

Chemical: Heptachlor, $C_{10}H_5Cl_7$

Sources: Heptachlor was a widely used pesticide prior to American and European industries voluntarily discontinuing its production, except for termite and fire ant control operations, in 1982. As one of the POPs on the EPA's "dirty dozen" list of the most significant ecotoxins, it is extremely persistent and now contaminates food webs throughout the biosphere due to its stability and bioavailability. Heptachlor was found in 13 of 27 subjects in an EWG study.[1][2] Large quantities of heptachlor are still produced by countries outside of the European Union and United States. A volatile organic compound, heptachlor is transported throughout the global atmospheric water cycle.

Transport Vectors: Heptachlor's widespread use and extreme persistence make it a commonly encountered chemical found in food, soil, in high concentrations in tap water near waste sites, in lower concentrations throughout the hydrosphere, and in the air in treated homes. Some organisms break down heptachlor to heptachlor epoxide, a similar toxin. Heptachlor epoxide has a half-life of at least 4 years in water. [4][1]

Treated Homes ($\mu g/m^3$):	Water:	Food Products:	Human Fat tissue:
Prior: 0.5 **During Treatment**: 5 **After 24 hours**: 2.0 **After 180 Days**: 2.0	Lower Mississippi Delta, 1997: 10 ng/g [6]	Organically grown Buenos Aires Tomatoes, 2003: <10 ng/g dry weight [3]	Victoria, Australia women's breast milk, 1997: .007 mg/kg [7]

Health Effects: An MRL of 0.0006 mg/kg/day has been derived for acute-duration oral exposure to heptachlor. Liver damage, excitability, and decreases in fertility have been observed in animals ingesting heptachlor. The effects are worse when the exposure levels were high or when exposure was prolonged. Although there is very little information on heptachlor epoxide, the ATSDR asserts it is likely that similar effects would also occur after exposure to this compound. Lifetime exposure to heptachlor resulted in liver tumors in animals. The International Agency for Research on Cancer (IARC) and the EPA have classified heptachlor as a possible human carcinogen and the EPA also considers heptachlor epoxide as a possible human carcinogen. [1]

Animals exposed to heptachlor during gestation and infancy are sensitive to heptachlor and heptachlor epoxide; changes in nervous system and immune function were found in these animals. Exposure to higher doses of heptachlor in animals can also result in decreases in body weight and death in newborn animals.

Citations:

1. Agency for Toxic Substances and Disease Registry. (2007). *Toxicological profile for heptachlor*. Atlanta: US Department of Health and Human Services; http://www.atsdr.cdc.gov/tfacts12.html#bookmark05

2. http://www.ewg.org/sites/humantoxome/chemicals/chemical.php?chemid=100326

3. Gonzalez M, Miglioranza KSB, Aizpun De Moreno JE, et al. 2003. Occurrence and distribution of organochlorine pesticides (OCPs) in tomato (*Lycopersicon esculentum*) crops from organic production. *Journal of Agricultural Food Chemistry*. 51. pg. 1353-1359.

4. Eichelberger, JW, Lichtenberg JJ. (1971). Persistence of pesticides in river water. *Environmental Science and Technology*. 5. pg. 541-544.

5. Kamble ST, Ogg CL, Gold RE, et al. 1992. Exposure of applicators and residents to chlordane and heptachlor when used for subterranean termite control. *AECT Publications*. 22. pg. 253-259.

6. Zimmerman LR, Thurman EM, Bastian KC. 2000. Detection of persistent organic pollutants in the Mississippi Delta using semipermeable membrane devices. *Science of theTotal Environment*. 248. pg. 169-179.

7. Sim M, Forbes A, McNeil J, et al. 1998. Termite control and other determinants of high body burdens of cyclodiene insecticides. *AECT Publications*. 53(2). pg. 114-121.

Category: Organochloride (Chlorinated Hydrocarbons)

Chemical: Hexachlorobenzene, C_6Cl_6

Sources: Hexachlorobenzene, a chemical on the EPA's "dirty dozen" list, was a seed-treatment fungicide, which is now banned under the Stockholm Convention. It is also produced as a byproduct of chemical production, including solvents and pesticides. It was also applied directly to seeds of onions, wheat, and other grains to prevent fungal infection and is used in the manufacture of various consumer products ranging from fireworks and ammunition to rubber.

Transport Vectors: Hexachlorobenzene is a long-lived ecotoxin, which accumulates in lipid tissue and is biomagnified in higher trophic levels of the food chain. It is not water soluble but is highly fat soluble (lipophilic). Its half-life in soil is 3-6 years and 0.63 to 6.28 years in air. While its production has been banned in the United States and European Union, annual average uptake is still estimated at 1 ug/kg via food and 0.01

ug/kg via inhalation. It was found in all 3,979 people in a 2002 CDC study. Populations who consume large amounts of seafood register the highest levels of this chemical in blood. [1][2] Hexachlorobenzene has been found in at least 84 of the 1,430 National Priorities List sites identified by the Environmental Protection Agency (EPA).

Water	Marine Life	Humans	At-risk groups
Uncontrolled Waste site, Bayou Baton Rouge, 1986: 8,100,000 ppb [3] **France Rainfall Rural, 1996:** 2.5 – 4.5 ng/L [4]	**Porpoise Blubber:** 223 – 1,070 ng/g wet [5] **Texas Sea Catfish:** 913 **Louisiana Catfish:** 202 [6]	**US National Survey:** 0 – 1,300 ppb **California womens' Breast tissue, ng/g:** 14 – 170 [9]	**Breast milk, ug/g in Quebec Inuits:**.136 **Quebec Whites:** .028 [7] **German pregnant women, active smokers, ng/mL:** .87 **Passive smokers:**.55 **Nonsmokers:** .46 [8]

Health Effects: Exposure to hexachlorobenzene occurs primarily from eating contaminated food. The main health effect from eating contaminated food is liver disease. The US Department of Health and Human Services (DHHS) has determined that hexachlorobenzene may reasonably be expected to be a carcinogen. Studies in animals show that eating hexachlorobenzene for a long time can damage the liver, thyroid, nervous system, bones, kidneys, blood, and immune and endocrine systems. [10]

Citations:

1. Agency for Toxic Substances and Disease Registry. (2002). *Toxicological profile for hexachlorobenzene*. Atlanta: US Department of Health and Human Services.

2. http://www.ewg.org/sites/humantoxome/chemicals/chemical.php?chemid=20001

3. Davis BD, Morgan RC. (1986). Hexachlorobenzene in hazardous waste sites. *IARC Science Publications No. 77*. pg. 23-30.

4. Chevreuil M, Garmouma M, Teil MJ, et al. (1996). Occurrence of organochlorines (PCBs, pesticides) and herbicides (*Triazines, phenylureas*) in the atmosphere and in the fallout form urban and rural stations of the Paris area. *Science of the Total Environment*. 182. pg. 25-37.

5. Becker PR, Mackey EA, Demiralp R, et al. (1997). Concentrations of chlorinated hydrocarbons and trace elements in marine mammal tissues archived in the U.S. National Biomonitoring Specimen Bank. *Chemosphere*. 34. pg. 2067-2098.

6. EPA. (1992b). *National study of chemical residues in fish; Guidance for assessing chemical contaminants data for use in fish advisories*. 2 Volumes: EPA-823-R-92-008a and -008b. U.S. Environmental Protection Agency, Office of Water, Office of Science and Technology. Washington, DC.

7. Dewailly E, Ayotte P, Bruneau S, et al. (1993). Inuit exposure to organochlorines through the aquatic food chain in arctic Quebec. *Environmental Health Perspectives*. 101(7). pg. 618-620.

8. Lackmann GM, Angerer J, Tollner U. (2000). Parental smoking and neonatal serum levels of polychlorinated biphenyls and hexachlorobenzene. *Pediatric Research*. 47(5). pg. 598-601.

9. Petreas M, She J, Visita P, et al. (1998). Levels of PCDD/PCDFs, PCBs and OC pesticides in breast adipose of women enrolled in a California breast cancer study. *Organohalogen Compounds*. 38. pg. 37-40.

10. http://www.atsdr.cdc.gov/tfacts90.html.

Category: Heavy Metals

Chemical: Lead, Pb

Sources: Lead is a naturally occurring heavy metal, averaging around 15-20 mg/kg throughout Earth's crust. Industrially, it is used primarily for battery production, soldering materials, as a gasoline additive, in the manufacture of pesticides, in numerous consumer products, such as Tonka toys, and in the production of corrosion-resistant building products, such as paint and pipes.

Transport Vectors: Lead does not degrade and is absorbed into soil; use of lead, a naturally-occurring ecotoxin in a wide variety of industrial and consumer product applications means that concentrations of lead will not decrease naturally over time. While atmospheric levels have dropped since the banning of leaded gasoline in the US, a great deal continues to be released from commercial and industrial sources. [1] The US produced just under two million tons of lead a year from 1999 to 2003. [2] It was found in 13,641 of 14,333 subjects in the CDC's biomonitoring studies. [5]

Source Points	Homes (ng/m^3)	Products	Humans (ug/dL, mean)
House Dust mean: 467.4 ug/g **Max:** 30,578 ug/g[3]	**Built before 1940:** 46 **1960-1979:** 13 [3]	**Hair Dye:** 2,300 – 6,000 ug/g [4]	**Mechanics:** 4.80 **Health Service:** 1.76 **Average:** 2.42

Health Effects: Elevated levels of lead in children are considered to be those over 10 ug (micrograms) per dL (deciliter) of blood. "No safe blood level in children has been identified." (CDC 2005).

Lead has been shown to cause a wide variety of health effects. Many of the effects have been known since ancient times, although some of the more subtle effects have been discovered only recently. [7]

Acute symptoms occur relatively soon after a serious exposure and can include: metallic taste, stomach pain, vomiting, diarrhea, and black stools. Severe exposure can cause nervous system damage, intoxication, coma, respiratory arrest, and even death.

Chronic effects take time to develop and are often attributed to low exposures that accumulate to a toxic level over a period of time. The symptoms often seen with significant long-term exposure include: loss of appetite, constipation, nausea, and stomach pain. Other symptoms can include: excessive tiredness, weakness, weight loss, insomnia, headache, nervous irritability, fine tremors, numbness, dizziness, anxiety, and hyperactivity. Because these symptoms are common to a variety of health problems, it is difficult to attribute the etiology of disorders, such as hyperactivity in children, to lead or another ecotoxin, such as methylmercury.

Citations:

1. Agency for Toxic Substances and Disease Registry. (2007). *Toxicological profile for lead.* Atlanta: US Department of Health and Human Services.

2. http://www.ilzsg.org/static/statistics.aspx?from=1

3. Bonanno LJ, Freeman NCG, Greenburg M, et al. (2001). Multivariate analysis on levels of selected metals, particulate matter, VOC, and household characteristics and activities from the midwestern states NHEXAS. *Applied Occupational Environmental Hygiene.* 16(9). pg. 859-874.

4. Cohen AJ, Roe FJC. (1991). Review of lead toxicology relevant to the safety assessment of lead acetate as a hair colouring. *Food and Chemical Toxicology.* 29(7). pg. 485-507.

5. http://www.ewg.org/chemindex/term/455

6. Yassin AS, Martonik JF, Davidson JL. (2004). Blood lead levels in U.S. workers, 1988-1994. *Journal of Occupational Environmental Medicine.* 46. pg. 720-728.

7. DOT Federal Highway Administration (FHWA); http://www.tfhrc.gov/hnr20/bridge/model/health/health.htm

---

Category: Heavy Metals

Chemical: Mercury, Hg

Sources: Mercury is found primarily in three forms, metallic (elemental) mercury, inorganic, and organic mercury. Elemental mercury is the familiar shiny liquid found in a number of consumer products such as thermometers. Inorganic mercury occurs when mercury is bound to an inorganic element, as is sometimes done for fertilizer production. Organic mercury (methylmercury) is produced by microbial or fungal action and is the most dangerous form to humans. Mercury occurs in small amounts naturally, but industrial activity, especially the burning of coal, releases large quantities of mercury sulfide, which is converted to methylmercury and now accounts for 33%-66% of Earth's ambient mercury levels.

Transport Vectors: 80% of anthropogenic mercury is released in the form of elemental mercury (mercuric sulfide) from fossil fuel combustion (including power plants), mining, and smelting, 5% is released in industrial wastewater, and 15% is derived from consumer products containing mercury (batteries, thermometers). It is then converted to methylmercury by microorganisms and bioaccumulates in the food chain. Seafood consumption greatly increases exposure and several states have issued warnings concerning mercury toxicity, especially in pregnant women. Historically, many consumer products have contained mercury and resulted in cases of mercury poisoning, including: over the counter lice medication, skin-lightening cosmetic cream, artists' paints, imported Chinese and Indian herbal remedies, and nasal drops. [1] Mercury was detected in every subject of the ACHM 2007 study; 15% were over the federal risk level for fetal brain development. [2] It was found in 70 of 71 of EWG's participants. [5]

Mercury was found in nearly 50 percent of tested samples of commercial high fructose corn syrup (HFCS), according to an article published in *Environmental Health*. A separate study by the Institute for Agriculture and Trade Policy (IATP) [10] detected mercury in nearly one-third of 55 popular brand name food and beverage products where HFCS is the first or second ingredient, including products by Quaker, Hershey's, Kraft and Smucker's.

HFCS has replaced sugar in many processed foods and is now found in sweetened beverages, breads, cereals, breakfast bars, lunch meats, yogurts, soups and condiments. On average, Americans consume about 12 teaspoons per day of HFCS. Consumption by teenagers and other high consumers can be up to 80 percent above average levels.

"Mercury is toxic in all its forms," said IATP's David Wallinga, M.D., and a co-author in both studies. [10] "Given how much high fructose corn syrup is consumed by children, it could be a significant additional source of mercury never before considered. We are calling for immediate changes by industry and the FDA to help stop this avoidable mercury contamination of the food supply."

Source Points	Fish, 1998	Umbilical cords	Human hair, 2007
Stauffer Chemical Alabama Site: 4.3 – 316 ppm [3] **Imported Chinese herbs:** 540,000 ppm [4]	**Largemouth Bass:** 0 - 8.94 ppm **Smallmouth Bass:** .08 – 5.0 **Perch:** 0 – 3.15 [8]	**1976 Iowa, mean:** 1.24 ppb [6] **1995, Canadian aborigines**: 21.8% >20 ppb, **high:** 224 ppb [7]	**Maine range:** 186 – 1180 ppb **median:** 396 ppb **National median:** 200 ppb [2]

Health Effects: The level of mercury concentration associated with neurodevelopmental effects in the fetus is 58 ug dL. Data from the *Third National Report* for the period 1999-2002 [11] show that all women of childbearing age had levels below 58 micrograms per liter (µg/L), a concentration associated with neurodevelopmental effects in the fetus. However, mercury levels in these women merit close monitoring because 5.7% of women of childbearing age had levels within a factor of 10 of those associated with neurodevelopmental effects. Defining safe levels of mercury in blood continues to be an active research area (see *Appendix O and P.*)

Acute (short-term) exposure to high levels of elemental mercury in humans results in central nervous system (CNS) effects such as tremors, mood changes, and slowed sensory and motor nerve function. Chronic exposure to elemental mercury in humans also affects the CNS, with effects such as increased excitability, irritability, excessive shyness, and tremors.

Acute exposure to inorganic mercury by the oral route may result in effects such as nausea, vomiting, and severe abdominal pain. The major effect from chronic exposure to inorganic mercury is kidney damage. Mercuric chloride (an inorganic mercury compound) exposure has been shown to result in forestomach, thyroid, and renal tumors in animal exposure studies.

Acute exposure of humans to very high levels of methylmercury results in CNS effects such as blindness, deafness, and impaired level of consciousness. Chronic exposure to methylmercury in humans also affects the CNS with symptoms such as paresthesia (a sensation of pricking on the skin), blurred vision, malaise, speech difficulties, and constriction of the visual field. Methylmercury exposure from ingestion has led to significant developmental effects. Infants born to women who ingested high levels of methylmercury exhibited mental retardation, ataxia, constriction of the visual field, blindness, and cerebral palsy. [9]

Citations:

1. Agency for Toxic Substances and Disease Registry. (1999). *Toxicological profile for mercury*. Atlanta: US Department of Health and Human Services.
2. Schmitt, Catherine, Mike Belliveau, Rick Donahue, and Amanda Sears. (2007). *Body of evidence - a study of pollution in Maine people*. Alliance for a Clean and Healthy Maine. Portland, ME.
3. Hayes LC, Rodenbeck SE. (1992). Developing a public-health assessment: Impact of a mercury contaminated discharge to surface-water. *Journal of Environmental Health*. 55(2). pg. 16-18.
4. Perharic L, Shaw D, Colbridge M, et al. (1994). Toxicological problems resulting from exposure to traditional remedies and food supplements. *Drug Safety*. 11(6). pg. 284-294.
5. http://www.bodyburden.org/chemicals/chemical.php?chemid=30002.
6. Pitkin RM, Bahns JA, Filer LJ, et al. (1976). Mercury in human maternal and cord blood, placenta, and milk. *Proceedings of the Society for Experimental Biology and Medicine*. 151. pg. 565-567.
7. Wheatley B, Paradis S. (1995). Exposure of Canadian Aboriginal people to methylmercury. *Water, Air, and Soil Pollution*. 80. pg. 3-11.

8. NESCAUM. (1998). *Northeast states and eastern Canadian provinces - mercury study - a framework for action.* Northeast States for Coordinated Air Use Management. Boston, MA.

9. EPA; http://www.epa.gov/ttn/atw/hlthef/mercury.html.

10. Institute for Agriculture and Trade Policy (IATP). (January 26, 2009). *Brand-name food products also discovered to contain mercury.*

11. Centers for Disease Control and Prevention. (2005). *Third national report on human exposure to environmental chemicals.* CDC, Atlanta, GA. http://www.cdc.gov/exposurereport/default.htm.

Category: Aromatic Hydrocarbons

Chemical: Polycyclic Aromatic Hydrocarbons (PAH)

Sources: PAHs are a group of various chemicals that result from combustion of carbon containing materials such as coal, oil, gasoline, wood, garbage, tobacco, and charbroiled food. They are an extremely widespread substance and also occur naturally. The ATSDR isolated a specific group of 17 considered more dangerous and prevalent than the rest.

Transport Vectors: Most PAHs enter the air from a source of combustion and are transported both as vapors and as particulates. They can also move into groundwater, are lipid soluble, and biomagnify readily up the food chain. Several types have been found in 100% of the EWG's test subjects. [1][2]

Avg. Soil, ug/kg*	Sediment	Seafood	Human Fat Tissue
Rural: 128.3 – 1674.7 **Urban:** 25,190 – 582,860 [1]	**1993 Boston Harbor:** 900 – 33,000 ug/kg dry[3]	**Mussels, over 3 years:** n/d – 15,200 [4]	**Cancer-Free corpses:** 11 – 2,700 ppb (ng/g) [5]

*composite of 17 PAHs

Health effects: PAHs may be attached to dust or ash, causing lung irritation and damage. Skin contact may cause redness, blistering, and peeling. The liver and kidneys can also be damaged by exposure to PAHs. Benzo(a)pyrene, a common PAH, is shown to cause lung and skin cancer in laboratory animals. Extracts of various types of smoke containing PAHs caused lung tumors in laboratory animals and the US Surgeon General long ago determined that cigarette smoke causes lung cancer. Reproductive problems and prenatal developmental problems have occurred in laboratory animals exposed to benzo(a)pyrene. [6]

Citations:

1. Agency for Toxic Substances and Disease Registry. (1995). *Toxicological profile for Polycyclic aromatic hydrocarbons.* US Department of Health and Human Services. Atlanta.

2. http://www.bodyburden.org/chemicals/chemical_classes.php?class=Polyaromatic+hydrocarbons+(PAHs).

3. Demuth S, Casillas E, Wolfe DA, et al. (1993). Toxicity of saline and organic solvent extracts of sediments from Boston Harbor, Massachusetts and the Hudson River-Raritan Bay Estuary, New York using the Microtox bioassay. *Archives of Environmental Contamination and Toxicology.* 25. pg. 377-386.

4. NOAA. (1989). A summary of data on tissue contamination from the first three years (1986-1988) of the Mussel Watch Program. NOAA Technical Memorandum NOS OMA 49. National Oceanic and Atmospheric Administration. Rockville, MD.

5. Obana H, Hori S, Kashimoto T, et al. (1981). Polycyclic aromatic hydrocarbons in human fat and liver. *Bulletin of Environmental Contamination and Toxicology.* 27. pg. 23-27.

6. Wisconsin Department of Health Services; http://dhs.wisconsin.gov/eh/chemfs/fs/PAH.htm.

Category: Brominated Flame Retardants

Chemical: PBDEs, $C_{12}H_{10-x}Br_xO$

Sources: The PBDE family of chemicals consists of 209 cogeners, each with differing numbers of attached bromine ions. These chemicals are used as flame retardants in a variety of consumer products ranging from plastic television cases, computers, cell phones, fax machines, printers, and scanners to textile fibers in consumer products, such as couches, cushions, mattresses, and pillows. In countries or states where only one type of PBDE is banned, others are generally used in its place. In the European Union, the use of all PBDEs is entirely banned. [2] Worldwide production of PBDEs from BFR has recently been reported as 203,790 tons. [2]

Transport Vectors: 49 million pounds of DecaPBDE was added to consumer products in North America in 2001. Products containing PBDEs slowly release them as particulates, including airborne particulates, where they are often converted to more toxic forms by sunlight. Due to their prevalence in consumer products, they are ubiquitous throughout America in quantities 10 to 40 times those found in Europe or Japan, illustrating the wide range of concentration in human subjects from region to region. A positive correlation has been shown between working with electronics disassembly and PBDE levels. Sweden's PBDE levels rose exponentially from 1972 to 1996, then began a gradual decline, possibly due to a voluntary ban in the early 1990's.

[3] None of these studies includes all of the PBDE cogeners and most generally focus on the most common cogeners, meaning that total PBDE levels in all samples are actually higher than indicated. PBDEs have low water and high lipid solubility. [2] BDE-17, a cogener not found at source points, was detected in the air on a Baltic Sea island, indicating that PBDEs are transformed as they move through the environment. Studies have found food is responsible for up to 93% of human exposure to PBDEs. [3]

Environmental	Food	House Dust	Humans
China: 4434-16088 ng/g [7] **Spain:** 30-14395 ng/g [6] **USA, Great Lakes:** 1.7-4 ng/g [5]	(US Levels) **Fish:** 8.5 – 3078 ppt **Meat:** .2 – 1373 ppt **Dairy:** .9 – 679 ppt [4]	**Singapore:** .11-13 ug/g [8] **USA (alt. study):** .78-30 ug/g [9]	**NYC, fat tissue, ng/g Mean:** 399 **Min/max:** 17-9,630 [10] **US Breast milk:** 9 – 1,078 ng/g [11]

Health Effects: Research has shown that low levels of exposure to PBDEs causes permanent disturbances in behavior, memory, and learning in young mice. PBDEs have also been shown to disrupt the thyroid hormone system, a crucial part of the development of the brain and body. [12] PBDEs are neurotoxic and persistent in humans and their environment. [13] They are known endocrine system disrupters, a probable cause of birth defects and developmental delays, and a possible cause of cancer and liver dysfunction. [13] A report conducted jointly by researchers at the EPA National Health and Environmental Effects Laboratory and Indiana University and published in Environmental Science & Technology suggests that many cases of feline hyperthyroidism are associated with exposure to environmental contaminants such as PBDEs. The feline hyperthyroidism epidemic began in the late 1970s, paralleling the introduction of PBDEs. Early cases were first noted in the United States, but other cases have been diagnosed in Canada, Australia, Japan, and Europe.

> The growing use of PBDEs in consumer products over the past 30 years has paralleled the increase in feline hyperthyroidism, and a preliminary study indicates that higher levels are found in cats stricken with this disease… In addition to PBDEs, hyperthyroidism in cats could be linked to the plastics chemical and potent endocrine disruptor BPA that is known to leach from the pop-top cat food lining. (www.ewg.org/node/26238)

Most pet cats are indoor animals and research suggests that their exposure to household toxins is similar to that experienced by a 2 year old child. Cats ingest PBDE in the dust that coats their fur while grooming. Cats are also small animals with a small blood volume; a moderate dose of PBDE dissolved in the blood volume of a cat results in the more rapid accumulation of ecotoxins than would occur in a larger animal, disrupting the normal production and metabolism of thyroid hormones, ultimately resulting in hyperthyroidism.

PBDE is also linked to malignancies and birth defects, and some current studies are showing an association between endocrine disruptors and autism in children. PBDEs and their many cogeners are common and widespread in the biosphere and one of the most biologically significant of all "emerging" environmental threats. Closely related to PCBs, the huge irretrievable inventories of PBDEs in the environment will be a threat to human health for centuries. [15]

> The world's oceans are a sink and reservoir for the most toxic POPs, including PBDEs. With increasing coastal development, environmental deterioration from land-to sea pollution by POPs has become a critical issue for the regulatory community worldwide. Marine mammals such as seals, dolphins, and whales accumulate large amounts of these compounds and store them in their ample layers of blubber. Not surprisingly, numerous health problems have surfaced in sea mammals since in the early 1970s when Baltic seals with high tissue levels of PCBs and DDE were found to have skeletal deformities and uterine stenosis (closure of the uterine canal), causing widespread infertility and population declines. Since that time, seals in polluted waters of Europe, Asia, and North America have suffered from a wide range of health problems including large-scale disease outbreaks and mass mortalities. At the top of the ocean food chain, marine mammals such as seals are important sentinels for the health of the ocean and for pollutant-related effects in people. Despite obvious species-related differences, seals are "real world" wildlife models for humans: they are long-lived, biologically similar to people, attain similar body weights as adults (~150-220 lbs), eat many of the same fish (major items in the seal diet are hake, herring, cod, redfish, butterfish), and breast-feed their pups. Over our lifetimes, we are exposed to similar dietary "cocktails" of toxic contaminants, pass them on in large quantities to our young in the womb and in breast milk, and show similar responses to toxic chemicals.[16]

A recent study conducted by the Marine Environmental Research Institute (MERI) research revealed that Atlantic coast harbor seals carry an average of 65 pollutants in their bodies including 55 organic chemicals and 10 heavy metals. PBDEs are one of the most biologically significant of these pollutants (www.meriresearch.org/PDF/Toxins.pdf).

"Together PCBs and PBDEs deliver a potent one-two punch… They lock onto thyroid hormone receptors, starving animals of thyroid hormone."[18]

Citations:

1. Schmitt, Catherine, Mike Belliveau, Rick Donahue, and Amanda Sears. (2007). *Body of evidence - a study of pollution in Maine people*. Alliance for a Clean and Healthy Maine. Portland, ME.

2. Teclechiel, Daniel. (2008). Synthesis and characterization of highly polybrominated diphenyl ethers. PhD diss., Stockholm University. http://www.diva-portal.org/su/theses/abstract.xsql?dbid=7410.

3. Washington State Department of Health. (2006). *Washington state polybrominated diphenyl ether (pbde) chemical action plan: final plan.* Department of Ecology Publications Distribution Office. Olympia, WA. www.cleanproduction.org/library/WA%20PBDE%20Chem%20Action%20Plan%202006.pdf.

4. Schector, Arnold; Päpke, Olaf; Tung, Kuang-Chi; Staskal, Daniele; and Birnbaum, Linda. (2004). Polybrominated Diphenyl Ethers Contamination of United States Food. *Environmental Science &. Technology.* http://pubs.acs.org/subscribe/journals/esthag-a/38/free/es0490830.html.

5. Song, W., Li, A., Ford, J.C., Sturchio, N.C., Rockne, K.J., Buckley, D.R., and Mills, W.J. (2005). Polybrominated Diphenyl Ethers in the Sediments of the Great Lakes. 2. Lakes Michigan and Huron. *Environmental Science &. Technology.* 39 pg. 3474-3479.

6. Eljarrat, E., Marsh, G., Labandeira, A., and Barcelo, D. (2007). Effect of sewage sludges contaminated with polybrominated diphenylethers on agricultural soils. *Chemosphere.*

7. Luo, Q., Cai, Z.W., and Wong, M.H. (2007). Polybrominated diphenyl ethers in fish and sediment from river polluted by electronic waste. *Science of the Total Environment.* 383. pg. 115-127.

8. Tan, J., Cheng, S.M., Loganath, A., Chong, Y.S., and Obbard, J.P. (2007). Polybrominated diphenyl ethers in house dust in Singapore. *Chemosphere.* 66. pg. 985-992.

9. Stapleton, H.M., Dodder, N.G., Offenberg, J.H., Schantz, M., and Wise, S.A. (2005). Polybrominated diphenyl ethers in house dust and clothes dryer lint. *Environmental Science &. Technology.* 39. pg. 925-931.

10. Johnson-Restrepo B. (2005). Polybrominated diphenyl ethers and polychlorinated biphenyls in human adipose tissue from New York. *Environmental Science and Technology.* 39(14). pg. 5177-5182.

11. Lunder, Sonya and Sharp, Renee. (2003). *Toxic fire retardants (PBDEs) in human breast milk.* Environmental Working Group; http://www.ewg.org/files/MothersMilk_final.pdf.

12. Friends of the Earth. (July 18, 1999). *Brominated flame retardants contaminate blood.*

13. EWG, Human Toxome Project; http://www.ewg.org/sites/humantoxome/chemicals/chemical_classes.php?class=Polybrominated+diphenyl+ethers+%28PBDEs%29.

14. http://autoimmunedisease.suite101.com/article.cfm/feline_hyperthyroidism.

15. Moore, Elaine. (8/18/07). *Flame retardant chemicals linked to Graves' Disease in cats*; http://autoimmunedisease.suite101.com/article.cfm/feline_hyperthyroidism.

16. Naidenko, Olga, Sutton, Rebecca and Houlihan, Jane. (2008). *High levels of industrial chemicals contaminate cats and dogs.* Environmental Working Group. http://www.ewg.org/reports/pets.

17. MERI. *Fact sheet. Toxic pollutants: The human-marine wildlife connection.* www.meriresearch.org/PDF/Toxins.pdf.

18. McGrath, Susan. Pandora's water bottle. (March-April 2010). *Audubon.* pg. 37. http://www.audubonmagazine.org/currents/currents1003.html.

Category: Organochlorines (Chlorinated Hydrocarbons)

Chemical: Pentachlorophenol, C_6HCl_5O

Sources: Since its synthesis in the 1930s, pentachlorophenol, or PCP, has been used as a pesticide, a disinfectant, and as a leather, wood, rope, and masonry preservative. Despite its ban for commercial use in the early 1980s, it is currently still used in railroad ties and utility poles. When incinerated or subject to other modes of combustion, as with other chlorinated hydrocarbons, the creation of dioxin-furan byproducts is a frequent occurrence.

Transport Vectors: PCP was one of the most widely used pesticides in the US and was discharged in the millions of pounds into the air and soil per year during peak use. PCP is a highly volatile ecotoxin and once released as a gas it is rapidly transported by the global atmospheric water cycle as a microcontaminant in rainfall events. It is lipid soluble and bioconcentrates, especially in aquatic life. Most human exposure is through ingestion, respiration, or direct contact; as a result, occupationally exposed populations show much higher contamination levels than populations not subject to chronic exposure. It was found in 794 of 5,022 of the CDC biomonitoring study's subjects. [1][2]

Source Points	Remote Regions	Seafood	Human blood, ug/L
Soil Sample at wood-treatment factory: 820 – 200,000 ug/kg [3]	Yellowknife, Canadian NW Territories: 0.43 – 3.68 ng/m³ [4]	Nanjing, China Market: 0-61 ug/kg wet weight [5]	Occupationally Exposed group: 1,273 Non-exposed: 15.3[6]

Health Effects: PCP has been shown to negatively affect organisms in soil and water at relatively low concentrations. Algae appear to be the most sensitive aquatic organisms; as little as 1 µg/liter can cause significant inhibition of the most sensitive algal species. Most aquatic invertebrates (annelids, mollusks, crustacea) and vertebrates (fish) are affected by PCP concentrations below 1 mg per liter in acute toxicity tests, with reproductive and juvenile stages being the most sensitive. Low levels of dissolved oxygen, low pH, and high temperature increase the toxic effects of PCP. Concentrations causing effects on fish are in the low µg/liter range; PCP contamination in many surface waters is in this range. PCP is bioaccumulated by aquatic organisms. Fresh-water fish show bioconcentration factors of up to 1000 compared to < 100 in oceanic fish; the amount of PCP taken up, either through the surrounding water or along the food chain, is probably species-specific. PCP taken up by terrestrial plants remains in the roots and is partly metabolized.

PCP enters humans by ingestion, respiration, and skin contact, after which it is distributed in the tissues. The highest concentrations have been observed in the liver and kidneys, with lower levels found in body fat, brain tissue, and muscle.

Some animal data indicate that there may be long-term accumulation and storage of small amounts of PCP in human beings. The fact that urine- or blood-PCP levels do not completely disappear in some occupationally exposed people, even long after exposure, seems to confirm this, though the biotransformation of hexachlorobenzene and related compounds provides an alternative explanation of this phenomenon. However, there is a lack of data concerning the long-term fate of low PCP levels in animals as well as in man and no data are available on the accumulation and effects of microcontaminants taken up by people along with PCP.

PCP is highly toxic, regardless of the route, length, and frequency of exposure. There is limited evidence that the most dangerous route of exposure to PCP is through the air as PCP is an irritant for exposed epithelial tissue, especially the mucosal tissues of the eyes, nose, and throat.

Several epidemiological studies from Sweden and the USA have indicated that occupational exposure to mixtures of chlorophenols is associated with increased incidence of soft tissue sarcomas, nasal and nasopharyngeal cancers, and lymphomas. In contrast, surveys from Finland and New Zealand have not detected such relationships. The major deficiency in all of these studies appears to be a lack of specific exposure data. [7]

Citations:

1. Agency for Toxic Substances and Disease Registry. (2001). *Toxicological profile for Pentachlorophenol*. US Department of Health and Human Services. Atlanta.

2. http://www.bodyburden.org/chemicals/chemical.php?chemid=20010

3. ATSDR. (1995). *Health assessment for American Creosote Works, Incorporated, (Winnfield Plant), Winnfield, Winn Parish, Louisiana, region 6. CERCLIS No. LAD000239814*. Government Reports Announcements & Index (GRA&I) No. 13.

4. Cessna AJ, Waite DT, Constable M. (1997). Concentrations of pentachlorophenol in atmospheric samples from three Canadian locations, 1994. *Bulletin of Environmental Contaminant Toxicology*. 58. pg. 651-658.

5. Jiachun Ge, Jianling Pan, Zhiliang Fei, Guanghong Wu, and Giesy, John P. (August 2007). Concentrations of pentachlorophenol (PCP) in fish and shrimp in Jiangsu Province, China. *Chemosphere*. Vol 69. (1). pg. 164-169.

6. Ferreira AG, Vieira DN, Marques EP, et al. (1997). Occupational exposure to pentachlorophenol: The Portuguese situation. Ann NY Acad Sci 837:291-299.International Programme on Chemical Safety (IPCS); www.inchem.org/documents/ehc/ehc/ehc71.htm#SectionNumber:1.5

Category: Perchlorates

Chemical: Perchlorate, HCLO$_4$

Sources: Perchlorates bound to magnesium, potassium, ammonium, sodium, and lithium are used for a variety of pyrotechnic and chemical synthesis applications, the most important of which is the worldwide production of solid fuels for military rockets and satellite launchings. It is also used in Chilean tobacco fertilizer. Due to the military applications of these compounds, most countries keep the production statistics secret so there is no clear way to determine how much is being produced today. [1]

Transport Vectors: Little is known about the environmental persistence of perchlorate, but it appears to have a long chemical and biological half-life and is now a ubiquitous contaminant in public drinking water in most communities in the United States, and in the tissues and blood of most Americans. A major spill of percolates occurred in Nevada in the 1990s, contaminating much of the San Joaquin Valley water supply. Most vegetables, including spinach, grown in the valley after this accident contained significant residues of percolate. It dissolves readily in water but not soil, causing it to be transported through both surface and groundwater. [3] It is also absorbed by plants where it bioaccumulates. It was found in 2,818 of 2,942 subjects in the CDC's monitoring project. [4][7]

Water	Plants	Products	Humans
1999 Sacramento groundwater high: 8,000 ug/L [2]	**1999 Texas Ammunition plant Crabgrass:** 1,060,000-5,557,000 ug/kg [4]	**2004 Romaine Lettuce Mean:** 11.9 ppb [5] **Plug Tobacco:** 2.3-149.3 ppb [6]	**Factory workers' blood, pre-exposure:** 0 - 2 ug/L **Post:** 358.9 – 838.4 ug/L [8]

Health Effects: In recent years, environmental agencies have increasingly found instances of perchlorate contamination in drinking water, groundwater, surface water, and soil. Because of the risks to public health and the environment posed by perchlorate releases, the California Legislature directed the Department of Toxic Substances Control (DTSC) to establish best management practices for the prevention of perchlorate contamination.

Perchlorates interfere with iodide uptake and thyroid hormone production, which may lead to developmental defects. Scientists consider pregnant women, children, infants, and individuals with thyroid disorders to be the populations most at risk from perchlorate exposure. [9] For additional information documenting perchlorates in milk, including breast milk, see *Appendix Z*.

Citations:

1. Agency for Toxic Substances and Disease Registry. (2005). *Draft Toxicological profile for Perchlorates*. US Department of Health and Human Services. Atlanta.
2. Okamoto HS, Rishi DK, Steeber WR, et al. (1999). Using ion chromatography to detect perchlorate. *Journal of the American Water Works Assoc*. 91(10). pg. 73-84.
3. Urbansky ET, Brown SK. 2003. Perchlorate retention and mobility in soils. *Journal of Environmental Monitoring*. 5. pg. 455-462.
4. Smith PN, Theodorakis CW, Anderson TA, et al. (2001). Preliminary assessment of perchlorate in ecological receptors at the Longhorn Army ammunition plant (LHAAP), Karnack, Texas. *Ecotoxicology*. 10. pg. 305-313.
5. FDA. 2004. Exploratory data on perchlorate in food. Food and Drug Administration. http://www.cfsan.fda.gov/~dms/clo4data.html.
6. Ellington JJ, Wolfe NL, Garrison AW, et al. (2001). Determination of perchlorate in tobacco plants and tobacco products. *Environment and Science Technology*. 35(15). pg. 3213-3218.
7. http://www.ewg.org/sites/humantoxome/chemicals/chemical.php?chemid=100377.
8. Braverman LE, He XM, Pino S, et al. (2005). The effect of perchlorate, thiocyanate, and nitrate on thyroid function in workers exposed to perchlorate long-term. *Journal of Clinical Endocrinology and Metabolism*. 90(2). pg. 700-706.
9. CA Department of Toxic Substances Control; www.dtsc.ca.gov/HazardousWaste/Perchlorate/upload/HWM_FS_Perchlorate_7-061.pdf.

Category: Perfluorochemicals

Chemical: Perfluorochemicals, C_xF_x

Sources: PFCs are a family of chemicals used in a number of medical procedures and as a nonstick component of commercial products such as tape and frying pans. Teflon is the most well known consumer product utilizing PFCs. The PFC family of ecotoxins is considered to be one of the most biologically significant emerging environmental chemical groups due to its ubiquitous use in consumer products, many cogeners and metabolites, and its worldwide distribution in all food webs.

Transport Vectors: PFCs present a serious risk as a greenhouse gas due to their extreme persistence, with CF4 having an atmospheric lifetime of 50,000 years. These chemicals proliferated rapidly in biotic media throughout all ecosystems due to the extent of their use in consumer products and the global nature of their distribution, contaminating virtually every species sampled. PFCs have been found in the blood of many species of wildlife around the world, including fish, bald eagles, and mink in the US Midwest, and nearly all people have some PFCs in their blood, regardless of age.

Some PFCs stay in the human body for many years. Multiple types of PFCs were found in all 1,591 subjects in the CDC's biomonitoring project and in all the subjects of the Maine *Body of Evidence* study. Perfluorooctanesulfonyl fluoride (POSF) is of particular concern due to its deleterious health effects on wildlife. [3][4]

Water	Bird eggs, ng/g ww	Seafood	Humans
Fayetteville, NC DuPont plant groundwater high: 765 ppb [6]	**POSF in Maine Kingfisher: 954.76 Eagle: 710.53** [5]	**Fish and Oyster High: 1100 ng/g** [7]	**2007 Blood serum mean: 27.96 ppb** [4] **DuPont Workers: 422 – 1,870 ppb** [6]

Health Effects: The PFC family of chemicals is relatively new and is the focus of active scientific research although there are not yet many studies of health effects in people. In laboratory animal studies, high concentrations of PFCs cause damage in the liver and other organs. Developmental delays and other problems have been seen in the offspring of rats and mice exposed to PFCs while pregnant. Both PFOA and PFOS in high concentrations over a long period of time also cause cancer in laboratory animals. PFBA is not suspected of causing cancer in animals.[8]

Citations:

1. http://www.bodyburden.org/chemicals/chemical_classes.php?class=Perfluorochemicals+(PFCs)
2. http://www.epa.gov/highgwp/scientific.html
3. www.cdc.gov/exposurereport/pdf/factsheet_pfc.pdf
4. Schmitt, Catherine, Mike Belliveau, Rick Donahue, and Amanda Sears. (2007). *Body of evidence - a study of pollution in Maine people*. Alliance for a Clean and Healthy Maine. Portland, ME.
5. Biodiversity Research Institute. (2007). *Preliminary findings of contaminant screening of Maine bird eggs: 2007 Field Season*. BioDiversity Research Institute. Gorham, ME.
6. http://www.cleanwateraction.org/mn/pfcnational.html
7. Chiao-Li Tseng, Li-Lian Liu, Chien-Min Chen and Wang-Hsien Ding. (February 10, 2006). Analysis of perfluorooctanesulfonate and related fluorochemicals in water and biological tissue samples by liquid chromatography–ion trap mass spectrometry. *Journal of Chromatography A*. 1105, (1-2). pg. 119-126.
8. Minnesota Department of Health. (January 2009). *Hazardous substances in Minnesota: perfluorochemicals and health.* www.health.state.mn.us/divs/eh/hazardous/topics/pfcshealth.html

Category: Aromatic Hydrocarbons

Chemical: Phenol, C_6H_5OH

Sources: Phenol and related chemicals, such as xylenol or dimethylphenol, cresol, alkylphenols like nonylphenol, or parabens, are widely used chemicals incorporated in numerous consumer products. Murder by phenol inhalation and/or injection was a common strategy used in the execution of a large percentage of the Jews and others at Auschwitz and other locations during the holocaust. [1][2] The annual production capacity for phenol in 2004 was 6.6 billion pounds. [3]

Transport Vectors: Phenol reaches the public consumer directly in the form of a number of products, including antiseptics, ointments, household cleaners, herbicides, and cough drops. It is also used in a number of industrial processes, which release the phenol family of ecotoxins in water, air, and soil. As a highly volatile organic compound, it is rapidly transported by the global atmospheric water cycle, but is not subject to significant bioaccumulation or biomagnification. It has a short half-life (less than 1 day in air, 2-5 days in soil) but constant production has kept ambient levels high, especially near landfills, waste sites and manufacturing plants. [1]

Source Points:	Products:	Cigarettes (in ug):	Other Sources:
NYC wastewater, 1989-'93: 8-490 ug/L[4]	Soy Curds: 450 – 6,000 ug/kg [5]	Unfiltered: 60-140 Filtered: 19-35 [6]	House Dust: 4 ug/g[6]

Health Effects: Phenols are suspected of causing disruptions in the endocrine system and long-term exposure to phenol has been associated with cardiovascular disease, but the workers studied were also exposed to other chemicals at the same time. Ingestion of liquid products containing concentrated phenol can cause serious gastrointestinal damage and even death. Application of concentrated phenol to the skin can cause severe skin damage. Short-term exposure to high levels of phenol has caused irritation of the respiratory tract and muscle twitching in animals. Longer-term exposure to high levels of phenol caused damaged to the heart, kidneys, liver, and lungs in animals. Drinking water with extremely high concentrations of phenol has caused muscle tremors, difficulty walking, and death in animals.[7]

Citations:

1. Agency for Toxic Substances and Disease Registry. (2006). *Draft toxicological profile for heptachlor*. US Department of Health and Human Services. Atlanta.
2. *Killing through lethal injection: Auschwitz - FINAL STATION EXTERMINATION.* Johannes Kepler University, Linz, Austria. Retrieved on 2006-09-29.
3. CMR. (May 23-29, 2005). Phenol: Chemical profile. *Chemical Market Reporter*. pg. 34-35.

4. Stubin AI, Brosnan TM, Porter KD, et al. (1996). Organic priority pollutants in New York City municipal wastewaters: 1989–1993. *Water Environment Research*. 68. pg. 1037-1044.

5. Chung HY. 1999. Volatile components in fermented soybean (*Glycine max*) curds. *Journal of Agricultural and Food Chemistry*. 47. pg. 26902697.

6. Nilsson A, Lagesson V, Bornehag C-G, et al. (2005). Quantitative determination of volatile organic compounds in indoor dust using gas chromatography-UV spectrometry. *Environment International*. 31. pg. 1141-1148.

7. Agency for Toxic Substances and Disease Registry. (September 2008). *Public health statement for phenol*. www.atsdr.cdc.gov/toxprofiles/phs115.html#bookmark05

Category: Organometals

Chemical: Tin & Tributyltin, $C_{12}H_{28}Sn$

Sources: Tributyltin and a similar organotin, triphenyltin, are used as antifouling agents in textiles and paints. Tin is a toxic, naturally-occurring element found in numerous products, including canned goods. Organotins and tin are also released into the environment by industrial processes, but the EPA does not require release reporting. [1]

Transport Vectors: Tin occurs in the Earth's crust naturally at 2-3 ppm. [2] Marine paints are the major source of coastal tributyltin pollution, though it can also leach out of landfills from consumer products. [3][4] It can be transported by particulates; human exposure occurs occupationally, through seafood, from drinking water distributed in PVC pipes (in the form of methyltin and butyltin), in household dust, and from contact with consumer products containing various plastics. In some marine creatures, tributyltin achieves high bioconcentration factors, including up to 350,000 in mollusks. [1][5]

Source Points	Temporal Data	Seafood	Human blood
Canadian drinking water max: 65.6 ng Sn/L (methyl- and butyltin)[6]	**Central Park Lake surface sediment:** 32 mg/kg **From period of incinerator operation:** 8,000 mg/kg [7]	**Squid:** 655 ng TBT/g **Canned food, Grapefruit:** 128 mg/kg **Tomatoes:** 84 mg/kg [8]	**1998 Michigan, in 70% of samples:** nd - .101 ug BT/mL[9]

Health Effects: Organotin compounds can be inhaled in commercial/industrial work environments where organotin compounds are produced or used. Some skin absorption may occur when direct contact with organotins takes place.

Brief contact with dialkyltin and trialkyltin compounds causes irritation of the skin and the respiratory tract. Acute intoxications can cause vomiting, headache, visual defects and abnormal electrical activity in the brain. Tributyltin compounds are moderately

50

toxic via both ingestion and dermal absorption. Tributyltin compounds may be strongly irritating to the skin and skin exposure may result in chemical burns in severe cases. Mucous membranes such as the eyes and nasal passages may also become irritated upon exposure. Shipyard workers occupationally exposed to dusts and vapors of tributyltin developed irritated skin, dizziness, difficulty in breathing, and flu-like symptoms. [10] Tributyltin, an endocrine disrupting chemical, has a potent effect on gene activity and has recently been associated with rapid increase in obesity and with irregularities in the sexual development of wildlife species. [11]

Citations:

1. Agency for Toxic Substances and Disease Registry. (2005). *Toxicological profile for Tin and Tin Compounds*. US Department of Health and Human Services. Atlanta.

2. Budavari S, ed. (2001). *The Merck index - An encyclopedia of chemicals, drugs, and biologicals, 13th edition*. Merck and Co., Inc., Whitehouse Station, NJ. pg. 1685.

3. Alzieu C. (2000). Environmental impact of TBT: The French experience. *Science of the Total Environment*. 258. pg. 99-102.

4. Fent K. (1996). Ecotoxicology of organotin compounds. *Critical Reviews in Toxicology*. 26. pg. 1-117.

5. Gomez-Ariza JL, Giraldez I, Morales E. (2001). Occurrence of organotin compounds in water, sediments and mollusca in estuarine systems in the southwest of Spain. *Water*, *Air*, and *Soil Pollution*. 126. pg. 253-279.

6. Sadiki A, Williams DT. (1996). Speciation of organotin and organolead compounds in drinking water by gas chromatography-atomic emission spectrometry. *Chemosphere* 32. pg. 1983-1992.

7. Park J, Presley BJ. (1997). Trace metal contamination of sediments and organisms from the Swan Lake area of Galveston Bay. *Environmental Pollution*. 98. pg. 209-221.

8. Biego GH, Joyeux M, Hartemann P, et al. (1999). Determination of dietary tin intake in an adult French citizen. *Archives of Environmental Contamination and Toxicology*. 36. pg. 227-232.

9. Kannan K, Senthilkumar K, Giesy JP. (1999). Occurrence of butyltin compounds in human blood. *Environmental Science Technology*. 33. pg. 1776-1779.

10. National Pollutant Inventory Substance Profile. http://www.npi.gov.au/database/substance-info/profiles/66.htm

11. Iguchi T, Katsu Y. (December 2008). Commonality in signaling of endocrine disruption from snail to human. *BioScience*. 58(11). pg. 1061-7. http://caliber.ucpress.net/doi/abs/10.1641/B581109.

Other Anthropogenic Ecotoxins of Interest
(Hotel California: Today's Menu Part 2: Appetizers, Soups, Salads, and Desserts)

This secondary list of ecotoxins includes many contaminants not included in the previous *Section III. Selected Major Anthropogenic Ecotoxins of Interest*, as well as specific listings or abbreviated definitions of ecotoxins already described in *Section II. The Chemistry of Ecotoxins*, including some of the most important POPs and their cogeners.

Acinetobacter baumannii: An important emerging antibiotic resistant bacteria (ABRB); as a Gram-negative bacteria, *A. baumannii* is one of several proliferating bacterial strains for which there are no antibiotic treatments whatsoever. See *Section IV. Health Effects Caused or Affected by Environmental Toxins* for an excerpt from *The New York Times* on this bacterium.

Acrylamide: A chemical that is produced naturally in some foods when they are cooked at high temperatures. It is also manufactured for use in the production of polyacrylamide gels, which are used, among other things, for treating wastewater and drinking water. Acrylamide is a known carcinogen in animals and can cause nerve damage in humans.

Agent Orange: During the Vietnam War, in an effort to clear vegetation where enemy troops often hid, the U.S. military employed aerial spraying to deliver about 19.5 million gallons of Agent Orange and other herbicides (http://www.braytonlaw.com/practiceareas/agentorange.htm). Agent Orange exposure has been linked to various cancers, including sarcoma, respiratory, lung, larynx, trachea, prostate, leukemia, non-Hodgkin's lymphoma, Hodgkin's disease, and to type 2 diabetes, multiple myeloma, damage to the peripheral nervous system, and severe skin diseases, such as chloracne and porphyria cutanea tarda. The children of soldiers exposed to Agent Orange in Vietnam had a heightened risk of spinal bifida, a birth defect in which the baby's spinal cord is not properly closed, which may result in fluid on the brain (hydrocephalus) and neurological disorders (Institute of Medicine 2005).

Aldrin: Aldrin is a potent organochlorine pesticide that was utilized to control grasshoppers, termites, and other soil pests from the 1950s until 1974, when its use was banned in the US except for termite control. In 1987, it was banned for all uses. When ingested by bacteria or under the influence of sunlight, aldrin undergoes a chemical change and becomes dieldrin. As one of the POPs banned by the Stockholm Convention in 2001, aldrin and dieldrin remain ubiquitous contaminants in the global atmospheric water cycle and thus in the food webs of the biosphere. Aldrin/dieldrin-induced convulsions and tremors have been observed in animals and humans. Organochlorine insecticides such as aldrin and dieldrin can act as GABA (gamma-aminobutyric acid) receptor antagonists, blocking the chloride ion channel in the central nervous system of both animals and humans, causing convulsions and tremors (Klaassen 1996).

Aliphatic hydrocarbons: A major category of petrochemicals, aliphatic hydrocarbons are characterized by open chains of carbon atoms typified by methane, ethane, propane, and butane. Subclasses are either saturated or unsaturated formations, which permit easy chemical transformation from simple, low carbon atom molecules such as olefins into more complex and often toxic petrochemicals.

Alkylphenols: A type of organic compound synthesized from phenol and used in innumerable commercial and industrial products, alkylphenol ethoxylates (APEs) are used as surfactants, mimicking estrogen, causing endocrine system disruption. "APE breakdown products are more toxic to aquatic organisms than their intact precursors are… throughout northern Europe, a voluntary…ban on APE use in household cleaning products began in 1995 and restrictions on industrial cleaning applications are set to follow in 2000." (Environmental Science and Technology 1997). Environmental concentrations of NPEs (nonylphenol ethoxylates) have been documented by the USGS in rivers throughout North America, the majority of which result from sewage plant discharge effluents.

Alkylphenol polyethoxylate: A household and industrial detergent that can function as an imposter hormone, binding to hormone receptors in cells, and as an endocrine disruptor. It can trigger estrogen mishaps such as feminization in males, birth defects, and other mutagenic consequences (Tulane/Xavier Center for Biomedical Research; http://www.cbr.tulane.edu/).

Aluminum: "Aluminum damages nervous systems in both infants and adults. Symptoms of aluminum toxicity include memory loss, learning difficulty, loss of coordination, disorientation, mental confusion, colic, heartburn, and headaches. In patients having Alzheimer's disease the brain is somewhat shrunken and, on postmortem examination, a definite loss of nervous tissue is noted. Examination of the brain tissues under a microscope reveals small bundles of material called senile plaques, scattered throughout the tissues. The more plaques that are present, the worse is the mental condition of the patient. Chemical analysis reveals the presence of the metal aluminum at the core of each plaque and within many of the cells found in the plaques. Evidence is accumulating to indicate that aluminum may be involved in the formation of the plaques, and it is therefore a prime suspect as the initial cause of the disease." (http://newsgroups.derkeiler.com/Archive/Alt/alt.politics/2008-02/msg03534.html).

Americium-241: Am_{241} is a manmade alpha gamma emitting radioisotope (atomic number 95, $\frac{1}{2} T = 432.7$ years) created during the production of plutonium-241 for nuclear weapons and during nuclear electricity production. Its daughter product is neptunium-237. The most important commercial use of americium is as the power source of smoke detectors; it is also used in fluid density gauges, distance sensing devices, aircraft fuel gauges, thickness gauges, and medical diagnostic devices. Americium-241 is potentially biologically significant if smoke detectors and other equipment in which it is used are improperly disposed of, lost, or incinerated. A typical smoke detector uses one microgram of americium-241; tens of millions of micrograms of americium-241 have now entered the environment, primarily through disposal in landfills.

Aromatic hydrocarbons: Compounds based around a benzene ring, or, in the case of polyaromatic hydrocarbons, rings that are generally the result of combustion and possess a sweet odor. The unique qualities of aromatic hydrocarbons include ease of manufacture and molecular stability, which help make them among the most useful of all petrochemicals, and ubiquitous ecotoxins. Their suffixes always end as "arenes."

Arsenic: As; Arsenic is a naturally occurring poisonous metalloid (atomic number 33) that is remobilized by anthropogenic activity and is now a widely distributed ecotoxin of concern. It is one of the heavy metal ecotoxins produced by the combustion of coal. As a mineral species, arsenic has many different allotropic forms and is now incorporated in a wide variety of pesticides, herbicides, insecticides, and various alloys. One of its first known uses was as an alloy in the production of bronze; arsenic-bronze preceded production of tin-bronze. It is known to cause skin, lung, and bladder cancer and is linked to an increased risk of other cancers including kidney, liver, prostate, and throat. Ingestion of low levels of arsenic can cause high blood pressure, cardiac arrhythmias, congestive heart failure, and immunological disorders. Arsenic can cross the placenta to the developing fetus and may cause low birth weights, spontaneous abortions, and other problems. Arsenic contamination of the food web at low levels of concentration due to human activity is now so ubiquitous that it is not normally subject to biological monitoring (National Pollutant Inventory Substance Profile; http://www.npi.gov.au/database/substance-info/profiles/66.htm). Arsenic-containing compounds are commonly used in animal feeds to induce faster weight gain and create the appearance of a healthy color in meat from chickens, turkeys, and hogs.

Asbestos: A fibrous silicate mineral frequently used for fireproofing, insulating and other industrial purposes. Asbestos is among the most toxic of all biologically significant, remobilized, naturally occurring toxins; the careless processing of asbestos products has resulted in hundreds of thousands of unnecessary cases of mesothelioma, a rare form of lung cancer.

Atrazine: An herbicide and notoriously persistent organic pollutant (POP), which has come into wide use in the United States in the early 21st century as other herbicides have been phased out. Banned by the European Union in 2004, atrazine is now possibly the most widely used herbicide in foreign countries outside of Europe as well as in the United States, where it is used to control weeds, especially in corn-growing regions of the US, and in conservation tillage. It also has many industrial uses, as in the manufacture of explosives and dyes. Atrazine is a known endocrine disruptor, a carcinogen, and a teratogen, and has been demonstrated to cause demasculinization in male frogs in low concentrations (Hayes 2003). Atrazine is among the thousands of ecotoxins now contaminating the global atmospheric water cycle.

Badge-4OH: A metabolite of BADGE, used to make metal food can linings. The EWG (Environmental Working Group) cites research suggesting strong evidence of hormone activity disruption by Badge-4OH with limited evidence of other health concerns. They also cite a study showing that in the human body, BADGE can break down into BPA, another ecotoxin of concern associated with the production of plastic consumer goods (National Pollutant Inventory Substance Profile; http://www.npi.gov.au/database/substance-info/profiles/66.htm).

Biocidal silver: An engineered material derived from the element silver that utilizes the unique properties of nanoparticles, such as relatively large surface areas, greater chemical reactivity, enhanced biological activity, and amplified catalytic behavior, which result in a much higher toxicological risk in comparison to larger particles of the same chemical composition due to their ability to penetrate biological membranes and impact cells, tissues, and organs, i.e. greater bioavailability. "Nanomaterials which measure less than 70 nm can even be taken up by our cells' nuclei, where they can cause major damage." (Senjen 2009, 4). See nanosilver and nanoparticle.

Bovine growth hormone: See growth hormones.

Carbon tetrachloride: A highly ecotoxic solvent for fats and oils used in dry cleaners; also formerly used in the production of fire extinguishers. Highly volatile, it is produced by chlorination of methane. Restricted production and use in western countries has reduced, but not eliminated, its presence as a water cycle contaminant.

Carbonate insecticides: Insecticides are available as four different chemical types -- carbonate, chlorinated hydrocarbons, pyrethrins, and organophosphorus compounds (OPs). The most common carbonates are carbaril and methomyl, which are often used to destroy worms and other insects by blocking the transmission of nervous impulses at nerve muscle junctions. Carbonate insecticides act similarly to organophosphorus compounds. The antidote is atropine sulfate by injection. All insecticides act on the nervous system and the general symptoms are similar, including muscle tremors, hyperactivity, convulsions, and death. OPs and similar carbonate insecticides affect the chemical transmitters that control the nerve endings in muscles, causing muscle hyperactivity and eventually continual spasms. The continual and excessive stimulation of the muscles in acute poisoning [*of pigs*] causes excessive salivation, the passing of feces and urine, and a stiff, awkward gait, often followed by vomiting, diarrhea, and muscle tremors of the face and body. Death from respiratory failure can occur within 1 to 4 hours. (The Pig Site, Pig Health; www.thepigsite.com/pighealth/article/476/insecticides).

Cesium-137: One of the most biologically significant of all anthropogenic isotopes (1/2 T = 30 years), cesium (Cs_{137}) follows the potassium cycle in the biosphere, providing a whole body dose upon uptake, primarily by the ingestion pathway. A ubiquitous component of weapons-testing-derived stratospheric fallout as well as Chernobyl-derived tropospheric fallout, cesium-137 is now a component of all food webs and most biological media, though not necessarily in biologically significant quantities. Cesium-137 is an indicator nuclide for evaluating the size and impact of nuclear accidents and releases. See *Appendices G, H,* and *I* for more information on cesium-137.

Cesium-137 chloride: Cesium-137 chloride was formerly a principal source of radiation used in food irradiation equipment, which typically contain 135,000 curies per unit. Most food irradiation equipment now uses cobalt-60 as its power source. Numerous older irradiation units are now missing.

Chlordane: A highly toxic insecticide and one of the most important and ubiquitous POPs. See *Section III. Selected Major Anthropogenic Ecotoxins of Interest.*

Chlorinated hydrocarbons: A class of chemicals derived from the use of chlorine to produce one form of the many halogenated hydrocarbons used as pesticides (insecticides and herbicides) by industrial agriculture after World War II. Most chlorinated hydrocarbons are POPs and were widely used by domestic consumers until their production in western countries was restricted after 1980. Important and widely used chlorinated hydrocarbons include chloro- and dichloromethane, chloroform, chlordane, chlorinated polyvinyl, methylene chloride, ethyl chloride, hexachloroethane, trichloropropane, and a wide variety of other useful industrial ecotoxins. A comprehensive survey of ecotoxic chlorinated hydrocarbons and their health physics impact would run into the thousands of pages.

Chlorofluorocarbons (CFCs): CFCs are compounds that contain only carbon, fluorine and chlorine and were used widely in refrigerants before being banned by the Montreal Protocol due to their ozone depleting properties. A single CFC molecule typically degrades between ten thousand and several million ozone molecules in its lifetime (http://www.epa.gov/fedrgstr/EPA-AIR/1995/October/Day-11/pr-1117.html).

Clotrimazole: Clotrimazole is an antifungal medication commonly used to treat fungal infections such as vaginal yeast infections, athlete's foot, jock itch, and ringworm in humans and animals.

Cotinine: A major metabolite of nicotine, the cotinine level in the blood is a bioindicator for exposure to environmental tobacco smoke (ETS) and chewing tobacco.

Deca-BDE: One of three forms of polybrominated diphenyl ethers utilized as flame retardants in computers and other electronic equipment; all are significant neurotoxins and endocrine system disrupters. High levels of this ecotoxin have been found in peregrine falcons. Significant contamination was noted in computer wipe samples from public building from eight US states in March of 2004. "Women in North America have the highest levels globally of these chemicals in their breast milk… these levels are doubling every two to five years in the North American population." (www.computertakeback.com). Incineration or combustion transforms all forms of PBDEs into brominated dioxins.

DEHA: di(2-ethylhexyl)adipate is a plasticizer (softener) in items such as cling wrap. Exposure is linked to negative effects on the liver, kidney, spleen, bone formation, and body weight.

Depleted Uranium (DU): The most common form of naturally occurring Uranium, DU is composed almost entirely of the U-238 isotope. It is a poisonous, radioactive heavy metal that has been implicated as a cause of Gulf War Syndrome due to its use in munitions. Roughly 95% of DU reserves are stored as Uranium Hexafluoride (UF_6), which is not only toxic but chemically unstable, reacting with water vapor in the air to yield UO_2F_2 + HF, both of which are highly toxic and transport easily. Several accidents involving the 480,000 tons of UF_6 in storage in the US have occurred; world supply is estimated at around 1.2 million tons (http://www.wise-uranium.org/eddat.html; http://web.ead.anl.gov/uranium/faq/health/faq30.cfm; Pellmar 1998).

Dethane: Decabromodiphenyl ethane is the "key ingredient in Firemaster 2100, a brominated flame retardant in the same chemical family as the developmental neurotoxin, Deca-BDE" (Andrews 2009, 5).

Dibutyltin: A chemical used in PVC plastics "that can interfere with the natural ability of human and animal cells to control important immune responses and inflammation." (Gumy 2008). One more among thousands of rapidly proliferating endocrine disrupting chemicals, the legacy of the age of chemical fallout.

Dichloroethane: $C_2H_4Cl_2$; commonly used as a solvent in many industrial operations, Dichloroethane is a highly volatile, colorless liquid, a microcontaminant in most food webs, and a highly ecotoxic source material for vinyl chloride production derived from tetrachloroethane, another biologically significant industrial ecotoxin.

Diclofenac: A nonsteroidal, anti-inflammatory drug (NSAID) used to reduce pain and inflammation, especially in arthritis. Diclofenac is sold worldwide under more than three dozen different brand names and is used in both human and veterinary medicine; it has recently achieved widespread notoriety due to the decimation of the vulture population (99% die-off rate) on the India subcontinent due to renal failure following the ingestion of the carcasses of cattle that had been treated with diclofenac (*Wall Street Journal* May 1, 2008).

Dieldrin: Among the dieldrin family of ecotoxins are diethylphosphate (DEP), diethylphosphorodithidate (DEDTP), diethylphosphorothidate (DEPT), dimethylphosphate (DMP), dimethylphosphorodithidate (DMPT), malathion dicarboxylic acid, and malathion mono-carboxylic acid. See aldrin/dieldrin for a sketch of their health effects. Though dieldrin production was banned by the Montreal Protocol, third world production continues. Due to it persistence, its large production quantities, and its numerous metabolites, cogeners, and daughter products, dieldrin residues are and will continue to be persistent ubiquitous contaminants in most food webs.

Dimethyl phthalate (DMP): A highly ecotoxic breakdown product of the phthalates used in the production of many plastic products.

Dioxin: $C_4H_4O_2$; an organic compound, which includes several groups of biologically significant POPs, each with dozens or hundreds of cogeners. See the survey of major contaminant groups for a sketch of several dioxin families and their health effects. Dioxins are carcinogenic and bioaccumulate in lipids in all biotic media (http://www.atsdr.cdc.gov/tfacts104.html). See *Section III. Selected Major Anthropogenic Ecotoxins of Interest*.

Endocrine disrupting chemical (EDC): "Endocrine disrupting compounds encompass a variety of chemical classes, including hormones, plant constituents, pesticides, compounds used in the plastics industry and in consumer products, and other industrial by-products and pollutants. Some are pervasive and widely dispersed in the environment. Some are persistent organic pollutants (POP's), and can be transported long distances across national boundaries and have been found in virtually all regions of the world. Others are rapidly degraded in the environment or human body or may be present for only short periods of time." (Wikipedia.org). Another take on endocrine disrupting chemicals, which were first noted in 1991 by the environmental health analyst Theo Colborn, is that of The Endocrine Society. "These are chemicals that can interfere with hormone action. These chemicals are designed, produced, and marketed largely for specific industrial purposes, (e.g. plasticizers, pesticides, etc). They are also found in some natural foods and may become further concentrated as foods are processed or can even contaminate foods during processing or storage… The endocrine disrupting potential of a compound extends far beyond actions at hormone receptors. It is… evident that EDCs need not bind to a hormone receptor in order to disrupt endocrine signaling in the exposed individual, her offspring, and subsequent generations." (The Endocrine Society 2009). Commonly encountered EDCs include bisphenol A (BPA), perchlorates, phthalates, PBDEs, persistent organic pollutants (POPs) such as DDT and PCBs, and the miscarriage prevention drug diethylstilbestrol (DES).

Endotoxin: A potent toxin released by the death of bacterial cells and a component of the etiology of typhoid fever and cholera, endotoxin is a powerful activator of macrophages, which may ingest other ecotoxins, incorporating them in living tissue. They are a possible tool of bioterrorism, especially if reformulated with the help of clever bioengineering.

Endrin: A potent biologically significant insecticide used on cotton and grain and to control rodents. It was one of the first POPs to be subject to restricted manufacture and use, but it is still persistent as a contaminant signal in biotic media.

Ethyl chloride: The older term for what is now known as Chloroethane, ethyl chloride is a flammable volatile liquid (C_2H_5Cl) formerly used as a key ingredient for the gasoline additive tetra-ethyl lead. It was also used as a a blowing agent for foam packaging , a refrigerant, in the manufacture of aerosol sprays, and as an anesthetic. It is still used in the manufacture of ethylcellulose, which is used as a thickening agent in paints, cosmetics, and similar products. As a chlorinated hydrocarbon, ethyl chloride is a persistent organic pollutant and a worldwide contaminant in most food webs.

Ethylbenzene: $C_6H_5CH_2CH_3$; a toxic and potentially carcinogenic petrochemical made from benzene and ethylene and used to produce styrene (ATSDR 2007).

Ethylene dichloride: An ecotoxin produced from the chlorination of ethylene; biologically significant daughter products include vinyl chloride and hydrogen chloride.

Ethylenedibromide: A biologically significant ecotoxin produced from the bromination of ethylene.

Event 32: The accidental release of an unapproved, genetically modified corn seed produced by Dow Agrosciences that contained significant amounts of ecotoxic pesticides, and was considered a threat to the genetic integrity of nearby corn species that it had contaminated.

Fluoride: "Fluoride, the active ingredient in many pesticides, is a powerful poison – more acutely poisonous than lead. Because of this, accidental over-ingestion of fluoride can cause serious toxic symptoms... Over the past ten years a large body of peer-reviewed science has raised concerns that fluoride may present unreasonable health risks, particularly among children, at levels routinely added to tap water in American cities... Boiling fluoridated tap water in an aluminum pan leached almost 200 parts per million (ppm) of aluminum into the water in 10 minutes; Leaching of up to 600 PPM occurred with prolonged boiling; Using non-fluoridated water showed almost no leaching from aluminum pans." (http://newsgroups.derkeiler.com/Archive/Alt/alt.politics/2008-02/msg03534.html).

Freon: A chlorofluorocarbon, Freon has many industrial uses, including air conditioning, refrigeration, and fire fighting applications. The Montreal Protocol on Substances that Deplete the Ozone Layer banned Freon-11 and Freon-12 in 1996; the extent of its continuing production is unknown.

Fullerenes: The allotropic form of carbon that characterizes nanotubes. Discovered in 1985, fullerenes are a form of carbon molecules that are the key component of superplasticity in metals (e.g. Damascus swords, nuclear power plant and commercial jet airplane engine turbines). The industrial production of fullerenes as a component of nanotechnology in the form of buckminsterfullerenes –"buckyballs" – for a wide variety of products is responsible for the creation of a group of newly emerging ecotoxins, many of whose health physics impact are unknown. See Nanoparticles.

Glycol ethers: Ecotoxic solvents used in the photo-resist step of the micro-chip etching process.

Glyphosate: "Glyphosate (N-(phosphonomethyl) glycine) is a broad-spectrum systemic herbicide used to kill weeds... Initially patented and sold by Monsanto Company in the 1970s under the tradename *Roundup*, its U.S. patent expired in 2000... Some crops have been genetically engineered to be resistant to it (i.e. Roundup Ready). Such crops allow farmers to use glyphosate as a post-emergence herbicide against both broadleaf and cereal weeds, but the development of similar resistance in some weed species is emerging as a costly problem." (http://en.wikipedia.org/wiki/Glyphosate).

Growth Hormones: Hormones can be steroids or proteins. Steroid hormones, such as birth control pills, become active in the body after ingestion. Protein hormones such as insulin are broken down and lose their ability to act in the body when eaten and are usually injected. Synthetic steroid hormones used as pharmaceutical drugs can increase the risk of cancer in users. Diethylstilbestrol (DES), a synthetic estrogen drug originally developed to fatten chickens and later used in human medicine, was banned when it was found to increase the risk of vaginal cancer in daughters of treated women. Its use in food production was phased out in the late 1970s. Lifetime exposure to the natural steroid hormone estrogen is also associated with an increased risk for breast cancer (Clark 1998). There are six different kinds of steroid hormones that are currently approved by FDA for use in food production in the US: estradiol, progesterone, testosterone, zeranol, trenbolone acetate, and melengestrol acetate. Estradiol and progesterone are natural female sex hormones; testosterone is the natural male sex hormone; zeranol, trenbolone acetate, and melengesterol acetate are synthetic, hormone-like growth promoters. Currently, federal regulations allow these hormones to be used in raising cattle and sheep, but not for poultry or hogs. In 1993, the Food and Drug Administration (FDA) approved the recombinant bovine growth hormone (rBGH), also known as bovine somatotropin (rBST) for use in dairy cattle to increase milk production. Recent estimates by Monsanto, the manufacturer of the hormone, indicate that 30% of the cows in the US may be treated with rBGH. Because of increased milking, hormone-treated cows become more prone to mastitis, an infection of the udders. This results in more antibiotics being used to treat the cows, in turn leading to more residues of antibiotics remaining in the milk. In 1989, the European Community (now European Union) issued a ban on all meat and milk products from animals treated with steroid growth hormones, which is still in effect. Countries within the European Union also do not allow the use of the protein hormone rBGH, for dairy cattle. The use of steroid hormones for beef cattle is permitted in Canada, but in 1999, the Canadian government refused approval for the sale of rBGH for dairy cattle, based on concerns about health effects including mastitis in treated animals (Gandhi 2000). See rBGH for additional information on its health effects.

Halogenated aromatic hydrocarbons: These are produced by the incorporation of one of the halogen molecules, fluorine, chlorine, bromine, iodine, or astatine, in conjunction with a preexisting benzene ring or rings in the naphthalene group of hydrocarbons, to form the highly stable pesticides so useful in industrial agriculture. Most halogenated aromatic hydrocarbons are classed as highly ecotoxic persistent organic pollutants (POPs).

Hexabromocyclododecane (HBCD): One of the worlds' most commonly produced flame retardants; HBCD is a POP, highly bioaccumulative, and a potent thyroid/endocrine disruptor that may play a role in the etiology of obesity and diabetes.

Hexachloroethane: C_2Cl_6; also known as perchloroethane (PCA), is a highly toxic chemical used in the military production of smoke bombs and by aluminum foundries to remove hydrogen. In a reductive atmosphere, it has a chemical half life of less than one hour; additional information about the toxicity and biological half life of its cogeners and other chemical forms is not available.

Hydrochlorofluorocarbons (CFC): A form of halogenated hydrocarbon and an important greenhouse gas, Hydrochlorofluorocarbons are a subspecies of haloalkanes, as are chlorofluorocarbons; both types were manufactured by DuPont and known by the trade name Freon, before being banned because of their ability to destroy the stratospheric ozone layer, which protects the Earth from harmful ultraviolet radiation.

Hydrophobic compounds: Compounds such as polycyclic aromatic hydrocarbons (PAHs) that are not water soluble.

Ketoprofen: A highly toxic arthritis drug with similarities to diclofenac; also widely utilized in livestock operations.

Lead: A naturally-occurring metal and one of the most ancient ecotoxins, lead is famous as the contaminant that leached out of Roman drinking and eating utensils and water pipes. Lead is omnipresent as an ecotoxin in a wide variety of industrial and consumer products and, along with methylmercury, is one of the most significant heavy metal health hazards of modern society, despite having been banned for most contemporary uses. The irretrievable amount of lead in the biosphere derived from anthropogenic activities is in the hundreds of millions of tons. See ecotoxin health issues and major anthropogenic ecotoxins of interest.

Melamine: The FDA began intensive testing of hundreds of food products for melamine after Chinese-made infant formula was found to be tainted with melamine in 2008. Chinese manufacturers had added the chemical to make it appear to contain higher levels of protein after diluting formula. More than 50,000 Asian infants were hospitalized, and at least four died. The FDA found melamine and cyanuric acid, a related chemical, in samples of baby formula made by major U.S. manufacturers. Melamine can cause kidney and bladder stones and, in the worst cases, kidney failure and death. If melamine and cyanuric acid combine, they can form round yellow crystals that can also damage kidneys and destroy renal function.

An FDA position paper asserts the "FDA is currently unable to establish any level of melamine and melamine-related compounds in infant formula that does not raise public health concerns." Agency scientists have maintained they could not set a safe level of melamine exposure for babies because they do not understand the effects of long-term exposure on a baby's developing kidneys. The problem is exacerbated because infant formula is a baby's only source of food for many months. Premature infants absorb an especially large dose of the chemical compared with full-term babies.

Mercury/methylmercury: The mercury derivative methylmercury is among the most toxic of all forms of biologically significant chemical fallout (BSCF). The skeleton in the closet of a pyrotechnic society that uses fossil fuels to fuel its heat engines, methylmercury is the product of bacterial action on mercuric sulfide after its release by coal-fired electricity generation, industrial power sources, and domestic use of coal for home heating and cooking, the latter of which is especially commonplace in far eastern countries with cold climates. Harmless in its naturally occurring form, mercuric sulfide undergoes transformation through bacterial action into highly toxic methylmercury, which is readily absorbed by living organisms, and is easily transported in pathways to human consumption. Increasing rates of the cross placental transfer of methylmercury from mother to child has been documented in the late 20th century and it is now a ubiquitous contaminant in most food webs. Methylmercury contaminant pulses in tuna fish, swordfish, and other species now occasionally exceed 1 ug/g (ppm). Widespread contamination of fresh water and food products, including some farm raised fish may play a much more important role in the rising rates of autism in children than methylmercury contamination from specific point sources, such as thimerosal, the antiseptic/antifungal once commonly used in children's vaccinations.

The US EPA has set a fish tissue criterion for MeHg at 0.3 ug/g under section 304(a) of the Clean Water Act. Important sources of Hg to the environment include electric utilities, incinerators, industrial manufacturing, wastewater treatment plants, and improper disposal of consumer products (e.g. batteries, fluorescent light bulbs, Hg switches) (Driscoll 2007). See the *Appendices* and *Section III. Selected Major Anthropogenic Ecotoxins of Interest.*

Methane: The third most potent greenhouse gas, methane gas is released as a result of permafrost melting, as a gas from domestic animals, especially cattle, from rice paddy farming, and from other anthropogenic activities such as petroleum production, coal mining, internal combustion engines, wastewater treatment, and other industrial activities. Methane has a global warming potential (GWP) of 62 over 20 years, 23 over 100 years, and 7 over 500 years; it is degraded to water and CO_2 by chemical reactions in the atmosphere. Methane is approximately 20 times more powerful in trapping heat in the atmosphere than is carbon dioxide.

Methyl bromide: CH_3Br; also called bromomethane, is a liquid with a boiling point of 4° C at which temperature it becomes a colorless gas. Methyl bromide has been widely used as a soil sterilizer and pesticide, especially on golf courses. It has also been extensively used to treat soils used in strawberry production. As a toxic substance, its health physics impact includes neurological, respiratory, and kidney damage. It is a significant ozone depleting chemical; while the Montreal Protocol called for its use to be curtailed, it continues in widespread use as a pesticide, fumigant, and soil sterilizer, particularly within the state of California. The state of California is now in the process of replacing methyl bromide with methyl iodide with respect to strawberry production. Unfortunately, methyl iodide may even be a more significant developmental neurotoxin and endocrine disrupter than methyl bromide.

Methyl ethyl ketone: A potent ecotoxin used in the manufacture of electronic equipment.

Methyl iodide: Methyl iodide is a toxic chemical derived from methyl bromide and used as a fumigant on crops like strawberries; highly volatile, it is a carcinogen and causes thyroid tumors. On October 5, 2007, the EPA approved this chemical for widespread use by US farmers (www.democracyinaction.org).

Methyl isocyanate: A highly toxic, water soluble, liquid, synthetic chemical that killed tens of thousands of area residents when it was spilled in 1984 at the Union Carbide pesticide plant in Bhopal, India.

Methyl tertiary butyl ether (MTBE): A highly toxic chemical compound that is added to gasoline to increase the oxygen level and raise the octane number; also used as an industrial solvent and in medical applications to dissolve gallstones. It is widely used as a substitute for tetraethyl lead and is now a ubiquitous contaminant in ground water and private and public water supplies, especially in the Northeast.

Methylene chloride: A potent ecotoxin used in manufacturing printed circuit boards for electronic equipment.

Mycoplasma fermentans incognitus: A controversial biological agent, potentially useful in biological warfare, that contains most, but not all, of the HIV envelope gene. The US Department of Defense has declined to confirm that Gulf War veterans have suffered from mycoplasma infections or to document the high death rates of infected troops.

Mirex: A highly toxic insecticide and one of the first POPs banned by the Stockholm Convention.

Nanosilver: Highly toxic silver nanoparticles are reported to be able to kill virtually any virus including HIV and are being considered as a treatment for Methicillin-resistant staphylococcus aureus (MRSA). Nanosilver is currently used as a topical antibiotic in medical settings (Free Market News Network 2005; www.freemarketnews.com/WorldNews.asp?nid=1401). The following products use nanosilver: health and food supplements, food packaging, household appliances, medical implants and devices, wound plasters, salves, paints, lacquers, textiles, shoes, cosmetics, personal care products, bedding, air and water filters, electronic articles, children's articles, and agricultural products.

Nanoparticles: "Currently we see the health, safety and environmental hazards of nanotechnologies as being restricted to discrete manufactured nanoparticles and nanotubes in a free rather than embedded form… The evidence that we have reviewed suggests that some manufactured nanoparticles in nanotubes are likely to be more toxic per unit mass than particles of the same chemicals at larger size and will therefore present a greater hazard… the difference comes largely from two size-dependent factors: the relatively greater surface area of nanoparticles, given equal mass, and their probable ability to penetrate cells more easily and in a different way. To pose a risk, these nanoparticles must come into contact with humans or the environment in a form and quantity that can cause harm… There is virtually no evidence available to allow the potential environmental impacts of nanoparticles and nanotubes to be evaluated… Nanoparticles and nanotubes that persist in the environment or bioaccumulate will present and increased risk and should be investigated… we have recommended that the release of free manufactured nanoparticles in to the environment for remediation (which has been piloted in the USA) be prohibited until there is sufficient information to allow the potential risks to be evaluated as well as the benefits." (The Royal Society & The Royal Academy of Engineering 2004, chapter 9).

Nitrates: Nitrate runoff from fertilizer use in the Mississippi River Basin, especially for corn production, for example, is a major cause of oceanic dead zones in the Gulf of Mexico and elsewhere. The nitrates facilitate extensive annual algae growth, which then die and decay, reducing the oxygen content of Gulf waters to near zero, resulting in significant die-offs of fish, shrimp, and other marine species. Nitrate contamination can also originate from sewage treatment facilities and urban environments. High nitrate levels in drinking water interrupt the normal body processes of some infants and may cause methemoglobinemia, a condition known as "blue baby." Nitrate becomes toxic when it is reduced to nitrite, a process that can occur in the stomach as well as in the saliva. Infants are especially susceptible because their stomach juices are less acidic and therefore more conducive to the growth of nitrate-reducing bacteria than are the stomach juices of adults (Illinois Department of Public Health 1999; http://www.idph.state.il.us/envhealth/factsheets/NitrateFS.htm).

Nitro and Polycyclic Musks: Linked to hormone disruption, and in some cases to cancer and reproductive problems, nitro and polycyclic musks are a common component of many popular health care products. In laboratory studies, some nitro musks have been linked to cancer (Maekawa, Matsushima, et al. 1990; Apostolidis, Chandra, et al. 2002). Studies of nitro musks in people suggest that high levels of some of these chemicals are associated with reproductive and fertility problems in women (Eisenhardt, Runnebaum, et al. 2001). Some also produce skin irritation and sensitization (Parker, Buehler, et al. 1986; Hayakawa, Hirose, et al. 1987). Growing concerns led the European Union to ban the use of some of these chemicals in cosmetics and personal care products. Laboratory studies suggest that polycyclic musks may also affect hormone systems. In the United States, all musk chemicals are unregulated, and safe levels of exposure have not yet been determined (Human Toxome

Project; http://www.bodyburden.org/chemicals/chemical_classes.php?class=Nitro-+and+polycylic-+musks).

Nitrogen trifluoride: NF_3; a greenhouse gas 17,000 times more potent than CO_2, with an atmospheric half life of 550 years. Due to its use in the fabrication of widescreen televisions and other electronic equipment for the plasma etching of silicon wafers, its annual production, 4,000 tons in 2007, is expected to reach 10,000 tons in 2010 (Hoag 2008).

Nitrous oxide: The fourth most potent greenhouse gas, nitrous oxide is popularly known as laughing gas from its use in medical and dental venues as an analgesic. It is a potent oxidizer used in rocket and racing car fuels and also released from tropical soils and oceanic sources; it has an atmospheric lifetime of 120 years and a global warming potential (GWP) of 296 over 100 years.

Nonylphenol: An ecotoxic xenoestrogen, which is a ubiquitous component in the manufacture of plastics.

Organochloride: An organic compound containing at least one chlorine atom. Most persistent organic pollutants (POPs), including DDT, heptachlor, and hexachlorobenzene, are organochlorides, also known as chlorinated hydrocarbons.

Organochlorine insecticides: Persistent organic pollutants (POPs) such as aldrin and dieldrin that have been halogenated using chlorine; also called organochlorides.

Organofluorine: An organic chemical containing at least one fluorine atom. A number of POPs are fluorinated hydrocarbons. See perfluorocarbons.

Organometals: A wide variety of metals where carbon atoms are linked to metal atoms, including the highly toxic alkali metals, the most notorious of which is lead tetraethyl, the recently banned (at least in Europe and the US) additive to gasoline.

Organophosphate (OP): Organophosphate is phosphoric acid with an alkyl group attached. Many natural and synthetic, biologically significant chemicals are organophosphates, typified by neurotoxic pesticides, such as malathion, parathion, and diazinon. OPs were developed initially in the 1940s and designed to kill insects by attacking the nervous system by poisoning the critical enzyme, acetylcholinesterase, which is essential for breaking down chemically delivered nerve impulses.

Ozone: O_3; the stratospheric ozone layer protects humans from the harmful effects of ultraviolet radiation; some progress has been made in reducing the world's output of ozone-depleting chlorofluorocarbons since the 1960s. Ozone is also a potent and biologically significant product of internal combustion engines and among the most ubiquitous of all tropospheric contaminants. As such, it is a respiratory irritant and the most common subject of air pollution alerts. Health effects include: "1) Irritation of the

66

respiratory system, causing coughing, throat irritation, and/or an uncomfortable sensation in the chest. 2) Reduced lung function, making it more difficult to breathe deeply and vigorously. Breathing may become more rapid and more shallow than normal, and a person's ability to engage in vigorous activities may be limited. 3) Aggravation of asthma… One reason this happens is that ozone makes people more sensitive to allergens, which in turn trigger asthma attacks. 4) Increased susceptibility to respiratory infections. 5) Inflammation and damage [to] the lining of the lungs. Within a few days, the damaged cells are shed and replaced much like the skin peels after a sunburn. Animal studies suggest that if this type of inflammation happens repeatedly over a long time period (months, years, a lifetime), lung tissue may become permanently scarred, resulting in permanent loss of lung function and a lower quality of life." (wikipedia.org; 2010).

PBDE: "PBDEs are brominated fire retardants, intentionally added to flexible foam furniture – primarily mattresses, couches, padded chairs, pillows, carpet padding and vehicle upholstery – and to electronic products." (Environmental Working Group; http://www.ewg.org). Note the large number of PBDE cogeners listed on the EWG website or discussed in *Section III. Selected Major Anthropogenic Ecotoxins of Interest*, under *Brominated Fire Retardants*. PBDEs are among the most biologically significant and widely distributed of all "emerging" environmental ecotoxins. Twenty four PBDE cogeners are also included in the *Appendices.*

Parabens: "(Alkyl-p-hydroxybenzoates) are one of the most widely and heavily used suites of antimicrobial preservatives in cosmetics (skin creams, tanning lotions, etc.), toiletries, pharmaceuticals, and even foodstuffs. Although the acute toxicity of these compounds is very low, Routledge, et al. report that these compounds (methyl through butyl homologs) display weak estrogenic activity in several assays. Although the risk from dermal application in humans is unknown, the probable continual introduction of these benzoates into sewage treatment systems and directly to recreational waters from the skin leads to the question of risk to aquatic organisms."

Parechovirus: A common virus contracted by most infants that displays few or no symptoms. Researchers at the Norwegian Institute of Public Health have recently suggested that parechovirus may serve as a trigger for Type 1 diabetes in people who are at genetic risk for the disease (MedIndia.com 2008).

Perchlorates: A common, highly water soluble contaminant in public water supplies, as well as in the water content of many vegetables grown in California's San Joaquin Valley and elsewhere. Perchlorate is naturally occurring as a salt derived from perchloric acid ($HClO_4$) and is manufactured for use as an oxidizer in explosives, rocket fuels, fireworks, submarine and space vehicle oxygen candles, airbags, and many other commercial products, and thus is a globally transported ecotoxin in the atmospheric, terrestrial, and public water cycles and systems. See *Section III. Selected Major Anthropogenic Ecotoxins of Interest*.

Perfluorodecanoic acid (PFDA): A "breakdown product of stain- and grease-proof coatings on food packaging, couches, carpets." (http://www.ewg.org/chemindex/chemicals/23244).

Perfluorooctanesulfonic acid (PFOS): A newly emerging ecotoxin as well as a proposed addition to the POPs, PFOS is a potent byproduct of PFCs and is biologically significant as an immunological disrupter. PFOS is a widely used chemical in semiconductor photolithography. One of the first attempts to document PFOS in biological media was the BioDiversity study of chemical ecotoxins in Maine birds (Goodale 2008, 35-6). See *Appendix M*.

Perfluorocarbon (PFC): A family of halogenated hydrocarbons using fluorine, PFCs have a multitude of medical, industrial, and consumer product uses, including as an anti-aging element in cosmetics, in fire fighting foams, as a replacement for CFCs in refrigeration, as stain repellents, as floor polish and ski wax, and in photography. One of their most important uses is for the purpose of etching semiconductors during the manufacture of computers. Some forms of PFCs, such as tetrafluoromethane and hexafluoroethane, are potent greenhouse gases with a greenhouse warming potential (GWP) thousands of times greater than CO_2 and a chemical lifespan of tens of thousands of years. The smelting of aluminum is an important secondary source of PFCs. The production of PFCs was curtailed by the Kyoto Protocol, which was initiated in 1997; production peaked in 1999. As an anthropogenic ecotoxin, PFCs and their many cogeners are found in abiotic and biotic media throughout the biosphere (http://en.wikipedia.org/wiki/Perfluorocarbon - cite_note-23).

Persistent organic pollutant (POP): Any organic compound of anthropogenic origin that resists degradation and accumulates in the food chain. POPs are among the most biologically significant ecotoxins; numerous specific POPs are discussed or documented throughout this publication. Most halogenated hydrocarbons are considered POPs. As lipophilic organic molecules, persistent organic pollutants are easily absorbed into the food chain in conjunction with their association with nutrient elements such as carbon, oxygen, hydrogen, nitrogen, potassium, calcium, phosphorous, and manganese. First noted by the UN's Stockholm Convention in 1995 as ecotoxins of extreme concern, they have since been listed as the "Dirty Dozen" by the EPA. The first chemicals to be listed as POPs are: dioxins, furans, polychlorinated biphenyls (PCBs), DDT, chlordane, heptachlor, toxaphene, hexachlorobenzene, aldrin, dieldrin, endrin, and mirex. Since the EPA compiled its list of the Dirty Dozen POPs, several hundred additional chemicals fitting the definition of a POP have either been documented as in

68

use or have been produced as chemically similar substitutes for POPs, which have been specifically banned for production and distribution. See *Section III. Selected Major Anthropogenic Ecotoxins of Interest*.

Pharmaceuticals: Among the principle ecotoxic forms of pharmaceuticals are: antibiotics, antidepressants, anti-cancer drugs, heart drugs, anticonvulsants, steroids, sex hormones, erectile dysfunction drugs, mood stabilizers, antimicrobials, detergent metabolites, antioxidants, and many other over-the-counter drugs. Many useful pharmaceuticals become biologically significant ecotoxins after entering water treatment plants, most of which lack the capacity for their removal. These pharmaceuticals then enter the global atmospheric water cycle as contaminants in surface water, quickly entering public water supplies and food webs as the legacy of sophisticated advanced modern medical biotechnologies. They or their metabolites have unanticipated health effects, many of which are just beginning to be documented, on humans, including pregnant women and children, who were not intended to be their targets or consumers. Recent laboratory research has found that small amounts of medication can affect human embryonic kidney cells, blood cells, and breast cancer cells. The cancer cells proliferated too quickly; the kidney cells grew too slowly; and the blood cells showed biological activity associated with inflammation. There are indications that pharmaceuticals in waterways are injuring wildlife worldwide. Male fish are becoming feminized, creating egg yolk proteins, usually an activity of females. Pharmaceuticals also are affecting sentinel species at the foundation of the pyramid of life, including earth worms in the wild and zooplankton in the laboratory (Donn 2008). Growth hormones in pathways to human consumption have the potential to cause unforeseen mutations, such as a lower age of puberty, altered DNA structure, increased risk of cancer, behavioral abnormalities, and other negative health effects. See health effects.

Phthalates (plasticizers): Phthalic acid composed of phthalates with an alkyl attached, are used to add flexibility to a number of plastic products and have been implicated as a cause of endocrine disruption and as possible carcinogens. See DEHP in *Section III. Selected Major Anthropogenic Ecotoxins of Interest*. Some of the most important phthalate cogeners are: mMEP, mEtP, mBuP, mBzP, mEHP, mEOHP, and mEHHP. See *Appendix L* and *Q*.

Plutonium: Pu_{238}, Pu_{239}, Pu_{241}; a manmade radioisotope that also, but rarely, occurs naturally; fissionable Pu-239 is the essential ingredient in atomic weapons. One of the most biologically significant components of stratospheric fallout as well as nuclear energy generation, Pu-239 (½T = 24,100 years) is a hard to detect, alpha-emitting radioisotope, which, if inhaled or ingested, is a potent lung and bone-seeking carcinogen. Pu-238 (½T = 87.4 years) is an alpha-emitting isotope, which in the form of blocks of plutonium oxide is used as a power source in RTG (radioisotope thermoelectric generator) satellites. Volumetrically, most of the plutonium released in nuclear explosions is Pu-241 ($\frac{1}{2}$T = 14.4 years). The daughter product of Pu-241 is Am-241 ($\frac{1}{2}$T = 432 years), a long term bioindicator of anthropogenic radiation in the environment. Due to nuclear weapons testing, the generation of nuclear electricity, nuclear accidents such as Chernobyl, and RTG satellite disintegration, anthropogenic plutonium and its daughter products are omnipresent contaminants in most abiotic and all biotic media. As an alpha-emitter, they are almost impossible to detect except in highly sensitive laboratory environments and are a common persistent contaminant of the world's food webs.

Polybrominated biphenyl (PBB): A close relative of polychlorinated biphenyls and a widely used fire retardant. A large quantity of PBBs was accidentally fed to farm animals in 1973 in Michigan, when the fire retardant *Firemaster*™ was accidentally substituted for the animal feed *Nutrimaster*™ and circulated to farms throughout Michigan, sickening thousands of farm animals. The accident also resulted in the undocumented contamination by ingestion and inhalation of hundreds of thousands of Michigan residents (Kletz 1993).

Polychlorinated Naphthalene (PCN): Organochloride chemicals with the formula $C_{10}H_{10-x}Cl_x$ used for cable insulation, wood preservatives, engine oil additives, capacitors, flame proofing, preservatives, and other commercial and industrial applications. Though currently banned due to their toxicity, particularly to aquatic organisms, several hundred million kilograms have been manufactured and these chemicals persist in ecosystems worldwide due to their stability and lipid solubility, which results in biomagnification. They have been detected in human tissue worldwide, cause liver damage, and are a suspected carcinogen (http://www.inchem.org/documents/cicads/cicads/cicad34.htm).

Polychlorinated Terphenyls (PCTs): Organochloride chemicals with the formula $C_{18}H_{14-x}Cl_x$. They are suspected to act similarly to PCBs in the environment and exhibit the same properties of bioaccumulation and biomagnification. Despite their extremely limited use, they are a ubiquitous contaminant in biotic and abiotic media (Jensen 1983).

Polycyclic aromatic hydrocarbon (PAH): A compound composed of fused aromatic (benzene) rings including only carbon and hydrogen. See major anthropogenic ecotoxins of interest.

Polyvinyl chloride (PVC): The most widely manufactured vinyl plastic used in an un-plasticized form in the production of water and sewer pipes with high chemical resistance; also used extensively in the production of computers, typically 20% by weight. If plasticized, it is used for cable covering, plastic sheeting, moldings, and as a fabric for clothing and furnishings. Relatively nontoxic until subject to combustion or incineration, at which time it becomes a point source for airborne emissions for dioxin-furan-PCB cogeners and their numerous chemical metabolites. Polyvinyl chloride is an anthropogenic industrial product that will play a major role in biocatastrophe in the centuries to come; one of the many skeletons in the closet of pyrotechnic global consumer society.

Pyrethroid insecticides: A group of synthetic insecticides commonly used around the home. 3-phenoxybenzoic acid is a common metabolite of these insecticides. "Very limited scientific information is available on potential human health effects of pyrethroid insecticides at levels presented for the U.S. population." (Centers for Disease Control and Prevention 2005).

Recombinant bovine growth hormone (rBGH): Bovine somatotropin is a hormone fed to cattle to increase milk production, produced synthetically with genetically engineered e. coli bacteria and, until recently, marketed exclusively through Monsanto as Posilac. The administration of the growth hormone rBGH to cattle results in the production of significant elevated levels of the naturally occurring hormone IGF-1 in the liver. The presence of excessive levels of IGF-1 in milk has been linked to a wide variety of health effects, including breast cancer, prostate cancer, and breast enlargement in men. The FDA currently does not regulate rBGH and Monsanto has repeatedly sued media outlets for suggesting it could have negative health impacts, citing studies (the most recent of which is over 16 years old) and indicating rBGH has no impact on dwarfism in humans (www.monsantodairy.com/about/human_safety/index.html). See IGF-1, *Health Effects*, and *Pharmaceuticals*.

Semi-volatile organic chemical (SVOC): Ubiquitous products of the petrochemical industry, semi-volatile organic chemicals are only slightly less volatile than the VOCs (see below). As ecotoxins, both VOCs and SVOCs are widely distributed in food webs and biotic media throughout the biosphere.

Silicon tetrachloride: A potent biologically significant waste product of polysilicon manufacturing resulting from the manufacture of photovoltaics (solar panels, thin film solar cells) and in the microelectronics industry for silicon semiconductors.

Squalene: $C_{30}H_{50}$; a highly ecotoxic, experimental adjuvant (immune enhancer) used by researchers experimenting with mice and other animals to trigger severe autoimmune reactions. Squalene is a potential contaminant in anthrax vaccines administered to veterans during the Gulf War and may have a link to Gulf War Syndrome.

Strontium-90: One of the most biologically significant components of weapons testing stratospheric fallout, strontium-90 ($\frac{1}{2}T = 30$ years) follows the calcium cycle in nature and is commonly ingested by humans after foliar deposition via the grass/cow pathway.

Styrene: C_8H_8; also called vinyl benzene, is used in synthetic rubber, fiberglass, insulation, and plastics. It is toxic, causing sensory impairment at high doses in humans, liver and reproductive damage as well as learning impairment in laboratory animals observed for long-term exposure effects, and is a possible carcinogen (ATSDR 2007).

Teflon: The three principle forms of Teflon are fluorinated ethylene propylene (FEP), perfluoroalkoxy (PFA), and poly(tetrafluoroethylene) (PTFE), all of which are ecotoxins now present in the global atmospheric water cycle and food webs, and constitute "emerging" environmental chemicals of interest. See *Section III. Selected Major Anthropogenic Ecotoxins of Interest*.

Tetrabromobisphenol A (TBBPA): A flame retardant used in printed circuit boards, which is highly toxic to aquatic organisms as well as a persistent organic pollutant (POP). If its chemical bonding to printed circuit boards is disrupted, TBBPA has the potential to become a significant long-lived ecotoxin when released into the environment.

Tetra-ethyl lead (TEL): A highly toxic anti-knock agent added to gasoline to raise the octane level, TEL was invented in the early 1920s by a General Motors chemist and was used in gasoline from 1924 to its final phase-out in 1986. The fine lead particulates emitted by gasoline exhaust resulted in elevated lead levels throughout the biosphere by the early 1970s. Since its phase-out, concentration levels of lead as an ecotoxin in biotic media have been significantly mitigated. Lead continues to be a heavy metal ecotoxin of concern as a ubiquitous product of global consumer society.

Tetrabromobisphenol A: A potent ecotoxin that is a component of the manufacturing of hard plastics. See bisphenol A.

Titanium dioxide: A potent ecotoxin used in many nano-sunscreens and nano-cosmetics that produces harmful free radicals in brain cells, which can result in brain cell damage.

Toxaphene: As a potent insecticide used on foods and to control ticks in livestock, significant quantities of toxaphene were incorporated in food consumed by humans.

Trenbolone: A growth steroid, which as an endocrine disrupting chemical (EDC), masculanizes fish (and humans?) producing females with male attributes.

Trichloroethane (TCA): $C_2H_3Cl_3$; a highly toxic chlorinated hydrocarbon solvent used in the production of semiconductors, TCA along with its sister trichloroethylene (TCE) are potent greenhouse and ozone layer-depleting gases. TCA also has the same chemical formula as 1,1,1-trichloroethane, also known as chlorothene, a potent solvent banned by the Montreal Protocol.

Trichloroethylene (TCE): A highly toxic chlorinated hydrocarbon solvent used in the production of semiconductors and in a wide variety of industrial applications. Along with trichloroethane, TCE is a potent greenhouse and ozone layer-depleting gas. TCE production in 1991 was 321 million pounds and appears to have a strong link with the etiology of Parkinson's disease (Barringer 2009).

Triclosan: Triclosan is an antibacterial agent used in a broad variety of common consumer products, including hospital and household liquid hand soap, detergents, other sanitizing products, toothpaste, cosmetics, plastic cutting boards, footwear (shoe insoles), and any plastic products labeled antibacterial. The popularity of antibacterial consumer products has led to increased use of Triclosan (Perencevich 2001; Tan 2002) and it has been detected in human breast milk and blood samples (Adolfsson-Erici 2002; Peters 2005), and in the urine of 61% of 90 girls age 6 to 8 tested in a recent study spearheaded by Mount Sinai School of Medicine (Wolff et al. 2007). "Triclosan has been found in 42 of the 49 people testing in EWG/Commonweal studies. It has also been found in 1,862 of the 2,514 people tested in CDC biomonitoring studies." (Human Toxome Project; www.bodyburden.org). "Scientists recently found triclosan in 58% of 85 streams across the U.S." (Kolpin, et al. 2002). Some strains of bacteria have already acquired reduced susceptibility to Triclosan (McMurry 1998; Chuanchuen, et al. 2001).

Trifluoromethane: HFC-23; now known as Fluoroform (CHF_3). It is an ozone-depleting greenhouse gas manufactured for use as a refrigeration gas, especially in developing nations.

Trihalomethanes (THMs): A complex family of biologically significant ecotoxins with hundreds of subtypes and cogeners produced during water and wastewater treatment when chlorine is added for disinfection. Highly volatile THMs are now a ubiquitous component of the global atmospheric water cycle and are present in all public and many private water supplies. THMs are associated with both acute and chronic health problems, including liver and kidney damage. Some THMs, such as bromodichloro-methane and bromoform, are also carcinogenic. THMs may interact with other chemical microcontaminants in public water supplies, creating new ecotoxins that have not yet been identified since they are not present in sufficient quantities to be detected, or do not constitute health threats that can be differentiated from those of other ecotoxins.

Uranium: A toxic, radioactive heavy metal, 92 on the periodic table of elements, various isotopes of which are used for nuclear power plants, armor-piercing munitions, and atom bombs. A large quantity of weapons-grade uranium (roughly 600 tons, or 40,000 warheads,) is stored in the former USSR in poorly guarded conditions and smuggling attempts are regularly intercepted by police (Gale Group 2003). Wind-blown dust associated with uranium mining operations and contaminated with uranium U_3O_8 has been a long standing environmental concern in impacted areas of the American southwest. See Depleted Uranium.

Vinclozolin: A fungicide, which as an endocrine disrupting chemical (EDC), interferes with thyroid and androgen hormones producing males with female characteristics and malformed organs.

Volatile organic compound (VOC): Ubiquitous products of the petrochemical industry, VOCs are broken down into the following categories: fumigants, gasoline, hydrocarbons, gasoline oxygenates, organic synthesis compounds, refrigerants, solvents, and trihalomethanes. For a USGS survey of selected VOCs in the nation's water supply, see *Appendix K*.

IV. Health Effects Caused or Affected by Environmental Toxins

This section on the dynamics of biocatastrophe is a brief sketch of some of the most important health physics issues raised by the spread of ecotoxins through the global atmospheric water cycle and throughout the food web of the biosphere. In no way do we intend this précis to constitute an adequate discussion of the health impact of the spread of thousands of ecotoxins through natural and human ecosystems. Suggestions, corrections, additional information, and other comments about any of the following topics or about any major health issues not listed are welcomed by the editors.

Table of Contents – Health Effects

Antibiotic Resistant Bacteria and Other Bacterial Infections

Acinetobacter baumannii: A highly virulent bacteria that occurs in hospitals and nursing homes; some strains are resistant to most currently-available drugs.

Acinetobacter baumannii is one of a category of bacteria that by some estimates is already killing tens of thousands of hospital patients each year. While Acinetobacter organisms do not receive as much attention as the one known as MRSA — for Methicillin-resistant *Staphylococcus aureus* — some infectious-disease specialists say they could emerge as a bigger threat. That is because there are several drugs, including some approved in the last few years, that can treat MRSA. For a combination of business reasons and scientific challenges, the pharmaceuticals industry is pursuing very few drugs for *Acinetobacter* and other organisms of its type, known as Gram-negative bacteria. Meanwhile, the germs are evolving and becoming ever more immune to existing antibiotics. "In many respects it's far worse than MRSA," said Dr. Louis B. Rice, an infectious-disease specialist at the Louis Stokes Cleveland V.A. Medical Center and at Case Western Reserve University. "There are strains out there, and they are becoming more and more common, that are resistant to virtually every antibiotic we have." The bacteria, classified as Gram-negative because of their reaction to the so-called Gram stain test, can cause severe pneumonia and infections of the urinary tract, bloodstream and other parts of the body. Their cell structure makes them more difficult to attack with antibiotics than Gram-positive organisms like MRSA. *Acinetobacter* came to wide attention a few years ago in infections of soldiers wounded in Iraq. Meanwhile, New York City hospitals, perhaps because of the large numbers of patients they treat, have become the global breeding ground for another drug-resistant Gram-negative germ, *Klebsiella pneumoniae*. According to researchers at SUNY Downstate Medical Center, more than 20 percent of the *Klebsiella* infections in Brooklyn hospitals are now resistant to virtually all modern antibiotics. And those supergerms are now spreading worldwide. Health authorities do not have good figures on how many infections and deaths in the United States are caused by Gram-negative bacteria. The Centers for Disease Control and Prevention estimates that roughly 1.7 million hospital-associated infections, from all types of bacteria combined, cause or contribute to 99,000 deaths each year. MRSA thus remains the single most common source of hospital infections; it is especially feared because it can also infect people outside the hospital. There have been serious, even deadly, infections of otherwise healthy athletes and school children. By comparison, the other drug-resistant Gram-negative germs for the most part threaten only hospitalized patients whose immune systems are weak. The germs can survive for a long time on surfaces in the hospital and enter the body through wounds, catheters and ventilators. What is most worrisome about the Gram-negatives is not their frequency but their drug resistance. "For Gram-positives we need better drugs; for Gram-negatives we need any drugs," said Dr. Brad Spellberg, an infectious-disease specialist at Harbor-U.C.L.A. Medical Center in Torrance, Calif., and the author of *Rising Plague*, a book about drug-resistant pathogens… In some cases, antibiotic resistance is spreading to Gram-negative bacteria that can infect people outside the hospital (Pollack 2010a).

Carbapenem-resistant *Enterobacteriaceae* **(CRE):** A highly antibiotic resistant superbug often referred to as a 'nightmare bacteria' have become much more common (Stein 2013).

Clostridium difficile: According to the Centers for Disease Control and Prevention (CDC), the rate of *C. difficile* infections has skyrocketed in recent years. An analysis by Michael Jhung, M.D., of the CDC showed that the number of people hospitalized for *C. difficile*-associated disease went from 148,900 in 2001 to 301,200 in 2005, an increase of 100%. The latest rate indicates a rapid increase over the preceding eight year period, during which there was a 74% increase from 85,700 to 148,900 annually. Among all *C. difficile*-associated discharges, there were 28,600 deaths in 2005 (Smith April 30, 2009, www.medpagetoday.com/InfectiousDisease/GeneralInfectiousDisease/9287). *C. difficile* infections often occur when healthy bacteria in the intestine are killed by the administration of antibiotics for treatment of another illness, allowing the previously controlled bacteria to multiply. Health effects include severe diarrhea, colon rupture, bowel perforation, kidney failure, blood poisoning, or death. While only about 3 to 5% of healthy, nonhospitalized adults are believed to carry *C. difficile* in their intestines, in hospitals and long-term care facilities, about 20% and 55% of patients respectively are carriers.

Cholera: Cholera is an intestinal infection that is spread through water and aggravated by crowded, unsanitary conditions. Although it is treatable with just fluids and tetracycline, when drug and/or personnel supply networks are disrupted, and/or the population has lowered immune response levels, the afflicted die of untreated diarrhea and dehydration. Cholera is endemic in many of the world's impoverished nations, but environmental conditions such as weather and social unrest periodically combine to initiate epidemics. In Zimbabwe, a cholera epidemic is estimated to have killed 1000 people in 2008.

Legionnaires' Disease: Legionnaires' Disease is the most severe form of Legionellosis infection. More than 90% of Legionellosis cases are caused by *Legionella pneumophila*, an aquatic organism that thrives in warm environments (77 to 113°F, with an optimum of approximately 95 °F). Legionnaires' Disease is a type of bacterial pneumonia that is transmitted by breathing in mist from water that contains the bacteria. The mist is usually spread through hot tubs, showers, or air-conditioning units in large buildings where it breeds in warm humid conditions. Symptoms of Legionnaires' Disease are similar to other types of pneumonia and include fever, chills, cough, and sometimes muscle aches and headaches. A chest x-ray is used to diagnose the pneumonia and lab tests can detect the specific bacteria that cause Legionnaires' Disease. See Norovirus.

Malaria: Malaria is a highly infectious water-borne disease transported by mosquitoes in the form of the human parasites *Plasmodium falciparum*, *Plasmodium vivax*, and *Plasmodium knowlesi*. Intensified rainfall events as a result of global warming are likely to increase the mosquito populations that spread malaria.

> **Malaria parasites**: There are four kinds of malaria parasites; *Plasmodium falciparum* is the most virulent type. *Plasmodium falciparum* enters the bloodstream through a mosquito bite; after incubating for about two weeks, it begins to multiply and take over red blood cells, causing fever, chills, headaches, nausea, and other symptoms. Left untreated, the infected cells can block blood vessels and fatally cut off blood supply to vital organs.

Methicillin-resistant staphylococcus aureus (MRSA): MRSA is a rapidly growing bacterial threat that, until recently, was confined to hospitals, nursing homes, and other health care facilities. In the last five years, it has become more and more resistant to antibiotics and has mutated into a community-acquired version. It has also been found in pork in Canada.

The National Institute of Allergy and Infectious Diseases provides the following history of MRSA development. The *Staphylococcus aureus* bacterium, commonly known as staph, was discovered in the 1880s when the infection commonly caused painful skin and soft tissue conditions (boils, scalded-skin syndrome, and impetigo). More serious forms of *S. aureus* infection can progress to bacterial pneumonia and bacteria in the bloodstream, both of which can be fatal. *S. aureus* acquired from improperly prepared or stored food can also cause a form of food poisoning. In the 1940s, medical treatment for *S. aureus* infections became routine and successful with the discovery and introduction of antibiotic medications such as penicillin. Since then, however, use of antibiotics, specifically misuse and overuse, has allowed natural bacterial evolution by helping the microbes become resistant to drugs designed to help fight these infections. Beginning in the late 1940s and throughout the 1950s, *S. aureus* developed resistance to penicillin, which led to the development of Methicillin, a semisynthetic penicillin-related antibiotic, to counter the increasing problem of penicillin-resistant *S. aureus*. In 1961, British scientists identified the first strains of *S. aureus* bacteria that resisted treatment with Methicillin. This was the "birth of MRSA." In 1968, the first human case of MRSA was reported in the United States. Since then, new strains of the bacteria have developed that can now resist Methicillin and most related antibiotics. MRSA is now resistant to an entire class of penicillin-like antibiotics called beta-lactams. This class of antibiotics includes Amoxicillin, Oxacillin, Methicillin, and others. In 2002, physicians in the United States documented the first *S. aureus* strains resistant to the antibiotic Vancomycin, which had been one of a handful of antibiotics of last resort for use against *S. aureus*. Though it is feared that this could quickly become a major issue in antibiotic resistance, thus far, Vancomycin resistant strains are still rare at this time.

Epidemiologists at the University of Iowa have discovered pigs and pig farmers who are carrying Methicillin-resistant Staphylococcus aureus (MRSA), the dangerous bacterium that is responsible for more U.S. deaths than AIDS. The study was the first to document MRSA in U.S. pigs and pig farmers, and the first to find the bacterial strain ST398, shown in Europe to move from pigs to humans, in the United States." (Union of Concerned Scientists. *Dangerous bacterium found in U.S. pigs*; http://www.ucsusa.org/food_and_agriculture/feed/feed-latest.html).

Head and neck MRSA infections: A study published in the January issue of *Archives of Otolaryngology Head & Neck Surgery* and reported in *The New York Times* suggests that the prevalence of pediatric head and neck infections with MRSA has risen alarmingly, documenting the need for "judicious use of antibiotic agents and increased effectiveness in diagnosis and treatment to reduce further antimicrobial resistance in pediatric head and neck infections." (Rabin 2009). A team led by a researcher from Emory University reviewed data from the Surveillance Network, a peer-reviewed national electronic microbiology database that collects strain-specific antimicrobial drug resistance test results from 300 US hospital-affiliated clinical laboratories. They examined 21,009 pediatric head and neck MRSA infections that occurred between January 2001 and December 2006 and found that rates of MRSA head and neck infections more than doubled during the study period, from 11.8 % in 2001 to 28.1% in 2006. The researchers also found that 47 percent of clinical MRSA isolates were resistant to the antimicrobial agent clindamycin (Naseri 2009).

New Delhi metallo-beta-lactamase (NDM-1): "New Delhi metallo-beta-lactamase (NDM-1) is an enzyme that makes bacteria resistant to a broad range of beta-lactam antibiotics." (http://en.wikipedia.org/wiki/New_Delhi_metallo-beta-lactamase).

Plague: Plague is caused by the *Yersinia pestis* microbe.

> **Bubonic plague**: A flea-born bacterial infection, commonly believed to be the disease known as Black Death or Black Plague that killed millions during the Middle Ages, has recently reappeared in Madagascar. Its most potent form, which has a death rate of 90-100% compared to 50% for the more common form, can be transmitted from person to person and is, along with anthrax, considered a potential bioterrorism agent.

Tuberculosis (TB): *Mycobacterium tuberculosis*, the bacterium of the *M. tuberculosis* complex, is the most common causative infectious agent of TB in humans and is now re-emerging as a threat to human health after almost being totally eliminated in the late 20th century. Potent new forms of antibiotic resistant TB are also now emerging, especially in eastern European communities and Africa, including **MDR-TB** (multi-drug-resistant tuberculosis), which is resistant to isoniazid and rifampicin. The most ominous strain is **XDR-TB** (extensively-drug-resistant tuberculosis), a family of TB that is currently incurable, due to its resistance to all known antibiotics.

> **TB/HIV co-infections**: An important emerging world threat to public health, HIV-TB co-infections involving the emerging forms of drug-resistant TB, MDA-TB, and XDR-TB are a disturbing example of the ability of the TB bacteria to mutate into drug-resistant forms, and are already appearing in Africa and Eastern Europe.

Vancomycin resistant enterococcus (VRE): An emerging form of antibiotic resistant bacterial infection, enterococci are bacteria that live in the human intestinal tract and have become resistant to the antibiotic Vancomycin and are now a commonplace ABRB in hospitals and other institutions.

Viruses

Bird flu: "Lab accidents involving bird flu and Ebola viruses have increased biosecurity fears in Europe… deadly H5N1 bird flu virus samples were mixed with seasonal flu samples at a Baxter International contracted laboratory in Austria… The World Health Organisation (WHO) fears that virus, which has killed 256 people since 2003, could trigger a deadly flu pandemic if it mutates and starts to spread more easily." (MacInnis March 29, 2009).

Chikungunya fever: "Chikungunya fever is a viral disease transmitted to humans by the bite of infected mosquitoes. Chikungunya virus was first isolated from the blood of a febrile patient in Tanzania in 1953, and has since been cited as the cause of numerous human epidemics in many areas of Africa and Asia, and most recently in a limited area of Europe." (http://www.cdc.gov/ncidod/dvbid/Chikungunya/CH_FactSheet.html).

Chytridiomycosis: A rapidly spreading fungal disease, "Chytridiomycosis, or chytrid, a deadly fungal disease that has driven at least 200 of the world's 6,700 amphibian species to extinction. One third of the world's frogs, toads and salamanders are threatened. Forty percent are declining. Chytrid's arrival has laid waste to the indigenous Sierra Nevada yellow-legged frog, Rana sierrae… Dr. Vredenburg and colleagues counted 512 populations scattered among the thousands of mountain lakes in the park in 1997. In 2009, 214 of these populations had gone extinct. A further 22 showed evidence of the disease. It is a far cry from the early 1900s, when frogs in the region were so common that lakeside visitors reported trampling them underfoot… Chytrid spreads in a linear wave across the landscape, an infection pattern like that of human epidemics. Infection levels start out light, then increase to very high. Then there is a mass die-off." (Rex 2010, D3).

Dengue fever: Dengue fever is a mosquito-transmitted disease caused by any of four closely related virus serotypes of the genus *Flavivirus*. Infection with one of the serotypes provides lifetime immunity to the infecting serotype only; a second dengue infection from a different serotype is possible, and second infections cause increased risk for dengue hemorrhagic fever (DHF), the more severe form of the disease. DHF is characterized by bleeding manifestations, thrombocytopenia (an abnormal drop in the number of blood cells [platelets] involved in forming blood clots), and increased vascular permeability that can lead to life-threatening shock. Many dengue infections are asymptomatic, and most ill persons likely do not seek medical attention

Hemorrhagic conjunctivitis is a rapidly progressing and contagious infection of the family Picornaviridae (picornaviruses) that affects the whites of the eyes. Hemorrhagic conjunctivitis usually begins as eye pain, followed quickly by red, watery, swollen eyes, light sensitivity, and blurred vision.

Hendra virus: "Hendra virus (HeV) is a rare, emerging zoonotic virus (a virus transmitted to humans from animals), that can cause respiratory and neurological disease and death in people. It can also cause severe disease and death in horses, resulting in considerable economic losses for horse breeders. Initially named Equine Morbilivirus, Hendra virus is a member of the genus *Henipavirus*, a new class of virus in the *Paramyxoviridae* family. It is closely related to Nipah virus. Although Hendra virus has caused only a few outbreaks, its potential for further spread and ability to cause disease and death in people have made it a public health concern." (http://www.who.int/mediacentre/factsheets/fs329/en/index.html).

Herpangina is an infection of the throat that causes red-ringed blisters and ulcers on the tonsils and soft palate. The blisters are usually accompanied by fever. Most of the many viruses that cause Herpangina are coxsackieviruses or other enteroviruses. See Hemorrhagic fevers.

HIV/AIDS (Human Immunodeficiency Virus/Acquired Immunodeficiency Syndrome): The greatest threat to world health to emerge in the late 20th century, AIDS is the disease that develops when humans are infected with HIV. Worldwide, 32,900,000 people are currently infected with HIV or AIDS and 2,000,000 died of the disease in 2007. In the US, 1,200,000 people are living with HIV/AIDS and 22,000 died in 2007 (UNAIDS 2008). A major world health issue in the 21st century is and will be the emergence and spread of HIV-TB co-infections, with the possible evolution of HIV-TB strains that are resistant to recently developed and increasingly effective AIDS treatment medications.

Israeli acute paralysis virus: A newly emerging virus, which, according to the EPA, may be a possible cause of the 2006/7 honey bee colony collapse disorder (www.epa.gov/pesticides/about/intheworks/honeybee.htm).

Newcastle disease virus: "Newcastle disease is a contagious bird disease affecting many domestic and wild avian species… Exposure of humans to infected birds (for example in poultry processing plants) can cause mild conjunctivitis and influenza-like symptoms, but the Newcastle disease virus (NDV) otherwise poses no hazard to human health. Interest in the use of NDV as an anticancer agent has arisen from the ability of NDV to selectively kill human tumour cells with limited toxicity to normal cells." (http://en.wikipedia.org/wiki/Newcastle_disease).

Nipah virus: A newly emerging virus, which first appeared in Malaysia in 1998; it is transmitted by fruit bats to pigs. Humans then get the virus from contact with infected pigs. Flu-like symptoms can lead to encephalitis and catatonia. Nipah is fatal in 40% of cases.

Norovirus: Noroviruses are a group of related, single-stranded RNA, non-enveloped viruses that cause acute gastroenteritis in humans. Norovirus was recently approved as the official genus name for the group of viruses provisionally described as "Norwalk-like viruses" (NLV). Also known as "cruise line" disease, norovirus usually occurs in environments where people gather in enclosed spaces. The illness is characterized by acute onset of nausea, vomiting, abdominal cramps, and diarrhea. Vomiting is relatively more prevalent among children, whereas a greater proportion of adults experience diarrhea. Primary cases result from exposure to a fecal-contaminated vehicle (e.g., food or water), whereas secondary and tertiary cases among contacts of primary cases result from person-to-person transmission (CDC, June 1, 2001, *Morbidity and Mortality Weekly Report*). See Legionnaires' Disease.

Novel H1N1 "Swine" influenza (April 2009): The new form of swine influenza that suddenly appeared in Mexico in late April of 2009 is, according to Dr. Ann Schuchat of the CDC, "found to be made up of genetic elements from four different flu viruses – North American swine influenza, North American avian influenza, human influenza, and swine influenza virus typically found in Asia and Europe – 'an unusually mongrelized mix of genetic sequences'." (http://en.wikipedia.org/wiki/Swine_flu). The outbreak appears to have originated in the manure lagoons of a Mexican pig farm, where viruses originally from Asia and Europe mixed with North American influenza viruses to form a new form of the H1N1 virus not previously seen before this outbreak. "This new strain appears to be a result of re-assortment of human influenza and swine influenza viruses, presumably due to superinfection in an individual human. Influenza viruses readily undergo re-assortment because their genome is split between eight pieces of RNA." This outbreak is the prototypical form of zoonosis in humans, i.e. an example of the transmission of a new mutated form of viral infection from animals to humans. The 1918-19 pandemic was also a form of swine flu. For an excellent bibliography of information sources on the current outbreak of swine influenza, see the footnotes on the Wikipedia page (April 28, 2009; http://en.wikipedia.org/wiki/Swine_flu).

Rinderpest: "In only the second elimination of a disease in history, rinderpest — a virus that used to kill cattle by the millions, leading to famine and death among humans — has been declared wiped off the face of the earth. Rinderpest, which means 'cattle plague' in German, does not infect humans, though it belongs to the same viral family as measles. But for millenniums in Asia, Europe and Africa it wiped out cattle, water buffalo, yaks and other animals needed for meat, milk, plowing and cart-pulling. Its mortality rate is about 80 percent — higher even than smallpox, the only other disease ever eliminated." (McNeil 2010).

Severe acute respiratory syndrome (SARS): SARS is a viral respiratory illness caused by a coronavirus – SARS-associated coronavirus (SARS-CoV). SARS was first reported in Asia in February 2003. Over the next few months, it spread to North and South America, Europe, and Asia before the global outbreak was contained. Although SARS has not been active worldwide since 2004, with a death rate of about 10%, in a world of rapidly increasing global trade and transportation, SARS could be one of the most dangerous emerging viral infections in the Age of Biocatastrophe. During the last widespread outbreak, there were 774 deaths among 8,096 cumulative victims (WHO; www.who.int/csr/sars/country/table2004_04_21/en/index.html).

Viral Hemorrhagic Fevers: Viral hemorrhagic fevers are a group of febrile illnesses caused by several distinct families of viruses, all of which are enveloped and have RNA genomes. These groups include Ebola and Marburg viruses, Lassa fever virus, the New World arenaviruses (Guanarito, Machupo, Junin, and Sabia), Rift Valley fever, dengue fever, and Crimean Congo hemorrhagic fever viruses (1-3). Although some types can be mild, many of these viruses can cause severe, life-threatening disease. Severe illness is characterized by vascular damage and increased permeability, multi-organ failure, and shock.

During recorded outbreaks of hemorrhagic fever caused by filovirus infections (e.g., Ebola), persons who care for (feed, wash, or prepare for burial ceremonies) or work very closely with infected persons are at highest risk for infection. Health care-associated transmission through contact with infectious body fluids has been an important factor in the spread of this disease.

The reservoir hosts for Ebola and Marburg viruses have not yet been identified. Outbreaks occur when a patient who has been exposed to that unknown reservoir species or an infected nonhuman primate transmits the virus to humans. The outbreak often becomes amplified in the health-care setting due to ineffective prevention and control procedures.

    Coxsackie virus: Coxsackie viruses are part of the enterovirus family of viruses (which also include polioviruses and hepatitis A virus) that live in the human digestive tract. Enteroviruses are extremely contagious, can live for several days, and spread from person to person, usually on unwashed hands and surfaces contaminated by feces. In most cases, Coxsackie viruses cause mild flu-like symptoms and go away without treatment, but they can lead to more serious infections such as viral meningitis (an infection of the three membranes, meninges, that envelop the brain and spinal cord), encephalitis (a brain infection), and myocarditis, (an inflammation of the heart muscle).

> **Hand, foot, and mouth disease**: A type of Coxsackie virus syndrome that causes painful red blisters in the throat and on the tongue, gums, hard palate, inside of the cheeks, the palms of hands, and the soles of the feet.

Lassa fever: An arenaviral disease caused by a virus transmitted from asymptomatically infected rodents to humans. Transmission of arenaviruses to humans can occur via inhalation of primary aerosols from rodent urine, by ingestion of rodent-contaminated food, or by direct contact of broken skin with rodent excreta. Rodent infestation, facilitated by inappropriate food storage, increases the risk of human infection. Person-to-person spread of Lassa and Machupo viruses has also been described, most notably by large droplets and contact transmission in hospital settings.

Rift Valley Fever (RVF): It is transmitted by the bites of mosquitoes, percutaneous inoculation, or exposure to aerosols from contaminated blood or fluids of infected animals. RVF virus is endemic to sub-Saharan Africa, where sporadic outbreaks occur in humans. An outbreak of RVF occurrFed in 2000 in southwestern Saudi Arabia and Yemen with a strain of RVF closely related to that of the 1997-1998 East African strain. This outbreak represented the first spread of the virus outside Africa, demonstrating its potential for spread to unaffected regions elsewhere in the tropics.

The worldwide threats posed by the now multiplying and spreading varieties of viral hemorrhagic fevers will be among the foremost of many challenges to maintaining human health in less affluent communities as the Age of Biocatastrophe unfolds. Declining public and NGO resources and increasing infection rates from emerging and remerging viral and bacterial infections will be greatly exacerbated if the global financial crisis is not quickly resolved. Cataclysmic climate change may also intensify the impact of emerging new forms of hemorrhagic fevers by supplying the warmer and moister conditions in which the viruses breed.

West Nile virus: West Nile virus is spread by mosquitoes and has been an emerging public health threat in the northeastern United States and many other locations for the last two decades. The appearance of dead crows and other birds often signals its presence.

Xenotropic murine leukemia virus-related virus (XMRV): First described in 2006, the xenotropic MLVs are a subspecies of the murine leukemia viruses (MLVs), which are both exogenous and endogenous. Recent research has associated the XMRV with both prostate cancer and chronic fatigue syndromes (CFS). Initially unconfirmed by other studies, a recent paper issued by researchers from the National Institutes of Health, Food and Drug Administration, and Harvard Medical School has reasserted the link of this virus with the etiology of chronic fatigue syndrome. As a result "some chronic fatigue patients are already trying HIV medications" (Tuller 2010, D6) in response to the possibility that CFS is caused by a retrovirus. Concern about MRV related viruses in the blood supply was also noted in *The New York Times* article on this subject.

Cancers

Due to space restrictions, the editors have decided that, aside from short notes based on well-established findings, which are included in *Section III. Selected Major Anthropogenic Ecotoxins of Interest*, no attempt will be made to open the Pandora's Box of links between anthropogenic ecotoxins and the many forms of cancer that have been connected to their growing presence in the biosphere. There is a vast body of existing literature on this subject.

However, two important observations should be made about cancer: its worldwide incidence is continuing to rise despite great advances in medical technology, and, it is almost impossible to attribute a specific type of cancer to groups of or specific ecotoxins. The link between cigarette smoke and lung cancer represents one of a small number of cases where such a correlation has been demonstrated.

Other Emerging or Reemerging Diseases

Fungal infections: Fungi (i.e. mushrooms, molds, and mildew) are actually primitive vegetables that live in air, in soil, on plants, in water, and in the human body. Only about half of all fungi are harmful. The most common problem-causing fungi are: **Candida albicans, Histoplasmosis, Sporotrichosis,** and **Microsporidia.** Candida albicans is a yeast microorganism that lives in the intestinal tract of every human being. The introduction of the mass use of antibiotics paralleled the rise in yeast infections. Initially a prescription for antibiotics directly preceded almost all yeast infections. By the 1970s, women and men began to have yeast infections even when they had not been taking antibiotics, and fungal infections, including yeast, are common today.

Histoplasmosis is caused by the fungus *Histoplasma capsulatum. H. capsulatum* grows in soil and other media contaminated with bat or bird droppings. The spores become airborne when contaminated soil is disturbed. Breathing the spores causes infection. **Sporotrichosis** is caused by *Sporothrix schenckii* and usually infects the skin. The fungus can be found in sphagnum moss, hay, other plant materials, and in the soil. **Microsporidia** is used as a general nomenclature for obligate intracellular protozoan parasites belonging to the phylum Microsporidia. More than 1,200 species in 143 genera have been identified. Human microsporidiosis represents an important and rapidly emerging opportunistic disease. The majority of cases occur in severely immunocompromised patients with AIDS, but also in persons not infected with HIV, as well as in immunocompetent persons (NIH – National Institute of Allergy and Infectious Diseases; http://www.nationalcandidacenter.com/; www.dpd.cdc.gov).

Cryptococcus gattii: "*Cryptococcus* is a genus of fungi… [which] primarily affects HIV-uninfected persons in tropical and subtropical regions." (DeBless 2010). First emerging on Vancouver Island and mainland British Columbia, this fungi infection has now spread to the United States where 12 cases were reported in Oregon in 2006-7. "By July 2010, a total of 60 human cases had been reported to CDC from four states (California, Idaho, Oregon, and Washington) in the Pacific northwest… Among 45 patients with known outcomes, nine (20%) died because of *C. gattii* infection... Physicians should consider *C. gattii* as a possible etiology of a cryptococcal infection among persons living in or traveling to the Pacific Northwest or traveling to other *C. gattii* -- endemic areas." (DeBless 2010).

***Haplosporidum nelsoni* (MSX)**: "*Haplosporidium nelsoni* is a pathogen of oysters, that originally caused oyster populations to experience high mortality rates in the 1950s and still is quite prevalent today. The disease caused by *H. nelsoni* is also known as MSX (multi-nucleated unknown)." (http://en.wikipedia.org/wiki/Haplosporidium_nelsoni). "It is not harmful to humans and can be present in small numbers without hurting oysters." (Deese 2011, 9).

Heart and skeletal muscle inflammation (HSMI): "HSMI is a frequently fatal disease of farmed Atlantic salmon… HSMI is associated with infection with piscine reovirus (PRV). PRV is a novel reovirus." (Palacios 2010).

Shell disease: Shell disease, also known as rust disease and by many other names, is characterized by a syndrome or group of signs and symptoms, in this case the progressive erosion of the shells of crustaceans. It is known to occur in blue crabs, king crabs, shrimps, prawns, crayfish, and lobster. Its direct cause is "bacteria that are capable of eroding the three innermost chitinous layers of the shell" (Lavalli 2003). "Lobster shell disease has been a contentious issue in Connecticut since the beginning of a massive lobster die-off in Long Island Sound in the late 1990s… The catch has plummeted, with catches falling to about one-sixth of their 1998 levels. Factors such as mosquito insecticides and warming temperatures have been implicated as potential risk factors for the disease, which creates dark lesions on the outside of the lobsters' shells that, over time, bore through the shell to the membranes underneath… [Hans Laufer] has found that by interfering with hormones crucial to young lobster growth, chemicals such as bisphenol A can slow the lobsters' molting patterns and interfere with regular development, leading to body deformations, susceptibility to disease, and potential death... He and his colleagues identified "hotspots" in the Sound where lobsters have high levels of alkylphenols – a group of chemicals derived from detergents, paints, and plastics – circulating in their bodies. The lobsters take in these chemicals from their food." (Buckley 2010). At the present time, shell disease has also been noted in the Gulf of Maine bioregion, but in a much smaller percentage of the lobster harvest than has occurred south of the Cape Cod Canal. This may relate to the fact that riverine water discharges into the Gulf of Maine contain significantly less concentrations of alkylphenols such as phthalates than do rivers draining the more industrialized regions of southern New England west of Buzzards Bay.

Emerging Tropical Diseases

Hundreds of thousands of low-income U.S. residents in inner cities, the Mississippi Delta, Appalachia, areas near the Mexico border, and tribal reservations remain undiagnosed and untreated for diseases that are prevalent in Africa, Asia and Latin America, according to an analysis published in *PLoS Neglected Tropical Diseases*.

The analysis, conducted in 2007 by Peter Hotez of the Global Network for Neglected Tropical Diseases found that residents in those areas are more likely to have mental retardation, heart disease, and epilepsy, among other conditions, caused by untreated tropical and other infectious diseases (http://gnntdc.sabin.org/). These diseases primarily affect women and children in those areas, according to the analysis (Sternberg 2008). The diseases include Chagas' disease, cysticercosis, and worm diseases, as well as dengue fever, syphilis, and cytomegalovirus (McNeil 2008). "If this were occurring among white mothers in the suburbs, you'd hear a tremendous outcry," Hotez said.

Stem Rust epidemics: The term stem rust epidemic refers to the reemergence of over 40 formerly well controlled fungal infections of wheat. The recent appearance of new strains of fungi, as well as the reappearance of old strains, has occurred in India, Yemen, and Pakistan, and has the potential to spread to US wheat crops as bioengineered varieties of wheat become increasingly susceptible to infection (Charles May 5, 2008, NPR). "In 1999, a new virulent race of stem rust emerged in Uganda called UG-99 that infects most current resistant wheat varieties by overcoming the most common resistance genes." (http://maswheat.ucdavis.edu/education/PDF/facts/rustfacts.pdf). "90 percent of the world's wheat has little or no protection against the Ug99 race of *P. graminis*. If nothing is done to slow the pathogen, famines could soon become the norm." (Koerner 2010).

Viral hemorrhagic septicemia (VHS) of fishes: A newly emerging and now pandemic systemic rhabdoviral infection of fresh water fish, including salmonoids (salmon), lake trout, steelhead trout, shad, and other fish. First noted as spreading rapidly in Pacific coastal states, VHS is now virulent in Great Lake fisheries and a worldwide threat to all lacustrine ecosystems. Having several serotypes and transmittable via fish urine, the spread of VHS may be connected to chemical fallout contaminants in the global atmospheric water cycle.

Other Syndromes and Conditions

Black mold: A common fungi, which can cause life threatening infections as well as warts, lesions, toenail infections, pneumonia, and sinusitis. Especially common in warm moist environments, such as Florida, black mold infections are antibiotic resistant. The 2012 meningitis outbreak from contaminated steroid injections was caused by Exserohilum, a type of black mold (Grady 2012). The CDC is investigating this multistate fungal meningitis outbreak.

Gulf War Syndrome/Illness: Gulf War Illness presents with a wide range of symptoms including: chronic fatigue, frequent vomiting, diarrhea, severe weight loss, coughing, joint pain, headaches, loss of memory and concentration, sleep disorders, rash, swollen lymph nodes, nervous system disruptions, night sweats, personality changes, and tumors. Originally linked to contaminants in an anthrax vaccine, other theories have been offered over the fifteen years in which it has been studied. In 2008, Gulf War Illness was accepted by the VA as an actual disease with benefits owed to those who develop it after military service (Silverleib 2008). The latest studies have determined that … there is an increase in 4 out of the 12 medical conditions reportedly associated with Gulf War Syndrome (fibromyalgia, chronic fatigue, eczema, and dyspepsia).

In 2008, the U.S. National Academy of Sciences published evidence suggesting that the large number of illnesses in Gulf War veterans can be explained in part by their exposure to acetylcholinesterase inhibitors. On November 17, 2008, the federally mandated Research Advisory Committee on Gulf War Veterans' Illnesses produced a 452-page report indicating that roughly 1 in 4 of the 697,000 veterans who served in the first Gulf War are afflicted with the disorder. Exposure to toxic chemicals was identified as the cause of the illness. The report states that "scientific evidence leaves no question that Gulf War illness is a real condition with real causes and serious consequences for affected veterans."

High fructose corn syrup (HFCS): "Since 1980, obesity rates in children have tripled. Today, 13 million children are obese, including 14 percent of all 6- to 11-year-olds, and 17 percent of adolescents. Over 70 percent of these children will be obese adults, with increased risks of diabetes, heart disease, and certain cancers… High-fructose corn syrup now represents 40 percent of the non-calorie-free sweeteners added to U.S. foods. It is virtually the only sweetener used in soft drinks. Because of subsidies, the cost of soft drinks containing HFCS has decreased by 24 percent since 1985, while the price of fruits and vegetables has gone up by 39 percent. By 2006, the average American child drank 132 calories of HFCS per day from sweetened beverages." (Landrigan 2010b).

Lead poisoning: Lead has been shown to cause a wide variety of health effects. Chronic effects often seen with significant long-term exposure include loss of appetite, constipation, nausea, and stomach pain. Symptoms also can include excessive tiredness, weakness, weight loss, insomnia, headache, nervous irritability, fine tremors, numbness, dizziness, anxiety, and hyperactivity. Because these symptoms are common to a variety of health problems, they can be easily missed.

Mercury (Hg) Poisoning: Mercury is a nerve poison and especially dangerous for unborn and developing babies. Evidence is building that mercury and PCBs together are much more toxic than either one individually. Some of the learning disabilities seen in Great Lakes children studies may be due to the multiplied health effect (synergism) of these chemicals together (Fox River Watch 2009). See the comments on mercury in both *Section III. Selected Major Anthropogenic Ecotoxins of Interest* and *Other Anthropogenic Ecotoxins of Interest*.

Obesity: "In the past 20 years, the rates of obesity have tripled in developing countries that have been adopting a Western lifestyle involving decreased physical activity and overconsumption of cheap, energy-dense food… In the developed world, 2 to 7% of total health care costs are attributable to obesity… The increase in the prevalence of type 2 diabetes is closely linked to the upsurge in obesity. About 90% of type 2 diabetes is attributable to excess weight… Consequently, diabetes is rapidly emerging as a global health care problem that threatens to reach pandemic levels by 2030; the number of people with diabetes worldwide is projected to increase from 171 million in 2000 to 366 million by 2030… overweight and obesity are contributing to a global increase in hypertension: 1 billion people had hypertension in 2000, and 1.56 billion people are expected to have this condition by 2025. This increase will have a disproportionate effect on developing countries, where the prevalence of hypertension is already higher than that in developed countries and where cardiovascular disease tends to develop earlier in affected persons." (Hossain 2007, 213).

Red tide: An algae bloom caused by a dinoflagellate phytoplankton *Alexandrium fundyense* that creates harmful toxins, "1,000 times more potent than cyanide" (Deese 2010) that contaminate shellfish, producing neurotoxic shellfish poisoning (NSP). A centuries-old seasonal phenomenon, red tide may be enhanced, especially in states with extensive agriculture and a warm climate, such as Florida, by industrial agricultural activities, which produce enhanced phosphates and nitrogen runoff. Due to its extensive coastline (3,478 miles) of enclosed estuaries, bays, and river basins, Maine shellfisheries are particularly vulnerable to red tide, as are the warmer inshore ocean waters south of Cape Cod.

Sjögren's ("SHOW-grins") **Syndrome**: A common autoimmune disorder as well as a new chronic systemic disease in which white blood cells attack the moisture-producing glands of the person afflicted (kidneys, gastrointestinal tract, blood vessels, lung, liver, pancreas, central nervous system). Sjögren's also causes debilitating fatigue and joint pain in many patients and is one of the most prevalent autoimmune disorders in the U.S., affecting an estimated four million Americans, nine out of ten of whom are women. In approximately 50% of cases, Sjögren's syndrome occurs alone (primary Sjögren's), but it also occurs concurrently with another connective tissue disease (secondary Sjögren's). The four most common disorders that co-exist with Sjögren's syndrome are rheumatoid arthritis, systemic lupus, systemic sclerosis (scleroderma) and polymyositis/dermatomyositis (Sjögren's Syndrome Foundation www.sjogrens.org).

The etiology of Sjögren's syndrome is not completely understood but may include hereditary, viral or bacterial infections. The disease etiology may also be impacted by the increasing intake of chemical fallout ecotoxins, which are neurotoxic or may cause other disruptions of the immune system.

Stevens-Johnson Syndrome (SJS): A life-threatening hypersensitivity complex affecting the mucus membranes and the skin, caused by medications such as dicloflex, penicillins, phenytoin, Nodafill, ibuprofen, etc. It is a form of toxic epidermal necrolysis.

Transthyretin (TTR): A protein, which is an essential component of normal thyroid function, and to which endocrine disrupting chemicals (EDCs) such as PBDEs, PBPs (pentabromo-phenols), and TBBPA (tetrabromobisphenol) bind. These EDCs, including "flame retardants, are similar in chemical form to natural human thyroid T4… [and] interfere with the natural binding of T4 to transthyretin thus potentially leading to disruption of the thyroid control over brain development." (Ilonka 2000).

Pharmaceuticals

There is the possibility that the agricultural, industrial, and consumer production of pharmaceutical drugs could result in the proliferation of pharmaceutical-derived ecotoxins that may equal or exceed the health physics impact of the more common persistent organic pollutants, which now number in the hundreds. The rapid expansion in production and distribution of pharmaceutical drugs for personal use, and agricultural production, including outdoor pharmaceutical crop production, has occurred during the last four decades. This proliferation of pharmaceutical drugs has produced contaminant pulses in the global atmospheric water cycle of hundreds of these pharmaceuticals as potential ecotoxins, which now contaminate most public water supplies and biotic media. The following series of excerpts provides a variety of commentary from several different perspectives on the health physics significance of the proliferation of pharmaceuticals in pathways to human consumption, i.e. these pharmaceuticals are now being recycled and are being ingested in tiny amounts by most of the human race. A key issue explored in the following observations is the long term synergistic interaction of these pharmaceuticals as ecotoxins in biotic media, including humans and animals.

Selected quotes from: *Pharmaceuticals and Personal Care Products (PPCP's) in the Environment: Overarching Issues and Overview* by Christian G. Daughton, Chief, Environmental Chemistry Branch, ESD/NERL, Office of Research and Development, EPA, Las Vegas, NV; www.epa.gov/ppcp/:

"The fact that PPCPs can be introduced on a continual basis to the aquatic environment via treated and untreated sewage essentially imparts a quality of 'persistence' to compounds that otherwise may not possess any inherent environmental stability -- simply because their removal/transformation (by biodegradation, hydrolysis, photolysis, etc.) is continually countered by their replenishment…"

"Perhaps more so than with any other class of pollutants, the occurrence of PPCPs in the environment highlights the intimate, inseparable, and immediate connection between the actions, activities, and behaviors of individual citizens and the environment in which they live. PPCPs, in contrast to other types of pollutants, owe their immediate origins in the environment directly to their worldwide, universal, frequent, highly dispersed, and individually small but cumulative usage by multitudes of individuals -- as opposed to the larger, highly delineated industrial manufacturing/usage of most high-volume synthetic chemicals… Their introduction to the environment has no geographic boundaries or climatic-use limitations as do many other synthetic chemicals. They are discharged to the environment wherever people… live or visit, regardless of season."

"Sewage and domestic wastes are the primary sources of PPCPs in the environment … These bioactive compounds are continually introduced to the environment (primarily via surface and ground waters) from human and animal use largely through sewage treatment works systems, failed septic fields, leaking underground sewage conveyance systems, and wet-weather runoff -- either directly by bathing/washing/swimming … or indirectly by excretion in the feces or urine of unmetabolized parent compounds. Bioactive metabolites (including reconvertible conjugates) are also excreted."

"Any chemical introduced to the aquatic domain can lead to continual exposure for aquatic organisms. Chemicals that are continually infused to the aquatic environment essentially become "persistent" pollutants even if their half-lives are short. Their supplies are continually replenished and this leads to life-long multi-generational exposures for aquatic organisms."

"Subtle, unnoticed effects could accumulate over time until any additional incremental burden imposed by a new, unrelated stressor could possibly trigger sudden collapse of a particular function or behavior across a population..."

Selections from India's Waterways a Toxic Stew of Pharmaceutical Chemicals Dumped from Big Pharma Factories by Mike Adams (2009):

"Under the Bush Administration, the U.S. Environmental Protection Agency refused to regulate pharmaceuticals as environmental hazards. It is still unclear whether the new administration will clamp down on pharmaceutical pollution."

"In many ways, Big Pharma's chemicals are far more dangerous than the industrial and household chemicals. HRT drugs, for example, are toxic at parts per billion, and they're now being found in public water supplies around the world."

"Municipal treatment facilities don't remove pharmaceutical chemicals from the water; they are passed through the water treatment centers which add more chemicals (fluoride and chlorine, typically) to the toxic brew. Citizens drinking public water supplies in India, the U.K., Canada and the US are now verifiably participating in a grand experiment involving mass medication of a population with low levels of untested pharmaceutical and pharmaceutical/other chemical combinations."

"So far, environmental regulators have done nothing to stop the dumping of drugs into public water supplies. This is true even in America, where hospitals, pharmacies, nursing homes, and doctors' offices routinely dispose of drugs by simply flushing them down the toilet (injecting them directly into the downstream water supply."

"Drugs that consumers swallow are also environmental pollutants. Many drugs pass right through the human body unaltered, where they are flushed back into the water supply. (Toilet water from one city becomes drinking water farther down the river.)"

"By allowing factories to dump drugs into local waterways, by tolerating a 'flush it' mentality at medical facilities, and by drugging consumers with an endless brew of vaccines, medications, and other toxic substances such as chemotherapy agents, the pharmaceutical industry has achieved the distinction as a major world polluter."

"When we destroy or disrupt the planet's delicate ecosystems through chemical contamination, we unleash a backlash of effects that put the entire human race in jeopardy: Outbreaks of infectious diseases, plummeting fish stocks in ocean waters, rising risks of superbugs across the population, and long-term disruptions in the food supply due to pharmaceutical contamination of food crops and soil microorganisms."

92

Selections from a second EPA report "*Pharmaceuticals and Personal Care Products in the Environment: Agents of Subtle Change?*" by Christian G. Daughton and Thomas A. Ternes, Environmental Sciences Division, U.S. EPA; http://www.epa.gov/ppcp/:

"Fragrances (musks) are ubiquitous, persistent, bioaccumulative pollutants that are sometimes highly toxic; amino musk transformation products are toxicologically significant. Synthetic musks comprise a series of structurally similar chemicals ... used in a broad spectrum of fragranced consumer items, both as fragrance and as fixative. Included are the older, synthetic nitro musks ... and a variety of newer, synthetic polycyclic musks that are best known by their individual trade names or acronyms. The polycyclic musks ... and especially the inexpensive nitro musks (nitrated aromatics accounting for about one-third of worldwide production) are used in nearly every commercial fragrance formulation (cosmetics, detergents, toiletries) and most other personal care products with fragrance; they are also used as food additives and in cigarettes and fish baits."

"The nitro musks are under scrutiny in a number of countries because of their persistence and possible adverse environmental impacts and therefore are beginning to be phased out in some countries…The human lipid concentration of various musks parallels that of other bioaccumulative pollutants such as PCBs."

"The specter of subtle, cumulative effects could make current toxicity-directed screening largely useless in any effort to test waste effluents for toxicologic end points."

"Abnormal behavior can masquerade as seemingly normal deviation within a natural statistical variation. Change can occur so slowly that it appears to result from natural events, with no reason to presume artificial causation. It is difficult to connect the issues of cause and ultimate effect, in part because of the ambiguous and subjective nature of subtle effects, but especially when these effects are confounded as aggregations of numerous, unrelated interactions."

Excerpts from *Meds Found in Water May be Affecting Humans* (CBS News 2008):

"A vast array of pharmaceuticals – including antibiotics, anticonvulsants, mood stabilizers and sex hormones – have been found in the drinking water supplies of at least 41 million Americans, an Associated Press investigation shows… The presence of so many prescription drugs – and over the counter medicines like acetaminophen and ibuprofen – in so much of our drinking water is heightening worries among scientists of long term consequences to human health… Even users of bottled water and home filtration systems don't necessarily avoid exposure. Bottlers, some of which simply repackage tap water, do not typically treat or test for pharmaceuticals… In the United States, the problem isn't confined to surface waters. Pharmaceuticals also permeate aquifers deep underground, source of 40% of the nation's water supply… There is evidence that adding chlorine, a common process in conventional drinking water treatment plants, makes some pharmaceuticals more toxic... Many concerns about chronic low level exposure focus on certain drug classes: chemotherapy that can act as a powerful poison, hormones, hormones that can hamper reproduction or development, medicines for depression and epilepsy that can damage the brain or change behavior, antibiotics that can allow human germs to mutate into more dangerous forms, pain relievers and blood-pressure diuretics... Some experts say medications may pose a unique danger because, unlike most pollutants, they were crafted to act on the human body... 'These are chemicals that are designed to have very specific effects at very low concentrations. That's what pharmaceuticals do. So when they get out to the environment it should not be a shock to people that they have effects.' Says zoologist John Sumpter at Brunel University in London." (http://cbs3.com/national/medicine.found.water.2.673206.html).

GMOs

Genetic engineering has become a major force in the agriculture industry in the past 10 years. Monsanto, one of the first corporations to sponsor the proliferation of for-profit manufacturing and distribution of genetically modified autotrophs, has been a leader in developing pesticide-resistant crops.

Unlike clones, which are supposed to be genetically identical to animals that already exist, genetically engineered plants and animals have DNA from other organisms, often other species, inserted into their genome. Milk and meat from cloned cows, pigs and goats received a blanket approval from the FDA in January 2008. Rapid advances in the science of bioengineering are a result of the proliferation of a wide variety of genetically modified food-producing animals such as cattle, chicken, and pigs. Genetically modified animals have been preceded by several decades of the genetic modification of basic crops such as soy beans, corn, cotton, wheat, and, more recently, tomatoes, carrots, apples, and a wide variety of other food crops.

Recent advances in genetic engineering have resulted in the development of numerous animals that have significant medical uses. In particular, pigs have been used to develop milk to treat a rare type of hemophilia; another strain of milk helps anemia patients produce more red blood cells. The somewhat famous Enviropig was developed in 2001, efficiently reducing pollution by excreting 50% less phosphorus than normal pigs. Also coming from the pig farm are transgenic pig parts, insulin-producing pig cells for diabetes patients and transgenic pig livers. Another well publicized innovation is the fluorescent zebra fish now utilized to detect water pollution. An FDA committee has also recently approved ATryn, a blood thinner made from the milk of genetically engineered goats. The European Commission already approved the drug in August

2006, and the FDA is expected to approve it for sale in the U.S. in early 2009. Of more universal efficacy are genetically engineered dairy cows that not only can produce more milk, but can resist mastitis, mad cow disease, and brucellosis (Minogue 2009). "Transgenic animals are probably going to be the drug stores of the future," said biologist Bryan Pickett, who works with genetically engineered zebra fish at Loyola University of Chicago. These are all examples of genetically modified organisms, which were produced as socially, medically, or environmentally useful entities. The long term genetic and biological impact of these genetically altered animals is unknown.

The Nature Institute maintains a website dedicated to tracking nontarget effects of genetic engineering at: http://natureinstitute.org/nontarget/. Some examples include: pigs fed GM canola with higher concentrations of undesirable substances had reduced daily weight gain; Atlantic salmon fed Bt corn had altered liver and intestinal enzyme activity and alterations in white blood cells; mice fed genetically modified, weevil-resistant peas showed immune reactions; pigs and chickens had reduced starch digestion; insulin-storing potatoes had higher alkaloid content and pigs fed on them had reduced daily weight gain; a diet containing glyphosate-resistant soybeans affected liver cells in mice. The Nature Institute asks, "Do we really know what we're doing?"

Rapid growth in adoption of genetically engineered crops in the U.S.

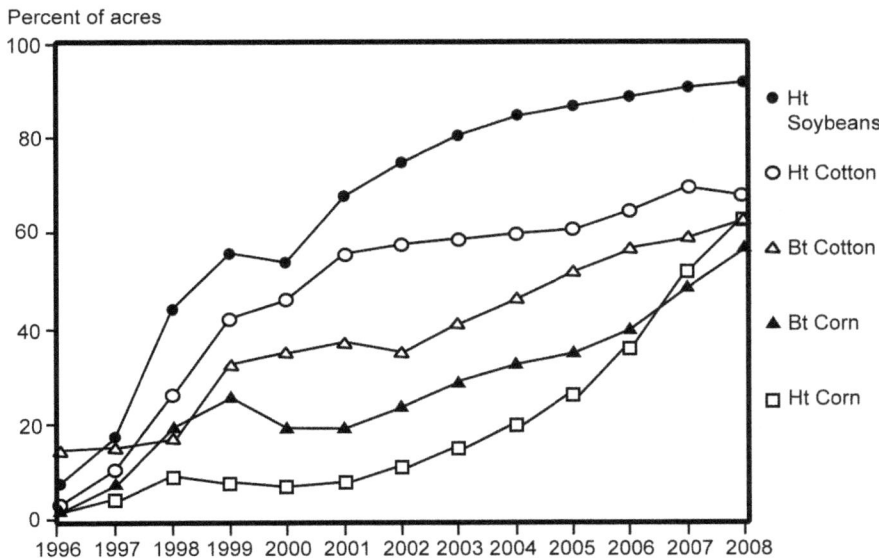

Data for each crop category includes varieties with both HT and Bt traits

Figure 5 Source: http://www.organicconsumers.org/gelink.cfm

95

The use of GMOs in the biosphere has resulted in chemical-resistant "superweeds" that even the GM industry admits need older, more dangerous control methods. Pesticide residues in GM animal feed are not monitored, so it is unknown what the animals eaten by consumers are actually eating. Of note is that the maximum legal limit in food for the weed killer Roundup was raised by 200% when GM Soya came on the market. Also of concern is the proliferation of genetically modified crops in other nations; native maize in Mexico is now contaminated with GM strains, even though growing genetically modified maize is banned there. The Austrian Government is taking steps to phase out GM animal feed. In Switzerland, biotech companies, not consumers, pay to segregate GM.

A recent article in *The New York Times* (Rosenthal 2010) highlighted one aspect of genetically modified crops, the MON810 corn seed, which is modified so that the corn produces a chemical that kills the larvae of the corn borer. Currently allowed by the European Union, but also banned in France, Austria, and Germany, the use of the seeds in Italy has resulted in action by Greenpeace activists, who "surreptitiously snipped off the stalks' tassels in the hope of preventing pollen from being disseminated." The use of genetically modified seeds to grow corn, soy, and wheat has resulted in the rapid expansion of GMO agriculture to all areas of the world; in spite of the obvious benefits of many GMO crops, their external costs and potential health effects have yet to be the subject of a comprehensive analysis.

A comprehensive overview of all the documented health risks derived from the use of genetically modified foods is contained in *Genetic Roulette* by Jeffery Smith (2007) who was also the author of *Seeds of Deception* (2003). The following synopsis of the chapter headings in *Genetic Roulette* provides a comprehensive overview of the downside of the genetically engineered crops that played a key role in the green revolution and the rapid expansion of the world's food supply in the late 20th and early 21st century.

1. Evidence of reactions in animals and humans

   - GM potatoes damaged rats
   - Rats fed GM tomatoes got bleeding stomachs, several died
   - Rats fed *Bt* corn had multiple health problems
   - Mice fed GM *Bt* potatoes had intestinal damage
   - Workers exposed to *Bt* cotton developed allergies
   - Sheep died after grazing in *Bt* cotton fields
   - Inhaled *Bt* corn pollen may have triggered disease in humans
   - Farmers report pigs and cows became sterile from GM corn
   - Twelve cows in Germany died mysteriously when fed *Bt* corn
   - Mice fed Roundup Ready soy had liver cell problems

- Mice fed Roundup Ready soy had problems with the pancreas
- Mice fed Roundup Ready soy had unexplained changes in testicular cells
- Roundup Ready soy changed cell metabolism in rabbit organs
- Most offspring of rats fed Roundup Ready soy died within three weeks
- Soy allergies skyrocketed in the UK, soon after GM soy was introduced
- Rats fed Roundup Ready canola had heavier livers
- Twice the number of chickens died when fed Liberty Link corn
- GM peas generated an allergic-type inflammatory response in mice
- Eyewitness reports: Animals avoid GMOs
- A GM food supplement killed about 100 people and caused 5,000-10,000 to fall sick

2. Gene insertion disrupts the DNA
   - Foreign genes disrupt the DNA at the insertion site
   - Growing GM crops using tissue culture can create hundreds or thousands of DNA mutations
   - Gene insertion creates genome-wide changes in gene expression
   - The promoter may accidentally switch on harmful genes
   - The promoter might switch on a dormant virus in plants
   - The promoter might create genetic instability and mutations
   - Genetic engineering activates mobile DNA, called transposons, which generate mutations
   - Novel RNA may be harmful to humans and their offspring
   - Roundup Ready soybeans produce unintentional RNA variations
   - Changes in proteins can alter thousands of natural chemicals in plants, increasing toxins or reducing phytonutrients
   - GM crops have altered levels of nutrients and toxins

3. The protein produced by the inserted gene may create problems
   - A gene from a Brazil nut carried allergies into soybeans
   - GM proteins in soy, corn, and papaya may be allergens
   - *Bt* crops may create allergies and illness
   - The *Bt* in crops is more toxic than the *Bt* spray
   - StarLink corn's built-in pesticide has a "medium likelihood" of being an allergen
   - Pollen-sterilizing barnase in GM crops may cause kidney damage
   - High lysine corn contains increased toxins and may retard growth
   - Cooking high lysine corn may create disease-promoting toxins
   - Disease-resistant crops may promote human viruses and other diseases

4. The foreign protein may be different than what is intended
   - GM proteins may be misfolded or have added molecules
   - Transgenes may be altered during insertion
   - Transgenes may be unstable and rearrange over time
   - Transgenes may create more than one protein
   - Weather, environmental stress, and genetic disposition can significantly change gene expression
   - Genetic engineering ignores and disrupts complex relationships in the DNA

5. Transfer of genes to gut bacteria, internal organs, or viruses
   - In spite of industry claims, transgenes survive the digestive system and can wander
   - Transgene design facilitates transfer into gut bacteria
   - Transgenes may proliferate in gut bacteria over the long-term
   - Transgene transfer to human gut bacteria is confirmed
   - GM foods might create antibiotic-resistant diseases
   - The promoter can also transfer and may switch on random genes or viruses
   - If *Bt* genes transfer, they could turn our gut bacteria into living pesticide factories
   - Genes may transfer to bacteria in the mouth or throat
   - Transfer of viral genes into gut microorganisms may create toxins and weaken viral defenses
6. GM crops may increase environmental toxins and bioaccumulate toxins in the food chain
   - Glufosinate-tolerant crops may produce herbicide "inside" our intestines
   - Herbicide-tolerant crops increase herbicide use and residues in food
   - Tiny amounts of herbicide may act as endocrine disruptors
   - GM crops may accumulate environmental toxins or concentrate toxins in milk and meat of GM-fed animals
   - Disease-resistant crops may promote new plant viruses, which carry risks for humans
7. Other types of GM foods carry risks
   - Milk from rbGH-treated cows may increase risk of cancer and other diseases
   - Milk from rbGH-treated cows likely increases the rate of twin births
   - Food additives created from GM microorganisms pose health risks
8. Risks are greater for children and newborns
   - Pregnant mothers eating GM foods may endanger offspring
   - GM foods are more dangerous for children than adults

V. Links to Information Sources

The information sources listed below represent those found most useful by the editors of this publication. Updates, comments, and additional suggestions about information sources relating to ecotoxins in the environment are welcome. Information sources are divided into the following groups. (WIP)

Table of Contents

International Government Organizations

United Nations (UN)
1st Ave & E 44th St
New York, NY 10017
Homepage: http://www.un.org/

Food and Agriculture Organization of the United Nations (FAO)
FAOSTAT
http://faostat.fao.org/

United Nations Economic Commission for Europe
Globally harmonized system of classification and labeling of chemicals (GHS):
http://www.unece.org/traans/danger/publi/ghs_welcome_e.html

Global Environment Monitoring System (GEMS)
http://www.gemswater.org/

International Labour Organization (ILO)
http://www.ilo.org/global/lang--en/index.htm

Joint Meeting on Pesticide Residues (JMPR)
http://www.fao.org/agriculture/crops/core-themes/theme/pests/pm/en/

Kyoto Protocol:
http://unfccc.int/kyoto_protocol/items/2830.php

The Joint *FAO*/WHO Committee on Food Additives
http://www.codexalimentarius.net/web/jecfa.jsp

Montreal Protocol
http://ozone.unep.org/Publications/MP_Handbook/Section_1.1_The_Montreal_Protocol/

United Nations Environment Programme
http://www.unep.org/

World Meteorological Organization
www.wmo.int/pages/index_en.html

World Health Organization (WHO)
Homepage: http://www.who.int/

World Health Report 1999
http://www.who.int/whr/1999/en/

Global Environment Monitoring System (GEMS/Food)
http://www.who.int/foodsafety/chem/gems/en/

Global Strategy on Diet, Physical Activity and Health
http://www.who.int/dietphysicalactivity/en/

International Programme on Chemical Safety
www.who.int/ipcs/food/active_chlorine/en/print.html

WHO Food Safety Program
http://www.who.int/foodsafety/en/

WHO Pesticide Evaluation Scheme (WHOPES)
http://www.who.int/whopes/en/

European Union (EU)
Homepage: http://europa.eu/

Environment
http://ec.europa.eu/environment/index_en.htm
European Climate Change Programme
http://ec.europa.eu/environment/climat/eccp.htm

European Commission Joint Research Center (JRC)
http://ec.europa.eu/dgs/jrc/index.cfm
Institute for Environment and Sustainability
http://ies.jrc.ec.europa.eu/index.php?page=home

European Fusion Development Agreement (EFDA)
www.efda.org
  Fusian Energy: Safety
  www.efda.org/fusion_energy/safety_and_the)environment.htm
European Food Safety Authority (EFSA)
www.efsa.europa.eu/EFSA/efsa_locale-1178620753812_home.htm
Human Epigenome Project
http://www.epigenome.org/
InnoMed PredTox
A Joint Industry and European Commission collaboration to improve drug safety.
http://www.innomed-predtox.com/
Questions and Answers on the RoHS
http://europa.eu/rapid/pressReleasesAction.do?reference=MEMO/08/763&format=H
TML&aged=0&language=EN&guiLanguage=en
Restriction of Hazardous Substances (RoHS)
http://eur-
lex.europa.eu/LexUriServ/LexUriServ.do?uri=OJ:L:2003:037:0019:0023:EN:PDF
Waste Electrical and Electronic Equipment Directive (WEEE Directive)
http://eur-
lex.europa.eu/LexUriServ/LexUriServ.do?uri=OJ:L:2003:037:0024:0038:EN:PDF

Global Health Reporting
Homepage: www.globalhealthreporting.org

Intergovernmental Panel on Climate Change (IPCC)
Homepage: http://www.ipcc.ch/
  General Assessment Reports
  http://www.ipcc.ch/ipccreports/assessments-reports.htm

  Special Topic Reports
  http://www.ipcc.ch/ipccreports/special-reports.htm

  Methodology Reports
  http://www.ipcc.ch/ipccreports/methodology-reports.htm

  Technical Papers
  http://www.ipcc.ch/ipccreports/technical-papers.htm

International Energy Agency
Homepage: http://www.iea.org/

International Organization for Standardization
Homepage: www.iso.org/iso/home.htm

Organization for Economic Co-operation and Development
Homepage: www.oecd.org

OSPAR Commission
Homepage: http://www.ospar.org/

Stockholm Convention
Homepage: http://chm.pops.int/

United States Federal Government Organizations

Armed Forces Health Surveillance Center
Homepage: http://www.afhsc.mil/home

Centers for Disease Control and Prevention
Homepage: http://www.cdc.gov/

Agency for Toxic Substances and Disease Registry
Homepage: www.atsdr.cdc.gov

FAQs
http://www.atsdr.cdc.gov/toxfaq.html

Toxic Substances Portal
http://www.atsdr.cdc.gov/toxicsubstances.html

Toxicalogical Profiles
http://www.atsdr.cdc.gov/toxprofiles/index.asp

Autism Spectrum Disorders
http://www.cdc.gov/ncbddd/autism/index.html

Avian Flu
www.cdc.gov/flu/avian/index.htm

Biomonitoring
www.cdc.gov/biomonitoring

Bovine Spongiform Encephalopathy (BSE) also known as "Mad Cow Disease"
http://www.fda.gov/oc/opacom/hottopics/bse.html

Chemical Agents, List of (by Category)
http://www.bt.cdc.gov/agent/agentlistchem-category.asp

Childhood Lead Poisoning Prevention Program
http://www.cdc.gov/nceh/lead/lead.htm

Comprehensive Environmental Response, Compensation, and Liability Act (CERCLA)
http://www.atsdr.cdc.gov/cercla/07list.html

Drug Resistance
http://www.cdc.gov/drugresistance/community/

Global Health – Global Disease Detection and Emergency Response (GDDER)
http://www.cdc.gov/globalhealth/gdder/

National Center for Health Statistics
http://www.cdc.gov/nchs

National Health and Nutrition Examination Survey
http://www.cdc.gov/nchs/nhanes.htm

NIOSH Pocket Guide to Chemical Hazards
http://www.cdc.gov/niosh/npg/npgd0000.html

Occupational Safety and Health (NIOSH), National Institute for Occupational Health and Safety Guidelines for Chemical Hazards
http://www.cdc.gov/niosh/81-123.html

Outbreak updates for international cruise ships
http://www.cdc.gov/nceh/vsp/surv/Glist.htm

Overweight and Obesity
http://www.cdc.gov/obesity/index.html

Pesticides and Public Health: Integrated Methods of Mosquito Management
http://www.cdc.gov/ncidod/eid/vol7no1/rose.htm

Pesticide Information
http://www.cdc.gov/nceh/hsb/pesticides/default.htm

Radiation Information
http://www.cdc.gov/nceh/radiation/default.htm

Registry of Toxic Effects of Chemical Substances (RTECS)
http://www.cdc.gov/niosh/rtecs

Third National Report on Human Exposure to Environmental Chemicals
http://www.cdc.gov/exposurereport/

Tobacco Information and Prevention Source
http://www.cdc.gov/tobacco

Consumer Product Safety Commission (CPSC)
Homepage: http://www.cpsc.gov

Defense Threat Reduction Agency & USSTRATCOM Center for Combating WMD
Homepage: http://www.dtra.mil/

Department of Agriculture
Homepage: http://www.usda.gov/wps/portal/usdahome

Forest Service Pesticide Fact Sheets
http://www.fs.fed.us/foresthealth/pesticide

Food Safety and Inspection Service
http://www.fsis.usda.gov

Department of Commerce

U. S. Commerce Department's Technology Administration

NIST (National Institute for Standards and Technology)

U. S. National Biomonitoring Specimen Bank
www.hml.nist.gov/PDF/Bankbibl-Website.pdf

NIST Chemistry WebBook
http://webbook.nist.gov/chemistry/

Department of Energy (DOE)
Homepage: http://www.energy.gov/

Annual Energy Review 2006
www.eia.doe.gov/emeu/aer/pdf/aer.pdf

International Energy Outlook 2007
www.eia.doe.gov/oeaf/ieo/index.html

National Energy Technology Laboratory
www.netl.doe.gov/

Official Energy Statistics
www.eia.doe.gov/

Office of Environment, Safety and Health
http://tis.eh.doe.gov/portal/home.htm

World Energy Reserves
www.eia.doe.gov/emeu/ieu/iea/res.htm

Department of Health and Human Services
Homepage: http://www.hhs.gov/

Department of Housing and Urban Development (HUD)
Homepage: http://www.hud.gov/

Office of Healthy Homes and Lead-Hazard Control:
http://www.hud.gov/offices/lead

Department of Labor, Occupational Safety and Health Administration (OSHA)
Homepage: http://www.osha.gov/index.html

Department of Transportation (DOT)
Homepage: http://www.dot.gov

Federal Highway Administration (FHWA), Lead:
http://www.tfhrc.gov/hnr20/bridge/model/health/health.htm

Hazardous Materials Emergency-Response Guidebook:
http://hazmat.dot.gov/pubs/erg/gydebook.htm

Environmental Health Policy Committee
Homepage: http://web.health.gov/environment

Environmental Protection Agency
Homepage: http://www.epa.gov/

Acid Rain Program
www.epa.gov/airmarkets/progsregs/arp/index.html

Endocrine Disruptors Research Initiative
http://www.epa.gov/endocrine/

Endocrine Disruptor Screening Program (EDSP)
http://epa.gov/endo/index.htm

Environmental Protection Community Right-to-know Act, effective January 20, 2009
http://www.epa.gov/emergencies/content/epcra/index.htm

Envirofacts
http://www.epa.gov/enviro/index_java.html

Fish Advisories
http://www.epa.gov/waterscience/fish/advice/volume2/index.html

Greenhouse gas emissions
http://epa.gov/climatechange/emissions/index.html
  Greenhouse gas inventory report
  http://epa.gov/climatechange/emissions/usinventoryreport.html

Integrated Risk-Information System (IRIS)
http://www.epa.gov/iris

Lead
http://www.epa.gov/OGWDW/dwh/c-ioc/lead.html

Mercury
http://www.epa.gov/mercury/

National Center for Environmental Research (NCER)
http://www.epa.gov/ncer/

National Pesticides Information Center (NPIC)
http://npic.orst.edu/index.html

National Primary Drinking Water Regulations
http://www.epa.gov/safewater/consumer/pdf/mcl.pdf

Nitrous oxide
http://www.epa.gov/nitrousoxide/sources.html

Office of Air and Radiation (OAR)
http://www.epa.gov/oar

Office of Environmental Information (OEI)
http://www.epa.gov/oei

Office of Pesticide Programs
http://www.epa.gov/pesticides

Office of Prevention, Pesticides, and Toxic Substances (OPPTS)
http://www.epa.gov/opptsmnt/index.htm

Office of Research and Development (ORD)
http://www.epa.gov/ORD

Office of Water (OW)
http://www.epa.gov/OW

Response to BP spill in the Gulf of Mexico
http://www.epa.gov/bpspill/dispersants.html#summary

Superfund
www.epa.gov/superfund/sites/npl/npl/htm

Toxicity and Exposure Assessment for Children's Health (TEACH)
http://www.epa.gov/teach/

Terms of Environment: Glossary, Abbreviations and Acronyms
www.epa.gov/OCEPAterms/

Toxic Release Inventory (TRI)
http://www.epa.gov/triexplorer/

Fish and Wildlife Service
Homepage: http://www.fws.gov/

Alaska Maritime National Wildlife Refuge
Homepage: http://alaskamaritime.fws.gov/

Food and Drug Administration (FDA)
Homepage: http://www.fda.gov/

Antibiotic Resistance
http://www.fda.gov/oc/opacom/hottopics/anti_resist.html

Bovine Spongiform Encephalopathy (BSE) Also Known as Mad Cow Disease
http://www.fda.gov/oc/opacom/hottopics/bse.html

Center for Devices and Radiological Health
http://www.fda.gov/cdrh

Center for Food Safety and Applied Nutrition
http://www.cfsan.fda.gov

National Center for Toxicological Research
http://www.fda.gov/nctr

National Academy of Sciences
Homepage: http://www.nasonline.org

National Atmospheric Deposition Program
Homepage: http://nadp.sws.uiuc.edu/

National Children's Study
Homepage: http://www.nationalchildrensstudy.gov/Pages/default.aspx

National Institute of Health
Homepage: http://www.nih.gov/

National Cancer Institute (NCI)
http://www.nci.nih.gov

National Institute of Allergies and Infectious Diseases (NIAD)
http://www3.niaid.nih.gov/

National Institute of Child Health and Human Development (NICHD)
http://www.nichd.nih.gov

National Institute for Environmental Health Sciences (NIEHS)
http://www.niehs.nih.gov

National Library of Medicine
http://www.nlm.nih.gov/

National Toxicology Program (NTP) Chemical Health and Safety Data
http://ntp.niehs.nih.gov/index.cfm?objectid=03610FA5-C828-304B-
FE31F1182E8F764C

National Toxicology Program (NTP) Report on Carcinogens
http://ntp.niehs.nih.gov/ntpweb/index.cfm?objectid=72016262-BDB7-CEBA-
FA60E922B18C2540

Center for the Evaluation of Risks to Human Reproduction (CERHR)
http://cerhr.niehs.nih.gov/

Chemical Carcinogenesis Research Information System
http://toxnet.nlm.nih.gov/cgi-bin/sis/htmlgen?CCRIS

Hazardous Substances Data Bank (HSDB®)
http://toxnet.nlm.nih.gov/cgi-bin/sis/htmlgen?HSDB

John E. Fogarty International Center for Advanced Study in the Health Sciences (FIC)
http://www.fic.nih.gov/

Mercury Abatement Program
http://orf.od.nih.gov/Environmental+Protection/Mercury+Free/

   Hatter's Library
   http://orf.od.nih.gov/Environmental+Protection/Mercury+Free/library.htm

   Hatter's Links - Other Mercury Related Sites
   http://orf.od.nih.gov/Environmental+Protection/Mercury+Free/Hatters+Links.htm

National Oceanic and Atmospheric Administration (NOAA)
Homepage: http://www.noaa.gov/

   Center for Coastal Monitoring and Assessment (CCMA)
   Monitoring data: Mussel Watch
   http://www8.nos.noaa.gov/cit/nsandt/download/mw_monitoring.aspx

   Climate.gov
   http://www.climate.gov/

   Earth System Research Laboratory (NOAA)
   www.esrl.noaa.gov/

   National Marine Mammal Tissue Bank
   http://www.nmfs.noaa.gov/pr/health/tissue/

   Ozone Hole
   www.theozonehole.com

Oak Ridge National Laboratory (ORNL)
Homepage: http://www.ornl.gov/

Office of Technology Assessment, Oceans & Environment Program
This office is no longer operational, but documents are available at:
http://www.gpo.gov/ota/

USA Spending
Homepage: http://www.usaspending.gov/

United States Geological Survey (USGS)
Homepage: http://www.usgs.gov/

Fact Sheet on pharmaceuticals, hormones, and other organic wastewater contaminants
http://toxics.usgs.gov/pubs/FS-027-02/

National Biomonitoring Specimen Bank
http://www.absc.usgs.gov/research/ammtap/nbsb.htm

North American Amphibians Monitoring Program
http://www.pwrc.usgs.gov/naamp/

STAMP (Seabird Tissue Archival and Monitoring Project)
http://www.absc.usgs.gov/research/ammtap/data.htm

Western Ecological Research Center
http://www.werc.usgs.gov/

Pacific Nearshore Project
http://www.werc.usgs.gov/project.aspx?projectid=221

U.S. State Government Organizations

Maine

Maine was the first state in the nation to issue guidelines for the consumption of fresh water fish. Maine state links are thus listed before those of any other state.

Maine Center for Disease Control and Prevention
Homepage: http://www.maine.gov/dhhs/boh/index.shtml
Public Health Data Reports:
http://www.maine.gov/dhhs/boh/phdata/environmental_health/reports_environmental_health.htm

Maine Department of Environmental Protection
Homepage: http://www.maine.gov/dep/index.shtml
Chemicals of High Concern List
http://www.maine.gov/dep/oc/safechem/highconcern/
Publications List:
http://www.maine.gov/dep/publications.htm

Science & Data:
http://www.maine.gov/dep/data.htm

Bureau of Land and Water Quality:
http://www.maine.gov/dep/blwq/monitoring.htm

Maine Department of Health and Human Services
Division of Environmental Health
Homepage: http://www.maine.gov/dhhs/

Environmental and Human Health Programs
Warnings on eating Fish and Game:
http://www.maine.gov/dhhs/eohp/fish/

Maine Family Fish Guide
http://www.maine.gov/dhhs/eohp/fish/documents/MeFFGuide.pdf

Bureau of Health Fish Tissue Action Levels
http://www.maine.gov/dhhs/eohp/fish/documents/Action%20Levels%20Writeup.pdf

Dioxins and furans in fish from Maine rivers
http://www.maine.gov/dhhs/eohp/fish/documents/FinalDraft_Eval_of_PCDD.pdf

Casco Bay Estuary Partnership
Homepage: http://www.cascobay.usm.maine.edu/index.html

Climate Change Institute
Homepage: http://www.climatechange.umaine.edu/

Friends of Casco Bay: Casco Baykeeper
Homepage: http://friendsofcascobay.org/default.aspx

Gulf of Maine Council on the Marine Environment
Homepage: http://www.gulfofmaine.org/

Census of Marine Life
http://www.gulfofmaine-census.org/

Gulf of Maine Times
http://www.gulfofmaine.org/times/fallwinter2008/index.php

Gulfwatch:
http://www.gulfofmaine.org/gulfwatch/

Other States

California
California Biomonitoring Program
Homepage: http://www.cdph.ca.gov/programs/Biomonitoring/Pages/default.aspx

California Institute of Technology
Jet Propulsion Laboratory, Orbiting Carbon Observatory (OCO)
http://oco.jpl.nasa.gov

State of California Ocean Protection Council
http://www.opc.ca.gov/

Resolution on Reducing and Preventing Marine Debris
http://www.opc.ca.gov/webmaster/ftp/pdf/agenda_items/20070208/0702COPC05_MarineDebris_Resolution.pdf

University of California Santa Barbara (UCSB) Marine Science Institute (MSI)
http://www.msi.ucsb.edu/

Florida
Emerging Pathogens Institute
Homepage: http://www.epi.ufl.edu/

Illinois
Illinois Department of Public Health
Homepage: http://www.idph.state.il.us/

Nitrates:
http://www.idph.state.il.us/envhealth/factsheets/NitrateFS.htm

Indiana
The Vulcan Project (U.S. fossil fuel CO_2 emissions)
http://www.purdue.edu/eas/carbon/vulcan/index.php

Massachusetts
Standards and Guidelines for Contaminants in Massachusetts Drinking Waters
http://www.mass.gov/dep/water/dwstand.pdf

New Hampshire
New Hampshire Department of Environmental Services
Homepage: http://des.nh.gov/index.htm

Monitoring programs: Gulfwatch
http://des.nh.gov/organization/divisions/water/wmb/shellfish/greatbay_s.htm

New York
New York State
Homepage: http://www.state.ny.us/

Environmental montoring
http://www.nysegov.com/citGuide.cfm?superCat=396&cat=403&content=main

Cornell University
Homepage: http://www.cornell.edu/

Rockefeller University
Homepage: http://www.rockefeller.edu/

Oregon

Office of Environmental Public Health
Homepage: http://www.oregon.gov/DHS/ph/oeph/index.shtml

Oregon Health & Science University
  Center for Research on Occupational and Environmental Toxicology (CROET)
  http://www.ohsu.edu/croet/index.cfm

Pennsylvania

Earth System Science Center (ESSC), Penn State
Homepage: http://www.essc.psu.edu/

Rhode Island

Rhode Island Sea Grant
Homepage: http://seagrant.gso.uri.edu/index.html

  9th Annual Ronald C. Baird Sea Grant Science Symposium
  http://seagrant.gso.uri.edu/baird/2010_diseases.html

Washington

State of Washington Department of Ecology
Homepage: http://www.ecy.wa.gov/ecyhome.html

Other Governmental Organizations

Australia

  Food Standards Australia New Zealand (FSANZ)
  Homepage: http://www.foodstandards.gov.au/

Canada

  Health Canada
  Homepage: http://www.hc-sc.gc.ca/index-eng.php

Antibiotic resistant germs
http://www.hc-sc.gc.ca/english/iyh/medical/antibiotic.html.

Denmark

Risø DTU National Laboratory for Sustainable Energy
Homepage: http://www.risoe.dk/?sc_lang=en

European Commission

Bio-safety Europe
Homepage: http://www.biosafety-europe.eu/index.html

ECOnetus
Homepage: http://econetus.polsl.pl/

REACH (Registration, Evaluation, Authorisation and Restriction of Chemical substances)
Homepage: http://ec.europa.eu/environment/chemicals/reach/reach_intro.htm

Finland

Finnish Centre for Radiation and Nuclear Safety
Helsinki, Finland

Germany

Jűlich Institute of Energy Research
Homepage: http://www.fz-juelich.de/portal/index.php?index=1159

India

Environmental Information Centre
Homepage: http://www.eicinformation.org/

Japan

Science Council of Japan
Homepage: www.scj.go.jp/en/

Netherlands, The

Netherlands Environmental Assessment Agency
Homepage: www.mnp.nl/en/index.html

Norway

The Fishery and Aquaculture Industry Research Fund
Homepage: www.fiskerifond.no

Sweden

Swedish National Food Administration
Homepage: www.slv.se/default.aspx?id=231&epslanguage=EN-GB

United Kingdom
Centre for Radiation, Chemical and Environmental Hazards
Homepage:
http://www.hpa.org.uk/webw/HPAweb&Page&HPAwebContentAreaLanding/Page/1
153822623782

Department for Environment Food and Rural Affairs
Homepage: www.defra.gov.uk/environment/nanotech

Department for Innovation, Universities & Skills
Homepage: http://www.dius.gov.uk/

Policy statement on nanotechnologies:
http://www.dius.gov.uk/policy/documents/statement-nanotechnologies.pdf

Food Climate Research Network
Homepage: www.fcrn.org.uk/

Food Standards Agency
Homepage: http://www.food.gov.uk/

Nanotechnology and Nanoscience
Homepage: www.nanotec.org.uk/

Royal Academy of Engineering
Homepage: www.raeng.org.uk/

Royal Society, The
6-9 Carlton House Terrace, London SW1Y 5AG
tel: +44 (0)20 7451 2500, fax: +44 (0)20 7930 2170
Homepage: www.royalsociety.org

Scottish Environment Protection Agency (SEPA)
Homepage: http://www.sepa.org.uk/

USSR/Russia
Bellona
Homepage: www.bellona.org

Non-Governmental Organizations

Research and Information Sources

Action Group on Erosion, Technology and Concentration (ETC Group)
Homepage: http://www.etcgroup.org

Alfred Wegener Institute for Polar and Marine Research (AWI)
Homepage: http://www.awi.de/en

Alliance for a Clean and Healthy Maine
Homepage: http://www.cleanandhealthyme.org/

Alliance for a Clean and Healthy Vermont
Homepage: http://www.alliancevt.org/
   Safer alternatives to Deca
   http://www.ppnne.org/site/DocServer/Deca_Alternatives_Fact_Sheet_-
   _FINAL.pdf?docID=5662

Alliance for Sustainable Built Environments
Homepage: http://www.greenerfacilities.org/

American Bird Conservancy
Homepage: http://www.abcbirds.org/

American College of Occupational and Environmental Medicine
Homepage: http://www.acoem.org

American Nuclear Society
Homepage: http://fire.pppl.gov/us_fusion50yr_dean.pdf

American Water Works Association
Homepage: http://www.awwa.org/index.cfm.

Amphibian Specialist Group (ASG)
Homepage: http://www.amphibians.org./ASG/Home.html

As You Sow
Homepage: http://www.asyousow.org/

Association of Occupational and Environmental Clinics
Homepage: http://www.aoec.org

Association of Public Health Laboratories
Homepage: http://www.aphl.org

Autism Society
Homepage: http://www.autism-society.org/site/PageServer

Autism Speaks
Homepage: http://www.autismspeaks.org/

Basel Action Network
Homepage: http://www.ban.org/

Basel Convention
Homepage: http://www.basel.int/

Belfer Center for Science and International Affairs
Homepage: http://belfercenter.ksg.harvard.edu/index.html

   Energy Technology Innovation Policy
   http://belfercenter.ksg.harvard.edu/project/10/energy_technology_innovation_policy.html

Biodiversity Research Institute
19 Flaggy Meadow Road
Gorham, ME 04038. Phone: (207) 839-7600
Homepage: www.BRIloon.org

Carbon Trade Watch
Homepage: http://www.carbontradewatch.org/

CATO Institute: Energy and Environment
Homepage: http://www.cato.org/energy-environment

Census of Marine Life (COML)
Homepage: http://www.coml.org/

Center for Biological Diversity
Homepage: http://www.biologicaldiversity.org/

Center for Economic and Policy Research (CEPR)
Homepage: http://www.cepr.net/

Center for Food Safety
Homepage: http://truefoodnow.org/

Center for Health, Environment and Justice
Homepage: http://www.chej.org/

116

Center for Public Integrity
Homepage: http://www.publicintegrity.org/

Centre for Research on Globalization
Homepage: http://www.globalresearch.ca/

Center for Whale Research (CWR)
Homepage: http://www.whaleresearch.com/thecenter/2008_Home.html

Channel Islands Marine and Wildlife Institute (CIMWI)
A California nonprofit public benefit corporation, 501(c)(3), committed to aiding and rescuing marine animals and wildlife affected by natural and unnatural causes.
Homepage: http://www.cimwi.org/stranded_domoic.html

Chemfinder
Homepage: http://www.chemfinder.com

Chemical Body Burden
Homepage: http://www.chemicalbodyburden.org/home.htm

   Chemicals in U.S. Population
   http://www.chemicalbodyburden.org/rr_cheminus.htm

Chemical Hazard and Alternatives Toolbox (ChemHAT)
Homepage: http://www.chemhat.org/

Children's Environmental Health Network
Homepage: http://www.cehn.org/

Cleantechnica
Homepage: http://cleantechnica.com/

Climate Progress
Homepage: http://climateprogress.org

Club of Rome
International Secretariat
Lagerhausstrasse 9
CH-8400 Winterthur (Canton Zurich)
Switzerland
Homepage: www.clubofrome.org

Collaborative on Health and the Environment
Homepage: http://www.healthandenvironment.org/
  http://www.cheforhealth.org/

Coming Clean Network
Homepage: http://www.come-clean.org/

A Compendium of On-Line Soil Survey Information
Homepage: http://www.itc.nl/~rossiter/research/rsrch_ss.html

Consumer Goods Forum
Homepage: http://www.ciesnet.com/
  Global Foods Safety Conference
  http://tcgffoodsafety.com/pro/fiche/quest.jsp;jsessionid=B6CF52DE9EDD5E145280
  B519C1483112.gl1

Convention on International Trade and Endangered Species (CITES)
Homepage: http://www.cites.org/

Corrosion Doctors
Homepage: http://www.corrosion-doctors.org/

Crop Science Society of America
Homepage: https://www.crops.org/

Davistown Museum
Homepage: www.davistownmuseum.org
  Department of Environmental History
  http://www.davistownmuseum.org/environment.html

Democracy in Action
Homepage: www.democracyinaction.org

Discovery Guides: Genetically Modified Foods: Harmful or Helpful?
Homepage: http://www.csa.com/discoveryguides/gmfood/overview.php

Downeast Institute for Applied Marine Research and Education (DEI)
Homepage: http://www.downeastinstitute.org/

EcoHealth Alliance
Homepage: http://www.ecohealthalliance.org/

Endocrine Disruption Exchange
see The Endocrine Disruption Exchange (TEDX)

Endocrine Society
Homepage: http://www.endo-society.org/
Endocrine disrupting chemicals
http://www.endo-society.org/advocacy/policy/upload/Endocrine-Disrupting-Chemicals-Position-Statement.pdf

Energy Watch Group
Homepage: www.energywatchgroup.org

Energy Watch Group Oil Report
www.energywatchgroup.org/fileadmin/global/pdf/EWG_Oilreport_10-2007.pdf

Environmental Ethics Journal
Homepage: http://www.cep.unt.edu/enethics.html

Environmental Health Strategy Center
Homepage: http://www.preventharm.org/

Environmental Research Foundation
Homepage: http://www.rachel.org/

Environmental Working Group (EWG)
Homepage: http://www.ewg.org/

EWG has all their reports online in an archive on PCBs, mercury, fire retardants, Teflon, arsenic, etc.

Farm Subsidies Database
http://farm.ewg.org/sites/farm/dp_text.php

Fire Retardants (PBDEs) in Breast Milk
http://www.ewg.org/reports/mothersmilk/

Fluorescent Light Bulbs: Green Lighting Guide
http://www.ewg.org/greenlightbulbs

National Drinking Water Database
http://www.ewg.org/tap-water/home

On-line Body Burden/ Community Monitoring Handbook
http://www.ewg.org/reports/mothersmilk/

PCB's in People of St. Lawrence Island
http://www.ewg.org/reports/mothersmilk/

Phthalates in Cosmetics
http://www.ewg.org/reports/mothersmilk/

Skin Deep: Searchable safety guide to cosmetics and personal care products
http://www.cosmeticsdatabase.com/

EXOTOXNET (Extension Toxicology Network)
http://extoxnet.orst.edu/
  Pesticide Information Profiles (PIPs)
  http://extoxnet.orst.edu/pips/ghindex.html

Fact Check
Homepage: http://factcheck.org/

Farmed Salmon Exposed
Homepage: http://farmedsalmonexposed.org/index.html

Federal Focus Inc.
Homepage: http://www.fedfocus.org/index.htm
  Center for the Study of Environmental Endocrine Effects
  http://www.fedfocus.org/science/cseee.html

Food First
Institute for Food and Development Policy
Homepage: http://www.foodfirst.org/

Food and Water Watch
Homepage: http://www.foodandwaterwatch.org

GAIA Foundation
Homepage: http://www.gaiafoundation.org/

GMO Compass
Homepage: http://www.gmo-compass.org/eng/home/

Global Atmosphere Watch
Homepage: www.wmo.ch/web/arep/gaw/gaw_home.html
  Station Information System
  http://gaw.empa.ch/gawsis

Global Crop Diversity Trust: A Foundation for Food Security
Homepage: http://www.croptrust.org/main/

Global Footprint Network: Advancing the Science of Sustainability
Homepage: http://www.footprintnetwork.org/en/index.php/GFN/

Global Health Reporting
Homepage: www.globalhealthreporting.org

Global Invasive Species Database
http://www.issg.org/database/welcome/

Global Viral
http://www.globalviral.org/

Global Warming Policy Foundation
Homepage: http://www.thegwpf.org/

Goldschmidt 2007 (Conference in August 2007, Cologne, Germany)
Homepage: http://www.the-conference.com/conferences/2007/gold2007/

Got Mercury?
Homepage: http://www.gotmercury.org/article.php?list=type&type=75

Green Facts
Homepage: http://www.greenfacts.org/en/index.htm

Groundwater Foundation
Homepage: http://www.groundwater.org/

Gulf Restoration Network
Homepage: http://healthygulf.org/

Harvard Medical School Center for Health and the Global Environment
Homepage: http://chge.med.harvard.edu/

Heal the Bay
Homepage: http://www.healthebay.org/

Healthy Stuff
Homepage: http://www.healthystuff.org

Heinrich Boll Foundation
Homepage: http://www.boell.org/

Human Toxome Project (HTP)
Homepage: www.bodyburden.org

IEA Greenhouse Gas R&D Programme
Homepage: http://www.co2captureandstorage.info/index.htm

International Network on Biofixation of CO_2 and Greenhouse Gas Abatement with Microalgae
www.co2captureandstorage.info/networks/Biofixation.htm

ISSP: The world's leading professional association of sustainability practitioners
Homepage: http://sustainabilityprofessionals.org/

Institute for Agriculture and Trade Policy
Homepage: http://www.iatp.org/
 Smart Plastics Guide: Healthier Food Uses of Plastics
 http://www.healthobservatory.org/library.cfm?refid=77083

Institute for Arctic and Alpine Research at the University of Colorado (INSTAAR)
Homepage: http://instaar.colorado.edu/

Institute for Energy and Environmental Research (IEER)
Homepage: www.ieer.org/
 Factsheets
 www.ieer.org/fctsheet/index.html

Institute of Medicine of the National Academies
Homepage: www.iom.edu

Interdisciplinary Sciences Complexity Wiki
Homepage: http://intersci.ss.uci.edu/wiki/index.php/Main_Page

International Center for Technology Assessment (ICTA)
Homepage: www.icta.org
 NanoAction
 http://www.nanoaction.org/nanoaction/index.cfm

International Chemical Safety Cards
Homepage:
http://www.ilo.org/public/english/protection/safework/cis/products/icsc/dtasht/index.ht
m

International Commission for the Conservation of Atlantic Tunas (ICCAT)
Homepage: http://www.iccat.int/en/

International POPs Elimination Network
Homepage: www.ipen.org/campaign/who/ipen.html

International Programme on Chemical Safety (IPCS)
Homepage: http://www.who.int/pcs

International Rice Research Institute (IRRI)
Homepage: http://irri.org/

International Union for Conservation of Nature (IUCN)
Homepage: http://www.iucn.org/

International Water Management Institute (IWMI)
Homepage: http://www.iwmi.cgiar.org/

International Maize and Wheat Improvement Center (CIMMYT)
Homepage: http://www.cimmyt.org/

Introduction to Endocrine Disrupting Chemicals
Homepage: http://website.lineone.net/~mwarhurst/

Johnson Foundation at Wingspread
Homepage: http://www.johnsonfdn.org/

Keep Antibiotics Working
Homepage: http://www.keepantibioticsworking.com

Kerr Center for Sustainable Agriculture
Homepage: http://www.kerrcenter.com/

Lead Zero
http://www.leadzero.org/

Learning Disabilities Association of Maine
Homepage: http://www.ldame.org/

Louisiana Environmental Action Network
http://www.leanweb.org/

Louisiana Universities Marine Consortium (LUMCON)
Homepage: http://www.lumcon.edu/

Maine Labor Group on Health
Homepage: http://www.mlgh.org/

Maine Organic Farmers and Gardeners Association (MOFGA)
Homepage: http://www.mofga.org/
   Position statement on genetically engineered organisms
   http://www.mofga.org/Default.aspx?tabid=439

Maine People's Resource Center
Homepage: http://www.mainepeoplesresourcecenter.org/

Maine Women's Policy Center
Homepage: http://www.mainewomen.org/

Marine Environmental Research Institute (MERI)
PO Box 1652
Blue Hill, ME, 04614
(207) 374-2135
Homepage: http://www.meriresearch.org/
  - An important information source for chemical fallout in Maine ecosystems, especially including PBDEs in marine mammals.

Marine Stewardship Council: Certified Sustainable Seafood
Homepage: http://www.msc.org/

Material Safety Data Sheets (MSDS)
http://www.hazard.com/msds

Mayo Clinic
Homepage: http://www.mayoclinic.com/

Mesothelioma and Asbestos Awareness Center (MAAC)
Homepage: http://www.maacenter.org/

Mesothelioma Resource Online
Homepage: http://www.mesotheliomasymptoms.com/

Midwest Renewable Energy Association (MREA)
Homepage: http://www.the-mrea.org/

Millennium Ecosystem Assessment
Homepage: www.millenniumassessment.org

Mount Desert Island Biological Laboratory (MDIBL)
  The Comparative Toxicogenomics Database
  http://ctd.mdibl.org/

National Audubon Society
Homepage: http://www.audubon.org/
  2007 Audubon Bird Watch List
  http://web1.audubon.org/science/species/watchlist/browsewatchlist.php

National Center for Ecological Analysis and Synthesis (NCEAS)
Homepage: http://www.nceas.ucsb.edu/

National Environmental Trust
Homepage: http://environet.policy.net/

National Fish Habitat Action Plan
Homepage: http://fishhabitat.org/

National Research Council (NRC)
Homepage: http://sites.nationalacademies.org/nrc/index.htm

National Vaccine Information Center
Homepage: http://www.nvic.org/

Natural Resources Council of Maine
Homepage: http://www.nrcm.org/

No Kid Hungry Campaign
Homepage: http://nokidhungry.org/

Northwest Coalition Against Pesticides
Homepage: http://www.pesticide.org/

Obesity in America
Homepage: http://www.obesityinamerica.org/

Obesity Society
Homepage: http://www.obesity.org/

Obesity Statistics
Homepage: http://www.annecollins.com/obesity/statistics-obesity.htm

Ocean Alliance
Homepage: http://www.oceanalliance.org/

Oceanic Society
Homepage: http://www.oceanicsociety.org/home

On Earth
Homepage: http://www.onearth.org

Organic Consumers Association
Homepage: www.organicconsumers.org

Organization for Autism Research (OAR)
Homepage: http://www.researchautism.org/

Oztoxics
Homepage: http://www.oztoxics.org/
  IPEN Body Burden Community Monitoring Handbook
  http://www.oztoxics.org/cmwg/index.html

Penobscot East Resource Center
Homepage: http://www.penobscoteast.org/default.asp

Perchlorate Information Bureau
Homepage: http://perchlorateinformationbureau.org/

Pesticide Action Network
Homepage: http://www.panna.org/
  Chemical Trespass: Report on Pesticide Body Burden Data
  http://www.panna.org/docsTrespass/chemicalTrespass2004.dv.html

Pew Center on Global Climate Change
Homepage: http://www.pewclimate.org/

Physicians for Social Responsibility (PSR)
Homepage: http://www.envirohealthaction.org/

Planned Parenthood
Homepage: http://www.plannedparenthood.org/

Political Instability Task Force
Homepage: http://globalpolicy.gmu.edu/pitf/

Pollution in People
Homepage: http://www.pollutioninpeople.org/

Prevent Cancer
Homepage: www.preventcancer.com
  Information on contaminants in milk
  www.preventcancer.com/consumers/general/milk.htm

Project on Emerging Nanotechnologies
Homepage: http://www.nanotechproject.org/inventories/consumer/

Pure Salmon Campaign: Raising the Standards for Farm-Raised Fish
Homepage: http://www.puresalmon.org/

Reality
Homepage: http://www.thisisreality.org

Rense
Homepage: http://rense.com/

Robert Wood Johnson Foundation
Homepage: http://www.rwjf.org

Rodale Institute
Homepage: http://www.rodaleinstitute.org

Safer Chemicals, Healthy Families
Homepage: http://www.saferchemicals.org

Save Our Gulf
Homepage: http://saveourgulf.org/

Science Communication Network
Homepage: http://www.sciencecommunicationnetwork.org/

Science and Environmental Health Network
Homepage: http://www.sehn.org/

Silicon Valley Toxics Coalition
Homepage: http://svtc.igc.org/
Lots of information on PBDEs available at this site

Sjogrens Syndrome Foundation
Homepage: www.sjogrens.org

Skin Deep – Cosmetic Safety Database
Homepage: www.cosmeticsdatabase.com/index.php?nothanks=1

Sky Truth
Homepage: http://www.skytruth.org/

Stockholm Environment Institute
Homepage: http://sei-international.org
  Stockholm Environment Institute – US Center
  Homepage: http://www.sei-us.org

Story of Stuff
Homepage: http://www.storyofstuff.com/

Sustainable Ocean Project (SOP)
Homepage: http://sustainableoceanproject.com/

The Endocrine Disruption Exchange (TEDX)
Homepage: http://www.endocrinedisruption.com

Toxic-Free Legacy Coalition
Homepage: http://toxicfreelegacy.org/

Toxicological Effects of Methylmercury
Homepage: http://books.nap.edu/books/0309071402/html/index.html

Toxics Action Center Campaigns
Homepage: http://www.toxicsaction.org/index.htm

Trust for America's Health
Homepage: http://healthyamericans.org/

Tulane Environmental Law Clinic
Homepage:
http://www.tulane.edu/~bfleury/envirobio/enviroweb/LawClinic/LawClinic.html

Tulane/Xavier Center for Biomedical Research
Homepage: http://www.cbr.tulane.edu/

Tyndall Centre for Climate Change Research
Homepage: www.tyndall.ac.uk/

US Green Building Council
LEED (Leadership in Energy and Environmental Design)
Homepage: www.usgbc.org/

Union of Concerned Scientists
Homepage: http://ucsusa.org/

University Corporation for Atmospheric Research
Homepage: http://www2.ucar.edu/
  National Center for Atmospheric Research
  http://www.ncar.ucar.edu/

Veta La Palma
Homepage: http://www.vetalapalma.es/index2.html

Water Encyclopedia
Homepage: http://www.waterencyclopedia.com/

Water Research Foundation
Homepage: http://www.waterresearchfoundation.org/
  Beta of new homepage
  http://www.waterrf.org/Pages/WaterRFHome.aspx

Waterkeeper Alliance
Homepage: http://www.waterkeeper.org/

Woodrow Wilson International Center for Scholars
Homepage: www.wilsoncenter.org

Woods Hole Oceanographic Institution (WHOI)
Homepage: http://www.whoi.edu/

World Health Organization, Switzerland
Homepage: www.euro.who.int

  Global Alert and Response (GAR)
  http://www.who.int/csr/en/

  World Health Organization/Convention Task Force on the Health Aspects of Air
  Pollution
  http://www.euro.who.int/air

World Energy Outlook
Homepage: www.worldenergyoutlook.org

World Business Council for Sustainable Development
Homepage: www.wbcsd.org

World Resources Institute
Homepage: www.wri.org

World Wildlife Fund
Homepage: www.worldwildlife.org

Yale Environment 360
Homepage: http://www.e360.yale.edu/

Zero Mercury Global Campaign
Homepage: http://www.zeromercury.org/index.html

Political Action Groups

Action for a Sustainable America
Homepage: http://www.asaseries.com/

Alliance for Climate Protection
Homepage: http://www.climateprotect.org/

American Rivers
Homepage: http://www.americanrivers.org/

Association for the Study of Peak Oil and Gas (ASPO International)
Homepage: http://www.peakoil.net/

Atrazine Lovers
Homepage: http://atrazinelovers.com/

Baby Milk Action
Homepage: http://www.babymilkaction.org/

Bonneville Environmental Foundation
Homepage: http://www.b-e-f.org/index.shtm

Bullitt Foundation
Homepage: http://bullitt.org/

CARE
Homepage: http://my.care.org/site/PageServer?pagename=Home

Calstart
Homepage: www.calstart.org/Homepage.aspx

Campaign for Safe Cosmetics
Homepage: http://www.safecosmetics.org/

Christian Children's Fund
Homepage: http://www.christianchildrensfund.org/

Climate Depot
Homepage: http://climatedepot.com/

Copenhagen Consensus Center
Homepage: http://www.copenhagenconsensus.com

Conservation International
Homepage: http://www.conservation.org/Pages/default.aspx

Consumer Federation of America
Homepage: www.consumerfed.org/

Consumer Policy Institute
Homepage: http://www.consumersunion.org/aboutcu/offices/CPI.html

Convention on Biological Diversity
Homepage: http://www.cbd.int/

E3G: Third Generation Environmentalism
Homepage: www.e3g.org/

Earth First
Homepage: http://www.earthfirst.org/

Earthworks
Homepage: http://www.earthworksaction.org/
  Halliburton Loophole
  http://www.earthworksaction.org/publications.cfm?pubID=398

Electronics Take Back Coalition
Homepage: http://www.electronicstakeback.com/index.htm

End of the Line
Homepage: http://endoftheline.com/film/

Environment Maine
Homepage: http://www.environmentmaine.org/

Environmental Defence
Homepage: www.environmentaldefence.ca
  Toxic nation
  http://www.toxicnation.ca/

Environmental Law & Policy Center: Protecting the Midwest's Environment and Natural Heritage
Homepage: http://elpc.org/

Friends of the Earth
Homepage: http://www.foe.org/

Gasland and Gasland 2 (A film by Josh Fox)
Homepage: http://gaslandthemovie.com/

Global Financial Integrity
Homepage: http://www.gfip.org/

Green Parties Worldwide
Homepage: http://www.greens.org/

Green Power Conferences
Homepage: http://www.greenpowerconferences.com

Green Thyself
Homepage: http://www.green.org/

Green Web
Homepage: http://home.ca.inter.net/~greenweb/

Greenpeace
Homepage: www.greenpeace.org/usa/
  Poisoning the unborn
  http://www.greenpeace.org/international/en/news/features/poisoning-the-unborn111/

Institute for Analysis of Global Security (IAGS)
Homepage: www.iags.org/

Oceana: Protecting the World's Oceans
Homepage: http://na.oceana.org/

Organic Consumers Association
6771 S Silver Hill Dr
Finland MN 55603
Homepage: http://www.organicconsumers.org/

Oxfam
Homepage: http://www.oxfam.org/

People for Puget Sound
Homepage: http://www.pugetsound.org/

Say No to GMOs
Homepage: http://www.saynotogmos.org/

Sierra Club
Homepage: http://www.sierraclub.org/

Skoll Global Threats Fund
Homepage: http://www.skollglobalthreats.org/

Source Watch
Homepage: http://www.sourcewatch.org
  Hydrofracking
  http://www.sourcewatch.org/index.php?title=Fracking

United Nations Convention to Combat Desertification
Homepage: http://www.unccd.int/

U. S. Climate Action Network
http://www.usclimatenetwork.org/

Wild Birds for the 21st Century, Inc.
Homepage: http://www.wildbirds.org/intro.htm

  U.S. Fish and Wildlife Service list of Threatened and Endangered Bird Species
  http://www.wildbirds.org/news/list.htm

World Wildlife Foundation
Homepage: http://www.worldwildlife.org

  Climate
  http://www.worldwildlife.org/climate/climatesavers2.html

Media
BBC Globe Scan
Homepage: http://www.globescan.com/news_center.htm

Democracy Now! The War and Peace Report
Homepage: http://www.democracynow.org/

Global Ideas Deutche Welle (DW.DE)
Homepage: http://www.dw.de/top-stories/environment/s-11798

Green Tech Media
Homepage: http://www.greentechmedia.com

Institutional Risk Analyst
Homepage: http://us1.institutionalriskanalytics.com/pub/IRAMain.asp

Journal of the American Medical Association (JAMA)
Homepage: http://jama.ama-assn.org/
   Occupational and Environmental Medicine
   http://jama.ama-assn.org/cgi/collection/occupational_and_environmental_medicine

Nature
Homepage: www.nature.com

   Malaria
   www.nature.com/nature/focus/malaria/index.html

ProPublica: Journalism in the Public Interest
Homepage: http://www.propublica.org/
   Natural gas drilling
   http://www.propublica.org/feature/natural-gas-drilling-what-we-dont-know-1231

Science News
Homepage: http://www.sciencenews.org/

The New York Times
Homepage: http://www.nytimes.com/
   Dealbook
   http://dealbook.blogs.nytimes.com/

Business

Bloom Energy
Homepage: http://www.bloomenergy.com

British Petroleum
Homepage: http://www.bp.com
   Statistical Review of World Energy
   www.bp.com/productlanding.do?categoryId=6929&contentId=7044622

Earth Shell
Homepage: http://www.earthshell.com/

Ecotel Certification
Homepage: http://concepthospitality.com/hotels-consultant/ecotel_certification.htm

First Solar
Homepage: http://www.firstsolar.com/en/index.php

Institutional Risk Analytics
Homepage: http://us1.institutionalriskanalytics.com/www/index.asp

Invenergy
Homepage: http://www.invenergyllc.com/

Jamie Oliver's Food Revolution
Homepage: http://www.jamieoliver.com/

Metabiota
Homepage: http://metabiota.com/

Monsanto
Homepage: http://www.monsanto.com
  Science behind Our Products
  http://www.monsanto.com/products/science.asp

Solar Energy Industries Association (SEIA)
Homepage: www.seia.org/

Sun Microsystems – Eco Responsibility
www.sun.com/aboutsun/media/presskits/ecoresponsibility/index.jsp

VI. Appendices

Appendix A - The CDC's Third National Report on Human Exposure to Environmental Chemicals

Centers for Disease Control, National Center for Environmental Health, Division of Laboratory Sciences. Atlanta, Georgia

NCEH Pub. No. 05-0570. As an introduction to the CDC's *Third National Report*, the authors make the following comment, "For more than three decades, laboratory scientists at NCEH [National Center for Environmental Health] have been determining which environmental chemicals enter people's bodies, how much of those chemicals are actually present, and how the amounts of those chemicals may be related to health effects." The report further states, "The National Report... provides an ongoing assessment of exposure of the US population to environmental chemicals used in biomonitoring." The Third Report contains new chemicals such as pyrethroid insecticides; the POPs Aldrin, Eldrin, and Dieldrin, and additional phthalate metabolites; pesticides; herbicides; and additional dioxins, furans, and polychlorinated byphenyls. What the CDC report does not contain is any discussion of the point sources where these chemicals originate, or their pathways to human consumption. Many of the chemicals listed are obviously well known pesticides and herbicides, some of which have been the subject of restricted use and mitigated environmental impact. As ubiquitous, persistent organic pollutants, they are nonetheless still present as ecotoxins in the biosphere, and thus included in the CDC survey of human blood, urine, or serum. Unfortunately, many of the ecotoxins listed in the *Third National Report* have much more obscure point sources, particularly as combustion products or other derivatives of the many hazardous chemicals used in the industrial and consumer products of modern society. Most of the following *Appendices* in this publication explore a few of the many labyrinths of the human tendency to create an endless variety of ecotoxins during the process of imposing increasing technometabolic human ecosystems upon natural ecosystems. The unfortunate result of this historical process is that in a biosphere with a human population approaching seven billion, the catastrophic impact of the rapid spread of a global consumer culture and its multiplicity of ecotoxins is now an unavoidable component of everyday life.

Appendix A has two components, a listing of the specific chemicals including heavy metals measured in the CDC's *Second and Third National Reports*, and a representative excerpt of one of the many ecotoxins monitored by the CDC. We have chosen to reproduce Tables 171 and 172, which display PCB 180, both for lipid adjusted serum concentrations and whole weight serum concentrations.

136

Part 1 – Chemical Groups Reported in the CDC Report, Except Heavy Metals

Polycyclic Aromatic Hydrocarbons:

1-Hydroxybenz[a]anthracene, 3-Hydroxybenz[a]anthracene,
9-Hydroxybenz[a]anthracene, 1-Hydroxybenzo[c]phenanthrene,
2-Hydroxybenzo[c]phenanthrene, 3-Hydroxybenzo[c]phenanthrene,
1-Hydroxychrysene, 2-Hydroxychrysene, 3-Hydroxychrysene,
4-Hydroxychrysene, 6-Hydroxychrysene, 3-Hydroxyfluoranthene,
2-Hydroxyfluorene, 3-Hydroxyfluorene, 9-Hydroxyfluorene,
1-Hydroxyphenanthrene, 2-Hydroxyphenanthrene, 3-Hydroxyphenanthrene,
4-Hydroxyphenanthrene, 9-Hydroxyphenanthrene, 1-Hydroxypyrene,
3-Hydroxybenzo[a]pyrene, 1-Hydroxynapthalene, 2-Hydroxynapthalene

Metals:

Antimony, Barium, Beryllium, Cadmium, Cesium, Cobalt, Lead, Mercury, Molybdenum,
Platinum, Tungsten, Thallium, Uranium

Tobacco Smoke:

Cotinine

Phytoestrogens:

Daidzein, Enterodiol, Enterolactone, Equol, Genistein, O-Desmethylangolensin

Polychlorinated Dibenzo-p-dioxins, Dibenzofurans, Coplanar and Mono-Ortho-Substituted Biphenyls:

1,2,3,4,6,7,8,9-Octachlorodibenzo-p- dioxin (OCDD)
1,2,3,4,6,7,8-Heptachlorodibenzo-p -dioxin (HpCDD)
1,2,3,4,7,8-Hexachlorodibenzo-p -dioxin (HxCDD)
1,2,3,6,7,8-Hexachlorodibenzo-p -dioxin (HxCDD)
1,2,3,7,8,9-Hexachlorodibenzo-p -dioxin (HxCDD)
1,2,3,7,8-Pentachlorodibenzo-p -dioxin (PeCDD)
2,3,7,8-Tetrachlorodibenzo-p -dioxin (TCDD)
1,2,3,4,6,7,8,9-Octachlorodibenzofuran (OCDF)
1,2,3,4,6,7,8-Heptachlorodibenzofuran (HpCDF)
1,2,3,4,7,8,9-Heptachlorodibenzofuran (HpCDF)
1,2,3,4,7,8-Hexachlorodibenzofuran (HxCDF)
1,2,3,6,7,8-Hexachlorodibenzofuran (HxCDF)
1,2,3,7,8,9-Hexachlorodibenzofuran (HxCDF)
1,2,3,7,8-Pentachlorodibenzofuran (PeCDF)
2,3,4,6,7,8-Hexachlorodibenzofuran (HxCDF)
2,3,4,7,8-Pentachlorodibenzofuran (PeCDF)
2,3,7,8-Tetrachlorodibenzofuran (TCDF)
2,4,4'-Trichlorobiphenyl (PCB 28)
2,3',4,4'-Tetrachlorobiphenyl (PCB 66)
2,4,4',5-Tetrachlorobiphenyl (PCB 74)
3,4,4',5-Tetrachlorobiphenyl (PCB 81)
2,3,3',4,4'-Pentachlorobiphenyl (PCB 105)
2,3',4,4',5-Pentachlorobiphenyl (PCB 118)

3,3',4,4',5-Pentachlorobiphenyl (PCB 126)
2,3,3',4,4',5-Hexachlorobiphenyl (PCB 156)
2,3,3',4,4',5'-Hexachlorobiphenyl (PCB 157)
2,3',4,4',5,5'-Hexachlorobiphenyl (PCB 167)
3,3',4,4',5,5'-Hexachlorobiphenyl (PCB 169)
2,3,3',4,4',5,5'-Heptachlorobiphenyl (PCB 189)

Non-dioxin-like Polychlorinated Biphenyls:
2,2',5,5'-Tetrachlorobiphenyl (PCB 52)
2,2',3,4,5'-Pentachlorobiphenyl (PCB 87)
2,2',4,4',5-Pentachlorobiphenyl (PCB 99)
2,2',4,5,5'-Pentachlorobiphenyl (PCB 101)
2,3,3',4',6-Pentachlorobiphenyl (PCB 110)
2,2',3,3',4,4'-Hexachlorobiphenyl (PCB 128)
2,2',3,4,4',5' and 2,3,3',4,4',6-Hexachlorobiphenyl (PCB 138&158)
2,2',3,4',5,5'-Hexachlorobiphenyl (PCB 146)
2,2',3,4',5',6'-Hexachlorobiphenyl (PCB 149)
2,2',3,5,5',6-Hexachlorobiphenyl (PCB 151)
2,2',4,4',5,5'-Hexachlorobiphenyl (PCB 153)
2,2',3,3',4,4',5-Heptachlorobiphenyl (PCB 170)
2,2',3,3',4,5,5'-Heptachlorobiphenyl (PCB 172)
2,2',3,3',4,5',6'-Heptachlorobiphenyl (PCB 177)
2,2',3,3',5,5',6-Heptachlorobiphenyl (PCB 178)
2,2',3,4,4',5,5'-Heptachlorobiphenyl (PCB 180)
2,2',3,4,4',5',6-Heptachlorobiphenyl (PCB 183)
2,2',3,4',5,5',6-Heptachlorobiphenyl (PCB 187)
2,2',3,3',4,4',5,5'-Octachlorobiphenyl (PCB 194)
2,2',3,3',4,4',5,6-Octachlorobiphenyl (PCB 195)
2,2',3,3',4,4',5,6' and 2,2',3,4,4',5,5',6-Octachlorobiphenyl (PCB196&203)
2,2',3,3',4,5,5',6-Octachlorobiphenyl (PCB 199)
2,2',3,3',4,4',5,5',6'-Nonachlorobiphenyl (PCB 206)

Other Pesticides
N,N-Diethyl-3-methylbenzamide, ortho –Phenylphenol, 2,5-Dichlorophenol
Carbamate Pesticides:
2-Isopropoxyphenol, Carbofuranphenol
Phthalates:
Mono-methyl phthalate, Mono-ethyl phthalate, Mono-n-butyl phthalate,
Mono-isobutyl phthalate, Mono-benzyl phthalate, Mono-cyclohexyl phthalate,
Mono-2-ethylhexyl phthalate, Mono-(2-ethyl-5-oxohexyl) phthalate,
Mono-(2-ethyl-5-hydroxyhexyl) phthalate, Mono-n-octyl phthalate,
Mono-(3-carboxypropyl) phthalate, Mono-isononyl phthalate
Organochlorine Pesticides:
Hexachlorobenzene, Beta-hexachlorocyclohexane,
Gamma-hexachlorocyclohexane, Pentachlorophenol, 2,4,5-Trichlorophenol, 2,4,6-
Trichlorophenol, p,p' –DDT, p,p' –DDE, o,p' –DDT, Oxychlordane,

trans-Nonachlor, Heptachlor epoxide, Mirex, Aldrin, Dieldrin, Endrin

Organophosphate Insecticides: Dialkyl Phosphate Metabolites:
Dimethylphosphate, Dimethylthiophosphate, Dimethyldithiophosphate, Diethylphosphate, Diethylthiophosphate, Diethyldithiophosphate

Organophosphate Insecticides: Specific Metabolites:
Malathion dicarboxylic acid, para –Nitrophenol,
3,5,6-Trichloro-2-pyridinol, 2-Isopropyl-4-methyl-6-hydroxypyrimidine,
2-(Diethylamino)-6-methylpyrimidin-4-ol/one,
3-Chloro-7-hydroxy-4-methyl-2H-chromen-2-one/ol

Herbicides:
2,4,5-Trichlorophenoxyacetic acid, 2,4-Dichlorophenoxyacetic acid,
2,4-Dichlorophenol, Alachlor mercapturate, Atrazine mercapturate, Acetochlor mercapturate,
Metolachlor mercapturate

Pyrethroid Pesticides:
4-Fluoro-3-phenoxybenzoic acid
cis-3-(2,2-Dichlorovinyl)-2,2-dimethylcyclopropane carboxylic acid
trans-3-(2,2-Dichlorovinyl)-2,2-dimethylcyclopropane carboxylic acid
cis-3-(2,2-Dibromovinyl)-2,2-dimethylcyclopropane carboxylic acid
3-Phenoxybenzoic acid

Part 2 –Tables 171 and 172 – 2,2',3,4,4',5,5'-Heptachlorobiphenol (PCB 180)

Table 171. 2,2',3,4,4',5,5'-Heptachlorobiphenyl (PCB 180) (lipid adjusted)

Geometric mean and selected percentiles of serum concentrations (in ng/g of lipid or parts per billion on a lipid-weight basis) for the U.S. population aged 12 years and older, National Health and Nutrition Examination Survey, 1999-2002.

	Survey years	Geometric mean (95% conf. interval)	Selected percentiles (95% confidence interval)				Sample size
			50th	75th	90th	95th	
Total, age 12 and older	99-00	*	< LOD	37.4 (32.7-41.8)	62.0 (56.6-66.6)	79.0 (72.1-89.2)	1924
	01-02	19.2 (17.4-21.1)	21.8 (19.0-24.6)	42.2 (38.3-47.5)	69.7 (63.7-75.9)	87.0 (83.3-93.0)	2302
Age group							
12-19 years	99-00	*	< LOD	< LOD	< LOD	< LOD	667
	01-02	*	< LOD	< LOD	12.2 (<LOD-14.8)	21.3 (14.2-26.0)	755
20 years and older	99-00	*	< LOD	41.0 (37.6-45.2)	65.5 (60.6-69.4)	83.8 (75.6-96.1)	1257
	01-02	23.0 (20.8-25.5)	26.4 (22.7-28.5)	46.7 (41.4-51.1)	74.0 (66.7-79.8)	90.7 (85.0-99.5)	1547
Gender							
Males	99-00	*	< LOD	40.5 (34.5-45.0)	65.1 (58.6-71.4)	83.8 (75.8-96.3)	919
	01-02	21.1 (18.8-23.7)	25.1 (21.2-30.0)	46.7 (39.8-51.8)	73.8 (63.2-79.8)	86.9 (78.0-99.4)	1073
Females	99-00	*	< LOD	34.4 (29.8-39.3)	56.7 (52.2-62.6)	74.6 (66.6-90.5)	1005
	01-02	17.5 (15.9-19.3)	18.3 (16.0-21.6)	39.6 (34.8-43.9)	64.0 (57.9-72.0)	87.9 (79.1-98.1)	1229
Race/ethnicity							
Mexican Americans	99-00	*	< LOD	< LOD	41.7 (33.2-50.5)	56.6 (49.7-63.8)	633
	01-02	*	< LOD	18.0 (11.4-22.2)	36.9 (28.2-45.7)	54.2 (42.2-60.0)	566
Non-Hispanic blacks	99-00	*	< LOD	39.1 (32.2-48.4)	78.4 (64.3-93.7)	117 (89.6-144)	414
	01-02	19.5 (16.5-23.1)	20.7 (16.9-23.8)	48.4 (36.7-57.9)	89.6 (73.7-101)	116 (96.1-167)	514
Non-Hispanic whites	99-00	*	< LOD	39.9 (34.5-45.3)	62.0 (56.5-68.6)	79.0 (71.3-91.4)	719
	01-02	21.4 (19.1-23.9)	24.8 (21.4-27.9)	45.6 (40.2-51.0)	72.3 (63.8-78.0)	87.9 (83.8-94.3)	1059

< LOD means less than the limit of detection, which may vary for some chemicals by year and by individual sample. See Appendix A for LODs.

* Not calculated. Proportion of results below limit of detection was too high to provide a valid result.

Table 172 2,2',3,4,4',5,5'-Heptachlorobiphenyl (PCB 180) (whole weight)

Geometric mean and selected percentiles of serum concentrations (in ng/g of serum or parts per billion) for the U.S. population aged 12 years and older, National Health and Nutrition Examination Survey, 1999-2002.

	Survey years	Geometric mean (95% conf. interval)	Selected percentiles (95% confidence interval)				Sample size
			50th	75th	90th	95th	
Total, age 12 and older	99-00	*	< LOD	.245 (.224-.266)	.414 (.373-.452)	.535 (.493-.604)	1924
	01-02	.118 (.106-.130)	.140 (.120-.155)	.278 (.253-.304)	.458 (.401-.508)	.605 (.541-.692)	2302
Age group							
12-19 years	99-00	*	< LOD	< LOD	< LOD	< LOD	667
	01-02	*	< LOD	< LOD	.061 (.053-.075)	.092 (.068-.135)	755
20 years and older	99-00	*	< LOD	.266 (.244-.296)	.441 (.407-.482)	.558 (.503-.617)	1257
	01-02	.146 (.131-.162)	.168 (.150-.188)	.302 (.275-.337)	.490 (.429-.555)	.637 (.572-.738)	1547
Gender							
Males	99-00	*	< LOD	.257 (.224-.300)	.459 (.365-.506)	.568 (.501-.648)	919
	01-02	.131 (.117-.147)	.165 (.137-.197)	.296 (.258-.339)	.479 (.405-.554)	.616 (.526-.716)	1073
Females	99-00	*	< LOD	.231 (.195-.266)	.377 (.344-.413)	.510 (.430-.557)	1005
	01-02	.106 (.096-.117)	.113 (.100-.135)	.261 (.228-.281)	.415 (.375-.470)	.604 (.519-.701)	1229
Race/ethnicity							
Mexican Americans	99-00	*	< LOD	< LOD	.275 (.198-.324)	.390 (.316-.452)	633
	01-02	*	< LOD	.110 (.079-.157)	.259 (.198-.307)	.353 (.279-.419)	566
Non-Hispanic blacks	99-00	*	< LOD	.245 (.188-.298)	.497 (.349-.656)	.727 (.534-1.04)	414
	01-02	.110 (.092-.131)	.119 (.091-.142)	.287 (.215-.345)	.560 (.436-.651)	.719 (.523-1.01)	514
Non-Hispanic whites	99-00	*	< LOD	.263 (.234-.299)	.430 (.377-.470)	.537 (.483-.604)	719
	01-02	.132 (.118-.148)	.160 (.142-.180)	.292 (.262-.327)	.471 (.405-.532)	.617 (.526-.730)	1059

< LOD means less than the limit of detection, which may vary for some chemicals by year and by individual sample. See Appendix A for LODs.

* Not calculated. Proportion of results below limit of detection was too high to provide a valid result.

Appendix B - Ridding the World of POPs: A Guide to the Stockholm Convention on Persistent Organic Pollutants

In 1995, the United Nations Environment Programme (UNEP), in response to a growing concern about the proliferation of toxic environmental chemicals, prepared an assessment of twelve notoriously persistent organic pollutants (POPs), which had been documented as globally transported ecotoxins of concern. After evaluation by the Intergovernmental Forum on Chemical Safety (IFCS) and the International Program for Chemical Safety (IPCS) the UNEP began negotiations to control eight organochlorine pesticides, two industrial chemicals (HCB and the PCB group), as well as their industrial byproducts, dioxins and furans, all of which are listed below. The Stockholm Convention on Persistent Organic Pollutants took effect May 17th 2004 and was ratified by 151 participating signatories. The convention has since been updated and other chemicals have been included or are being considered. These are: hexabromobiphenyl, octaBDE, pentaBDE, pentachlorobenzene, short-chained chlorinated paraffins, lindane, α- and β-hexachlorocyclohexane, dicofol, endosulfan, chlordecone, and PFOS (UNECE http://chm.pops.int/).

The first 12 POPs

Aldrin – A pesticide applied to soils to kill termites, grasshoppers, corn rootworm, and other insect pests.

Chlordane – Used extensively to control termites and as a broad-spectrum insecticide on a range of agricultural crops.

DDT – Perhaps the best known of the POPs, DDT was widely used during World War II to protect soldiers and civilians from malaria, typhus, and other diseases spread by insects. It continues to be applied against mosquitoes in several countries to control malaria.

Dieldrin – Used principally to control termites and textile pests, dieldrin has also been used to control insect-borne diseases and insects living in agricultural soils.

Dioxins – These chemicals are produced unintentionally due to incomplete combustion, as well as during the manufacture of certain pesticides and other chemicals. In addition, certain kinds of metal recycling and pulp and paper bleaching can release dioxins. Dioxins have also been found in automobile exhaust, tobacco smoke, and wood and coal smoke.

Endrin – This insecticide is sprayed on the leaves of crops such as cotton and grains. It is also used to control mice, voles, and other rodents.

Furans – These compounds are produced unintentionally from the same processes that release dioxins, and they are also found in commercial mixtures of PCBs.

Heptachlor – Primarily employed to kill soil insects and termites, heptachlor has also been used more widely to kill cotton insects, grasshoppers, other crop pests, and malaria-carrying mosquitoes.

Hexachlorobenzene (HCB) – HCB kills fungi that affect food crops. It is also released as a byproduct during the manufacture of certain chemicals and as a result of the processes that give rise to dioxins and furans.

Mirex – This insecticide is applied mainly to combat fire ants and other types of ants and termites. It has also been used as a fire retardant in plastics, rubber, and electrical goods.

Polychlorinated Biphenyls (PCBs) – These compounds are employed in industry as heat exchange fluids, in electric transformers and capacitors, and as additives in paint, carbonless copy paper, sealants and plastics.

Toxaphene – This insecticide, also called camphechlor, is applied to cotton, cereal grains, fruits, nuts, and vegetables. It has also been used to control ticks and mites in livestock.

Appendix C - EPA *List of Lists* and the Clean Water Act Priority Pollutants

The United States' Environmental Protection Agency has produced a complex labyrinth of listings of the most important environmental pollutants encountered in air, food, water, industrial processes, and consumer products. These listings, which begin with a generic listing of pollutants (40 CFR 401.15), the last listing in this appendix, was followed by a compilation of "priority pollutants" (40 CFR 423), also listed in this appendix. The EPA's frequently updated *List of Lists* of chemicals pertaining to the Clean Air Act is the first of their listings discussed in this appendix. A broad survey of hazardous chemicals is contained in 40 CFR 502.4 from which are extracted the extremely hazardous chemicals listed in *Appendix D* (40 CFR 355). *Appendix E* contains a listing of chemicals in food subject to EPA action as well as exempt chemicals (40 CFR 480). Chemicals of concern to the EPA in water are listed in *Appendix F*. The overall significance of these multiple listings, of which the contents of these appendices are only the tip of the ecotoxin information iceberg, includes the complexity of the chemical fallout crisis, our lack of a comprehensive guide to these chemicals, their broad distribution throughout the biosphere via the atmospheric water cycle, and, since 1976 and the establishment of the EPA, the rapid expansion of the large number of chemicals emerging as biologically significant ecotoxins.

United States Environmental Protection Agency. (2006). *List of lists: Consolidated list of chemicals subject to the Emergency Planning and Community Right-to-Know Act (EPCRA) and Section 112(r) of the Clean Air Act*. EPA 550-B-01-003. Chemical Emergency Preparedness and Prevention Office, Office of Solid Waste and Emergency Response, EPA. www.epa.gov/emergencies/docs/chem/title3_Oct_2006.pdf.

A one page specimen of the EPA's *List of Lists* pertaining to the Community Right-to-Know Act (EPCRA) and Section 112 of the Clean Air Act follows the introductory remarks to the EPA *List of Lists* quoted below. The page chosen for reproduction is one that includes PCBs as hazardous substances that might be produced in reportable quantities by an American manufacturer. The EPA's *List of Lists* has two distinct categories, hazardous and extremely hazardous chemicals. Both hazardous and extremely hazardous chemicals are contained in 40 Code of Federal Regulations (CFR) Section 302.4. Section 302.4 is supplemented by 40 CFR 355, which provides a specific listing of the extremely hazardous chemicals listed in 302.4. 40 CFR 355 is further divided into two appendices, the first of which (*Appendix A*) provides an alphabetical listing of what the EPA defines as extremely hazardous substances; 355 also includes their Threshold Planning Quantities (TPQs). A commentary on the significance and relationship of these various listings follows our one page excerpt from the *List of Lists* and will introduce the reader to the significance of *Appendix D*. Under the EPCRA legislation, any company that "receives or produces the substance on the site at or above EHS's TPQ" must notify "local emergency planning committees (LEPCs)," who must develop emergency response plans pertaining to accidents at these manufacturing sites. The following quotations are selected excerpts from the EPA introduction to the *List of Lists*. The EPA definition of basic terms such as Threshold Planning Quantity (TPQ) and Reportable Quantities (RQs) of the EPA's two categories -- hazardous and extremely hazardous substances -- are contained in the following excerpts.

Emergency Planning and Community Right-to-Know Act (EPCRA) Introduction

"This consolidated chemical list includes chemicals subject to reporting requirements under the EPCRA, also known as Title III of the Superfund Amendments and Reauthorization Act of 1986 (SARA), and chemicals listed under section 112(r) of the Clean Air Act (CAA). This consolidated list has been prepared to help firms handling chemicals determine whether they need to submit reports under section 302, 304, or 313 of EPCRA and, for a specific chemical, what reports may need to be submitted…"

EPCRA Section 302

…The presence of EHSs in quantities at or above the Threshold Planning Quantity (TPQ) requires certain emergency planning activities to be conducted. The extremely hazardous substances and their TPQs are listed in 40 CFR Part 355, *Appendices A* and *B*. For section 302 EHSs, Local Emergency Planning Committees (LEPCs) must develop emergency response plans and facilities must notify the State Emergency Response Commission (SERC) and LEPC if they receive or produce the substance on site at or above the EHS's TPQ. Additionally if the TPQ is met, facilities with a listed EHS are subject to the reporting requirements of EPCRA section 311 (provide material safety data sheet of a list of covered chemicals to the SERC, LEPC, and local fire department) and section 312 (submit inventory form – Tier I or Tier II). The minimum threshold for section 311-312 reporting for EHS substances is 500 pounds or the TPQ, whichever is less.

TPQ (threshold planning quantity). The consolidated list presents the TPQ (in pounds) for section 302 chemicals in the column following the CAS number. For chemicals that are solids, there may be two TPQs given (e.g., 500/10,000). In these cases, the lower quantity applies for solids in powder form with particle size less than 100 microns, or if the substance is in solution or in molten form. Otherwise, the 10,000 pound TPQ applies.

EHS (extremely hazardous substance) RQ (Reportable Quantities). Releases of reportable quantities of EHSs are subject to state and local reporting under section 304 of EPCRA. EPA has promulgated a rule (61 FR 20473, May 7, 1996) that adjusted RQs for EHSs without CERCLA RQs to levels equal to their TPQs. The EHS RQ column lists these adjusted RQs for EHSs not listed under The

144

Comprehensive Environmental Response, Compensation, and Liability Act (CERCLA), commonly known as Superfund (CERCLA) and the CERCLA RQs for those EHSs that are CERCLA hazardous substances…

Releases of CERCLA hazardous substances, in quantities equal to or greater than their reportable quantity (RQ), are subject to reporting to the National Response Center under CERCLA. Such releases are also subject to state and local reporting under section 304 of EPCRA. CERCLA hazardous substances, and their reportable quantities, are listed in 40 CFR Part 302, Table 302.4. Radionuclides listed under CERCLA are provided in a separate list, with RQs in Curies.

RQ. The CERCLA RQ column in the consolidated list shows the RQs (in pounds) for chemicals that are CERCLA hazardous substances. Carbamate wastes under RCRA that have been added to the CERCLA list with statutory one-pound RQs are indicated by an asterisk ("*") following the RQ (United States Environmental Protection Agency 2006).

NAME	CAS/313 Category Codes	Section 302 (EHS) TPQ	Section 304 EHS RQ	CERCLA RQ	Section 313	RCRA CODE	CAA 112(r) TQ
Asbestos (friable)	1332-21-4			1	313		
Hydrogen	1333-74-0						10,000
Sodium bifluoride	1333-83-1			100			
Lead subacetate	1335-32-6			10	313c	U146	
Hexachloronaphthalene	1335-87-1				313		
Ammonium hydroxide	1336-21-6			1,000	313		
PCBs	1336-36-3			1	X		
Polychlorinated biphenyls	1336-36-3			1	313^		
Methyl ethyl ketone peroxide	1338-23-4			10		U160	
Naphthenic acid	1338-24-5			100			
Ammonium bifluoride	1341-49-7			100			
Aluminum oxide (fibrous forms)	1344-28-1				313		
Antimycin A	1397-94-0	1,000/10,000	1,000				
Dinoterb	1420-07-1	500/10,000	500				
2,2'-Bioxirane	1464-53-5	500	10	10	X	U085	
Diepoxybutane	1464-53-5	500	10	10	313	U085	
Trichloro(chloromethyl)silane	1558-25-4	100	100				
Carbofuran phenol	1563-38-8			1*		U367	
Carbofuran	1563-66-2	10/10,000	10	10	313	P127	
Benezeneamine, 2,6-dinitro-N,N-dipropyl-4-(trifluoromethyl)-	1582-09-8			10	X		
Trifluralin	1582-09-8			10	313^		
Mercuric acetate	1600-27-7	500/10,000	500		313c		
Hydrazine, 1,2-diethyl-	1615-80-1			10		U086	
Ethanesulfonyl chloride, 2-chloro-	1622-32-8	500	500				
Methyl tert-butyl ether	1634-04-4			1,000	313		
Aldicarb sulfone	1646-88-4			1*		P203	
1,2-Dichloro-1,1-difluoroethane	1649-08-7				313		
HCFC-132b	1649-08-7				X		
3,5-Dibromo-4-hydroxybenzonitrile	1689-84-5				X		
Bromoxynil	1689-84-5				313		
Bromoxynil octanoate	1689-99-2				313		
Octanoic acid, 2,6-dibromo-4-cyanophenyl ester	1689-99-2				X		
1,1-Dichloro-1-fluoroethane	1717-00-6				313		
HCFC-141b	1717-00-6				X		
2,3,7,8-Tetrachlorodibenzo-p-dioxin (TCDD)	1746-01-6			1	313!^		
Acetone thiosemicarbazide	1752-30-3	1,000/10,000	1,000				
Ammonium thiocyanate	1762-95-4			5,000			
Benzene, 2,4-dichloro-1-(4-nitrophenoxy)-	1836-75-5				X		
Nitrofen	1836-75-5				313		
Benfluralin	1861-40-1				313		
N-Butyl-N-ethyl-2,6-dinitro-4-(trifluoromethyl) benzenamine	1861-40-1				X		
Ammonium benzoate	1863-63-4			5,000			
Hexachloropropene	1888-71-7			1,000		U243	
1,3-Benzenedicarbonitrile, 2,4,5,6-tetrachloro-	1897-45-6				X		
Chlorothalonil	1897-45-6				313		
Paraquat dichloride	1910-42-5	10/10,000	10		313		
6-Chloro-N-ethyl-N'-(1-methylethyl)-	1912-24-9				X		

The columns in the above EPA *List of Lists* table are:

- Column 1 – CAS/313: Chemical Abstract Services Registry Number

- Column 2 – Section 302: Extremely Hazardous Substance Threshold Planning Quantity Reporting Requirements: If a TPQ is noted in this column "certain emergency planning activities to be conducted." The chemical is not considered extremely hazardous if there are no data in this column.

- Column 3 – Section 4 EHS RQ – Reportable Quantities of extremely hazardous chemicals if released at the site of manufacturing or storage

- Column 4 – Reportable Quantities under the Comprehensive Environmental Response, Compensation, and Liability Act of 1980 (CERCLA) of hazardous and, in some cases, extremely hazardous substances

- Column 5 – Section 313: The listing of chemicals that, as part of the provisions of the CERCLA (40 CFR Part 372), must be reported as "emissions, transfers, and waste management data" Some of these chemicals have codes, such as X, which indicates that "the same chemical with the same CAS number appears on another list with a different chemical name.

- Column 6 – RCRA Code: Lists chemicals regulated by the Resource Conservation and Recovery Act

- Column 7 – Clean Air Act (CCA): Lists chemicals regulated under the Clean Air Act that might be generated by a manufacturing facility

A number of important observations need to be made about the chemicals reported in the EPA's *List of Lists*. Using PCBs as an example, it can be noted that this famous persistent organic pollutant is not considered an extremely hazardous substance if manufactured in the US, but does have a very low reporting quantity under the CERCLA legislation. PCBs, once subject to combustion or other forms of chemical change, are components of hundreds of ecotoxins, many of which are specifically listed in *Appendix A*, the CDC's *Third National Report on Human Exposure to Environmental Chemicals*. Furthermore, additional forms of PCBs are listed both in 40 CFR Section 302.4, Hazardous Substances, and 40 CFR Part 355, Extremely Hazardous Substances. PCBs are only one example of the chemicals listed as hazardous substances that are incorporated into a wide variety of industrial and consumer products. While disposal of these products is not generally regulated within the boundaries of the US, *Appendix J*, which displays waste electrical and electronic equipment generated in Europe, is a detailed listing of the many consumer products requiring special disposal. Many of these products contain PCBs as microcontaminants.

An inescapable component of the Age of Chemical Fallout is the now well-documented proliferation of a wide variety of persistent organic pollutants throughout the biosphere. PCB-like substances have been known since 1865 and the first synthetic PCB was produced in 1881. The Anniston (AL) Ordnance Co. (later known by the founder's name, Swann Chemical) started producing PCBs commercially in 1927. By 1933, 23 out of 24 workers at the plant were reporting symptoms such as lesions and pustules. In 1935, the Monsanto Co. purchased Swann Chemical and produced PCBs at the Alabama plant and in Sauget, Illinois until they licensed the product to other producers in 1977. By the time Miller and Beng published their famous text on chemical fallout in 1970, PCBs, with the help of the biogeochemical cycles of the biosphere, especially the global atmospheric water cycle, were ubiquitous contaminants in Antarctic sea birds. The Waste Electrical and Electronic Equipment (WEEE) listing, in view of the well documented human body burdens of PCB-related chemicals listed in *Appendix A*, suggests that the biomonitoring of the future should include tracing the pathways of persistent organic pollutants (POPs) and other ecotoxins from the electrical and electronic equipment tracked by European governments to their eventual repositories in biotic media including humans.

The introduction to the EPA's *List of Lists* contains a particularly significant footnote, which provides revealing commentary on the widespread distribution of chemicals catalogued for toxicity in both the *List of Lists* and the 40 CFR Sections 302.4 and 355.

> "This consolidated list does not include all chemicals subject to the reporting requirements in EPCRA sections 311 and 312. These hazardous chemicals, for which material safety data sheets (MSDS) must be developed under the Hazard Communication Standard (29 CFR 1910.1200), are identified by broad criteria, rather than by enumeration. There are over 500,000 products that satisfy the criteria..."

The significance of this footnote is that, utilizing the thousands of chemicals listed on the EPA's *List of Lists*, hundreds of thousands of industrial materials and consumer products containing ecotoxins are produced in America by its robust manufacturing industries. The introduction to the EPA's *List of Lists* also documents the growing knowledge of the wide variety of ecotoxins produced by combustion and other forms of chemical change. Of the many thousands of hazardous chemicals listed in 40 CFR 302.4, many have highly toxic byproducts, cogeners, and metabolites. All governmental listings of toxic chemicals are now expanding, but do not yet include some of the emerging ecotoxins documented in *Appendices L, M, Q, and R*. The EPA *List of Lists* introduction contains the following information.

> "Diisocyanates, Dioxins and dioxin-like Compounds, and PACs. In the November 30, 1994 expansion of the section 313 list, 20 specific chemicals were added as members of the diisocyanate category, and 19 specific chemicals were added as members of the polycyclic aromatic compounds (PAC) category. In October 1999, EPA added a category of dioxins and dioxin-like compounds that includes 17 specific chemicals. These chemicals are included in the CAS order listing on this consolidated list. The symbol "#" following the "313" notation in the section 313 column identifies

diisocyanates, the symbol "!" identifies the dioxins and dioxin-like compounds, and the symbol "+"identifies PACs, as noted in the Summary of Codes. Chemicals belonging to these categories are reportable under section 313 by category, rather than by individual chemical name."

Dioxins are not, unlike target-specific agricultural chemicals such as the POP Mirex, deliberately produced as a component of industrial activities. Their high degree of toxicity is reflected in the TRI threshold listing also contained in the EPA's introduction to the *List of Lists*.

"For most TRI chemicals [reported under EPCRA 313], the thresholds are 25,000 pound manufactured or processed or 10,000 pound otherwise used. EPA has recently lowered the reporting thresholds for certain chemicals and chemical categories that meet the criteria for persistence and bioaccumulation."

The listing that follows for dioxin has a reporting threshold of only .1 gram, and includes the following caveat:

"Dioxin and dioxin-like compound category (manufacturing and processing or otherwise use of dioxin and dioxin-like compounds if they are present as contaminants in a chemical and if they were created during the manufacture of that chemical."

A review of 40 CFR Part 355, the alphabetical listing of Extremely Hazardous Substances, does not include the word "dioxin," though it does include "furan," a chemical almost always associated with the dioxins family. This is a tip-off that our listing of EHS does not include the many cogeners and metabolites that are their progeny. The fact that modern global consumer culture is producing ecotoxins with the potency of the dioxin-furan family of compounds, and that they have infiltrated food webs and accumulated in biotic media including humans on a worldwide basis, are compelling reasons for much more detailed biomonitoring programs. Standing as obstacles to the documentation of anthropogenic chemical fallout are a lack of public resources, vested economic interests, political and religious ideology and sectarianism, and a general denial of the significance of the biological and economic significance of ecotoxins by a narcissistic consumer culture living on the edge of mass functional illiteracy.

Priority Pollutants

The Priority Pollutants are a set of chemical pollutants EPA regulates, and for which EPA has published analytical test methods. This list is published in 40 CFR 423, Appendix A. It is also online at http://www.epa.gov/waterscience/methods/pollutants.htm.

Acenaphthene	Benzene	Chlorobenzene
Acrolein	Benzidine	1,2,4-trichlorobenzene
Acrylonitrile	Carbon tetrachloride	Hexachlorobenzene

1,2-dichloroethane	Dichlorobromomethane	Aldrin
1,1,1-trichloreothane	REMOVED	Dieldrin
Hexachloroethane	REMOVED	Chlordane
1,1-dichloroethane	Chlorodibromomethane	4,4-DDT
1,1,2-trichloroethane	Hexachlorobutadiene	4,4-DDE
1,1,2,2-tetrachloroethane	Hexachlorocyclopentadiene	4,4-DDD
Chloroethane	Isophorone	Alpha-endosulfan
REMOVED	Naphthalene	Beta-endosulfan
Bis(2-chloroethyl) ether	Nitrobenzene	Endosulfan sulfate
2-chloroethyl vinyl ethers	2-nitrophenol	Endrin
2-chloronaphthalene	4-nitrophenol	Endrin aldehyde
2,4,6-trichlorophenol	2,4-dinitrophenol	Heptachlor
Parachlorometa cresol	4,6-dinitro-o-cresol	Heptachlor epoxide
Chloroform	N-nitrosodimethylamine	Alpha-BHC
2-chlorophenol	N-nitrosodiphenylamine	Beta-BHC
1,2-dichlorobenzene	N-nitrosodi-n-propylamine	Gamma-BHC
1,3-dichlorobenzene	Pentachlorophenol	Delta-BHC
1,4-dichlorobenzene	Phenol	PCB–1242 (Arochlor 1242)
3,3-dichlorobenzidine	Bis(2-ethylhexyl) phthalate	PCB–1254 (Arochlor 1254)
1,1-dichloroethylene	Butyl benzyl phthalate	PCB–1221 (Arochlor 1221)
1,2-trans-dichloroethylene	Di-N-Butyl Phthalate	PCB–1232 (Arochlor 1232)
2,4-dichlorophenol	Di-n-octyl phthalate	PCB–1248 (Arochlor 1248)
1,2-dichloropropane	Diethyl Phthalate	PCB–1260 (Arochlor 1260)
1,2-dichloropropylene	Dimethyl phthalate	PCB–1016 (Arochlor 1016)
2,4-dimethylphenol	benzo(a) anthracene	Toxaphene
2,4-dinitrotoluene	Benzo(a)pyrene	Antimony
2,6-dinitrotoluene	Benzo(b) fluoranthene	Arsenic
1,2-diphenylhydrazine	Benzo(b) fluoranthene	Asbestos
Ethylbenzene	Chrysene	Beryllium
Fluoranthene	Acenaphthylene	Cadmium
4-chlorophenyl phenyl ether	Anthracene	Chromium
4-bromophenyl phenyl ether	Benzo(ghi) perylene	Copper
	Fluorene	Cyanide, Total
Bis(2-chloroisopropyl) ether	Phenanthrene	Lead
	Dibenzo(,h) anthracene	Mercury
Bis(2-chloroethoxy) methane	Indeno (1,2,3-cd) pyrene	Nickel
Methylene chloride	Pyrene	Selenium
Methyl chloride	Tetrachloroethylene	Silver
Methyl bromide	Toluene	Thallium
Bromoform	Trichloroethylene	Zinc
	Vinyl chloride	2,3,7,8-TCDD

Toxic Pollutants

The Clean Water Act references the following list of toxic pollutants at §307(a)(1) (also labeled §1317(a)(1)). The list appears in the Code of Federal Regulations at 40 CFR 401.15. It may be found online at http://ecfr.gpoaccess.gov/cgi/t/text/text-idx?c=ecfr&rgn=div8&view=text&node=40:28.0.1.1.2.0.1.6&idno=40

Title 40: Protection of Environment
PART 401—GENERAL PROVISIONS

§ 401.15   Toxic pollutants.

The following comprise the list of toxic pollutants designated pursuant to section 307(a)(1) of the Act:

1. Acenaphthene
2. Acrolein
3. Acrylonitrile
4. Aldrin/Dieldrin[1]
[1] Effluent standard promulgated (40 CFR part 129).
5. Antimony and compounds[2]
[2] The term *compounds* shall include organic and inorganic compounds.
6. Arsenic and compounds
7. Asbestos
8. Benzene
9. Benzidine[1]
10. Beryllium and compounds
11. Cadmium and compounds
12. Carbon tetrachloride
13. Chlordane (technical mixture and metabolites)
14. Chlorinated benzenes (other than dichlorobenzenes)
15. Chlorinated ethanes (including 1,2-dichloroethane, 1,1,1-trichloroethane, and hexachloroethane)
16. Chloroalkyl ethers (chloroethyl and mixed ethers)
17. Chlorinated naphthalene
18. Chlorinated phenols (other than those listed elsewhere; includes trichlorophenols and chlorinated cresols)
19. Chloroform
20. 2-chlorophenol
21. Chromium and compounds
22. Copper and compounds
23. Cyanides
24. DDT and metabolites[1]
25. Dichlorobenzenes (1,2-, 1,3-, and 1,4-dichlorobenzenes)
26. Dichlorobenzidine
27. Dichloroethylenes (1,1-, and 1,2-dichloroethylene)
28. 2,4-dichlorophenol
29. Dichloropropane and dichloropropene
30. 2,4-dimethylphenol
31. Dinitrotoluene
32. Diphenylhydrazine
33. Endosulfan and metabolites
34. Endrin and metabolites[1]
35. Ethylbenzene
36. Fluoranthene
37. Haloethers (other than those listed elsewhere; includes chlorophenylphenyl ethers, bromophenylphenyl ether, bis(dichloroisopropyl) ether, bis-(chloroethoxy) methane and polychlorinated diphenyl ethers)
38. Halomethanes (other than those listed elsewhere; includes methylene chloride, methylchloride, methylbromide, bromoform, dichlorobromomethane
39. Heptachlor and metabolites
40. Hexachlorobutadiene
41. Hexachlorocyclohexane
42. Hexachlorocyclopentadiene
43. Isophorone
44. Lead and compounds
45. Mercury and compounds
46. Naphthalene
47. Nickel and compounds
48. Nitrobenzene
49. Nitrophenols (including 2,4-dinitrophenol, dinitrocresol)

50. Nitrosamines
51. Pentachlorophenol
52. Phenol
53. Phthalate esters
54. Polychlorinated biphenyls (PCBs)[1]
55. Polynuclear aromatic hydrocarbons (including benzanthracenes, benzopyrenes, benzofluoranthene, chrysenes, dibenz-anthracenes, and indenopyrenes)
56. Selenium and compounds
57. Silver and compounds
58. 2,3,7,8-tetrachlorodibenzo-p-dioxin (TCDD)
59. Tetrachloroethylene
60. Thallium and compounds
61. Toluene
62. Toxaphene[1]
63. Trichloroethylene
64. Vinyl chloride
65. Zinc and compounds

[44 FR 44502, July 30, 1979, as amended at 46 FR 2266, Jan. 8, 1981; 46 FR 10724, Feb. 4, 1981]

Appendix D – 40 CFR Appendix A Part 355 – The List of Extremely Hazardous Substances and their Threshold Planning Quantities

The following is a list of chemicals, the manufacture of which must be reported to the EPA above specified reportable quantities and threshold planning quantities, as noted in the following table. This listing includes +/- 350 chemicals, all of which are included in the 40 CFR 502.4 listing of hazardous chemicals. This list contrasts with the restricted categories of FDA *Action Levels for Unavoidable Pesticide Residues in Food and Feed*, which are generic categories of ecotoxins, many of which are in the list. The FDA Action Levels for many of the chemicals in their listing are in reporting units of ppm (ug/g), typically .1 and .2 ug/g. The compelling significance of these reporting levels is the fact that the FDA and other government surveillance programs do not require intervention for hundreds of chemicals that are often present in the food web and biotic media at levels ranging one to two orders of magnitude less than the FDA Action Level. In all cases, these small contaminant pulse signals have no observable health effects, but the synergistic effect of hundreds and possibly thousands of such pulses from multiple food, feed, and water sources are now a major public health consideration, both for individuals and large populations.

For commentary on the significance of this list and its relationship to the CDC's *Third National Report on Human Exposure to Environmental Chemicals* in *Appendix A* and see the essay and related EPA excerpts in *Appendix C*.

Appendix B to Part 355—The List of Extremely Hazardous Substances and Their Threshold Planning Quantities

CAS No.	Chemical name	Notes	Reportable quantity* (pounds)	Threshold planning quantity (pounds)
75–86–5	Acetone Cyanohydrin	10	1,000	
1752–30–3	Acetone Thiosemicarbazide	1,000	1,000/10,000	
107–02–8	Acrolein	1	500	
79–06–1	Acrylamide	f	5,000	1,000/10,000
107–13–1	Acrylonitrile	f	100	10,000
814–68–6	Acrylyl Chloride	d	100	100
111–69–3	Adiponitrile	f	1,000	1,000
116–06–3	Aldicarb	b	1	100/10,000
309–00–2	Aldrin	1	500/10,000	
107–18–6	Allyl Alcohol	100	1,000	
107–11–9	Allylamine	500	500	
20859–73–8	Aluminum Phosphide	a	100	500
54–62–6	Aminopterin		500	500/10,000

CAS No.	Chemical name	Notes	Reportable quantity* (pounds)	Threshold planning quantity (pounds)
78–53–5	Amiton		500	500
3734–97–2	Amiton Oxalate		100	100/10,000
7664–41–7	Ammonia	f	100	500
300–62–9	Amphetamine		1,000	1,000
62–53–3	Aniline	f	5,000	1,000
88–05–1	Aniline, 2,4,6-Trimethyl-		500	500
7783–70–2	Antimony Pentafluoride		500	500
1397–94–0	Antimycin A	b	1,000	1,000/10,000
86–88–4	ANTU		100	500/10,000
1303–28–2	Arsenic Pentoxide		1	100/10,000
1327–53–3	Arsenous Oxide	d	1	100/10,000
7784–34–1	Arsenous Trichloride		1	500
7784–42–1	Arsine		100	100
2642–71–9	Azinphos-Ethyl		100	100/10,000
86–50–0	Azinphos-Methyl		1	10/10,000
98–87–3	Benzal Chloride		5,000	500
98–16–8	Benzenamine, 3-(Trifluoromethyl)-		500	500
100–14–1	Benzene, 1-(Chloromethyl)-4-Nitro-		500	500/10,000
98–05–5	Benzenearsonic Acid		10	10/10,000
3615–21–2	Benzimidazole, 4,5-Dichloro-2-(Trifluoromethyl)-	c	500	500/10,000
98–07–7	Benzotrichloride		10	100
100–44–7	Benzyl Chloride		100	500
140–29–4	Benzyl Cyanide	d	500	500
15271–41–7	Bicyclo[2.2.1]Heptane-2-Carbonitrile, 5-Chloro-6-(((((Methylamino)Carbonyl)Oxy)Imino)-, (1s-(1-alpha,2-beta,4-alpha,5-alpha,6E))-		500	500/10,000
534–07–6	Bis(Chloromethyl) Ketone		10	10/10,000
4044–65–9	Bitoscanate		500	500/10,000
10294–34–5	Boron Trichloride		500	500
7637–07–2	Boron Trifluoride		500	500
353–42–4	Boron Trifluoride Compound With Methyl Ether (1:1)		1,000	1,000
28772–56–7	Bromadiolone		100	100/10,000
7726–95–6	Bromine	f	500	500
1306–19–0	Cadmium Oxide		100	100/10,000
2223–93–0	Cadmium Stearate	b	1,000	1,000/10,000
7778–44–1	Calcium Arsenate		1	500/10,000

CAS No.	Chemical name	Notes	Reportable quantity* (pounds)	Threshold planning quantity (pounds)
8001–35–2	Camphechlor		1	500/10,000
56–25–7	Cantharidin		100	100/10,000
51–83–2	Carbachol Chloride		500	500/10,000
26419–73–8	Carbamic Acid, Methyl-, O-(((2,4-Dimethyl-1, 3-Dithiolan-2-yl)Methylene)Amino)-		100	100/10,000
1563–66–2	Carbofuran		10	10/10,000
75–15–0	Carbon Disulfide	f	100	10,000
786–19–6	Carbophenothion		500	500
57–74–9	Chlordane		1	1,000
470–90–6	Chlorfenvinfos		500	500
7782–50–5	Chlorine		10	100
24934–91–6	Chlormephos		500	500
999–81–5	Chlormequat Chloride	d	100	100/10,000
79–11–8	Chloroacetic Acid		100	100/10,000
107–07–3	Chloroethanol		500	500
627–11–2	Chloroethyl Chloroformate		1,000	1,000
67–66–3	Chloroform	f	10	10,000
542–88–1	Chloromethyl Ether	d	10	100
107–30–2	Chloromethyl Methyl Ether	b	10	100
3691–35–8	Chlorophacinone		100	100/10,000
1982–47–4	Chloroxuron		500	500/10,000
21923–23–9	Chlorthiophos	d	500	500
10025–73–7	Chromic Chloride		1	1/10,000
62207–76–5	Cobalt, ((2,2'-(1,2-Ethanediylbis (Nitrilomethylidyne)) Bis(6-Fluorophenolato))(2-)-N,N',O,O')-		100	100/10,000
10210–68–1	Cobalt Carbonyl	d	10	10/10,000
64–86–8	Colchicine	d	10	10/10,000
56–72–4	Coumaphos		10	100/10,000
5836–29–3	Coumatetralyl		500	500/10,000
95–48–7	Cresol, o-		100	1,000/10,000
535–89–7	Crimidine	100	100/10,000	
4170–30–3	Crotonaldehyde		100	1,000
123–73–9	Crotonaldehyde, (E)-		100	1,000
506–68–3	Cyanogen Bromide		1,000	500/10,000
506–78–5	Cyanogen Iodide		1,000	1,000/10,000
2636–26–2	Cyanophos		1,000	1,000
675–14–9	Cyanuric Fluoride		100	100

CAS No.	Chemical name	Notes	Reportable quantity* (pounds)	Threshold planning quantity (pounds)
66–81–9	Cycloheximide		100	100/10,000
108–91–8	Cyclohexylamine	f	10,000	10,000
17702–41–9	Decaborane(14)		500	500/10,000
8065–48–3	Demeton		500	500
919–86–8	Demeton-S-Methyl		500	500
10311–84–9	Dialifor		100	100/10,000
19287–45–7	Diborane		100	100
111–44–4	Dichloroethyl ether		10	10,000
149–74–6	Dichloromethylphenylsilane		1,000	1,000
62–73–7	Dichlorvos		10	1,000
141–66–2	Dicrotophos		100	100
1464–53–5	Diepoxybutane		10	500
814–49–3	Diethyl Chlorophosphate	d	500	500
71–63–6	Digitoxin	b	100	100/10,000
2238–07–5	Diglycidyl Ether		1,000	1,000
20830–75–5	Digoxin	d	10	10/10,000
115–26–4	Dimefox		500	500
60–51–5	Dimethoate		10	500/10,000
2524–03–0	Dimethyl Phosphorochloridothioate		500	500
77–78–1	Dimethyl sulfate		100	500
75–78–5	Dimethyldichlorosilane	d	500	500
57–14–7	Dimethylhydrazine		10	1,000
99–98–9	Dimethyl-p-Phenylenediamine		10	10/10,000
644–64–4	Dimetilan		1	500/10,000
534–52–1	Dinitrocresol		10	10/10,000
88–85–7	Dinoseb		1,000	100/10,000
1420–07–1	Dinoterb		500	500/10,000
78–34–2	Dioxathion		500	500
82–66–6	Diphacinone		10	10/10,000
152–16–9	Diphosphoramide, Octamethyl-		100	100
298–04–4	Disulfoton		1	500
514–73–8	Dithiazanine Iodide		500	500/10,000
541–53–7	Dithiobiuret		100	100/10,000
316–42–7	Emetine, Dihydrochloride	d	1	1/10,000
115–29–7	Endosulfan		1	10/10,000
2778–04–3	Endothion		500	500/10,000
72–20–8	Endrin		1	500/10,000
106–89–8	Epichlorohydrin	f	100	1,000

CAS No.	Chemical name	Notes	Reportable quantity* (pounds)	Threshold planning quantity (pounds)
2104–64–5	EPN		100	100/10,000
50–14–6	Ergocalciferol	b	1,000	1,000/10,000
379–79–3	Ergotamine Tartrate		500	500/10,000
1622–32–8	Ethanesulfonyl Chloride, 2-Chloro-		500	500
10140–87–1	Ethanol, 1,2-Dichloro-, Acetate		1,000	1,000
563–12–2	Ethion		10	1,000
13194–48–4	Ethoprophos		1,000	1,000
538–07–8	Ethylbis(2-Chloroethyl)Amine	d	500	500
371–62–0	Ethylene Fluorohydrin	b, d	10	10
75–21–8	Ethylene Oxide	f	10	1,000
107–15–3	Ethylenediamine		5,000	10,000
151–56–4	Ethyleneimine		1	500
542–90–5	Ethylthiocyanate		10,000	10,000
22224–92–6	Fenamiphos		10	10/10,000
115–90–2	Fensulfothion	d	500	500
4301–50–2	Fluenetil		100	100/10,000
7782–41–4	Fluorine	e	10	500
640–19–7	Fluoroacetamide		100	100/10,000
144–49–0	Fluoroacetic Acid		10	10/10,000
359–06–8	Fluoroacetyl Chloride	b	10	10
51–21–8	Fluorouracil		500	500/10,000
944–22–9	Fonofos		500	500
50–00–0	Formaldehyde	f	100	500
107–16–4	Formaldehyde Cyanohydrin	d	1,000	1,000
23422–53–9	Formetanate Hydrochloride	d	100	500/10,000
2540–82–1	Formothion		100	100
17702–57–7	Formparanate		100	100/10,000
21548–32–3	Fosthietan		500	500
3878–19–1	Fuberidazole		100	100/10,000
110–00–9	Furan		100	500
13450–90–3	Gallium Trichloride		500	500/10,000
77–47–4	Hexachlorocyclopentadiene	d	10	100
4835–11–4	Hexamethylenediamine, N,N'-Dibutyl-		500	500
302–01–2	Hydrazine		1	1,000
74–90–8	Hydrocyanic Acid		10	100
7647–01–0	Hydrogen Chloride (gas only)	f	5,000	500
7664–39–3	Hydrogen Fluoride		100	100
7722–84–1	Hydrogen Peroxide (Conc > 52%)	f	1,000	1,000

CAS No.	Chemical name	Notes	Reportable quantity* (pounds)	Threshold planning quantity (pounds)
7783–07–5	Hydrogen Selenide		10	10
7783–06–4	Hydrogen Sulfide	f	100	500
123–31–9	Hydroquinone	f	100	500/10,000
13463–40–6	Iron, Pentacarbonyl-		100	100
297–78–9	Isobenzan		100	100/10,000
78–82–0	Isobutyronitrile	d	1,000	1,000
102–36–3	Isocyanic Acid, 3,4-Dichlorophenyl Ester		500	500/10,000
465–73–6	Isodrin		1	100/10,000
55–91–4	Isofluorphate	b	100	100
4098–71–9	Isophorone Diisocyanate	g	500	500
108–23–6	Isopropyl Chloroformate		1,000	1,000
119–38–0	Isopropylmethyl-pyrazolyl Dimethylcarbamate		100	500
78–97–7	Lactonitrile		1,000	1,000
21609–90–5	Leptophos		500	500/10,000
541–25–3	Lewisite	b, d	10	10
58–89–9	Lindane		1	1,000/10,000
7580–67–8	Lithium Hydride	a	100	100
109–77–3	Malononitrile		1,000	500/10,000
12108–13–3	Manganese, Tricarbonyl Methylcyclopentadienyl	d	100	100
51–75–2	Mechlorethamine	b	10	10
950–10–7	Mephosfolan		500	500
1600–27–7	Mercuric Acetate		500	500/10,000
7487–94–7	Mercuric Chloride		500	500/10,000
21908–53–2	Mercuric Oxide		500	500/10,000
10476–95–6	Methacrolein Diacetate		1,000	1,000
760–93–0	Methacrylic Anhydride		500	500
126–98–7	Methacrylonitrile	d	1,000	500
920–46–7	Methacryloyl Chloride		100	100
30674–80–7	Methacryloyloxyethyl Isocyanate	d	100	100
10265–92–6	Methamidophos		100	100/10,000
558–25–8	Methanesulfonyl Fluoride		1,000	1,000
950–37–8	Methidathion		500	500/10,000
2032–65–7	Methiocarb		10	500/10,000
16752–77–5	Methomyl	d	100	500/10,000
151–38–2	Methoxyethylmercuric Acetate		500	500/10,000
80–63–7	Methyl 2-Chloroacrylate		500	500
74–83–9	Methyl Bromide	f	1,000	1,000

158

CAS No.	Chemical name	Notes	Reportable quantity* (pounds)	Threshold planning quantity (pounds)
79–22–1	Methyl Chloroformate	d	1,000	500
60–34–4	Methyl Hydrazine		10	500
624–83–9	Methyl Isocyanate		10	500
556–61–6	Methyl Isothiocyanate	a	500	500
74–93–1	Methyl Mercaptan	f	100	500
3735–23–7	Methyl Phenkapton		500	500
676–97–1	Methyl Phosphonic Dichloride	a	100	100
556–64–9	Methyl Thiocyanate		10,000	10,000
78–94–4	Methyl Vinyl Ketone		10	10
502–39–6	Methylmercuric Dicyanamide		500	500/10,000
75–79–6	Methyltrichlorosilane	d	500	500
1129–41–5	Metolcarb		1,000	100/10,000
7786–34–7	Mevinphos		10	500
315–18–4	Mexacarbate	d	1,000	500/10,000
50–07–7	Mitomycin C		10	500/10,000
6923–22–4	Monocrotophos		10	10/10,000
2763–96–4	Muscimol		1,000	500/10,000
505–60–2	Mustard Gas	d	500	500
13463–39–3	Nickel Carbonyl		10	1
54–11–5	Nicotine	b	100	100
65–30–5	Nicotine Sulfate		100	100/10,000
7697–37–2	Nitric Acid		1,000	1,000
10102–43–9	Nitric Oxide	b	10	100
98–95–3	Nitrobenzene	f	1,000	10,000
1122–60–7	Nitrocyclohexane		500	500
10102–44–0	Nitrogen Dioxide		10	100
62–75–9	Nitrosodimethylamine	d	10	1,000
991–42–4	Norbormide		100	100/10,000
	Organorhodium Complex (PMN–82–147)		10	10/10,000
630–60–4	Ouabain	b	100	100/10,000
23135–22–0	Oxamyl		100	100/10,000
78–71–7	Oxetane, 3,3-Bis(Chloromethyl)-		500	500
2497–07–6	Oxydisulfoton	d	500	500
10028–15–6	Ozone		100	100
1910–42–5	Paraquat Dichloride		10	10/10,000
2074–50–2	Paraquat Methosulfate		10	10/10,000
56–38–2	Parathion	b	10	100
298–00–0	Parathion-Methyl	b	100	100/10,000

CAS No.	Chemical name	Notes	Reportable quantity* (pounds)	Threshold planning quantity (pounds)
12002–03–8	Paris Green		1	500/10,000
19624–22–7	Pentaborane		500	500
2570–26–5	Pentadecylamine		100	100/10,000
79–21–0	Peracetic Acid		500	500
594–42–3	Perchloromethylmercaptan		100	500
108–95–2	Phenol		1,000	500/10,000
4418–66–0	Phenol, 2,2'-Thiobis(4-Chloro-6-Methyl)-		100	100/10,000
64–00–6	Phenol, 3-(1-Methylethyl)-, Methylcarbamate		10	500/10,000
58–36–6	Phenoxarsine, 10,10'-Oxydi-		500	500/10,000
696–28–6	Phenyl Dichloroarsine	d	1	500
59–88–1	Phenylhydrazine Hydrochloride		1,000	1,000/10,000
62–38–4	Phenylmercury Acetate		100	500/10,000
2097–19–0	Phenylsilatrane	d	100	100/10,000
103–85–5	Phenylthiourea		100	100/10,000
298–02–2	Phorate		10	10
4104–14–7	Phosacetim		100	100/10,000
947–02–4	Phosfolan		100	100/10,000
75–44–5	Phosgene	f	10	10
13171–21–6	Phosphamidon		100	100
7803–51–2	Phosphine		100	500
2703–13–1	Phosphonothioic Acid, Methyl-, O-Ethyl O-(4-(Methylthio) Phenyl) Ester		500	500
50782–69–9	Phosphonothioic Acid, Methyl-, S-(2-(Bis(1Methylethyl)Amino)Ethyl) O-Ethyl Ester		100	100
2665–30–7	Phosphonothioic Acid, Methyl-, O-(4-Nitrophenyl) O-Phenyl Ester		500	500
3254–63–5	Phosphoric Acid, Dimethyl 4-(Methylthio)Phenyl Ester		500	500
2587–90–8	Phosphorothioic Acid, O,O-Dimethyl-S-(2-Methylthio) Ethyl Ester	b, c	500	500
7723–14–0	Phosphorus	a, d	1	100
10025–87–3	Phosphorus Oxychloride		1,000	500
10026–13–8	Phosphorus Pentachloride	a	500	500
7719–12–2	Phosphorus Trichloride		1,000	1,000
57–47–6	Physostigmine		100	100/10,000
57–64–7	Physostigmine, Salicylate (1:1)		100	100/10,000
124–87–8	Picrotoxin		500	500/10,000
110–89–4	Piperidine		1,000	1,000
23505–41–1	Pirimifos-Ethyl		1,000	1,000

160

CAS No.	Chemical name	Notes	Reportable quantity* (pounds)	Threshold planning quantity (pounds)
10124–50–2	Potassium Arsenite		1	500/10,000
151–50–8	Potassium Cyanide	a	10	100
506–61–6	Potassium Silver Cyanide	a	1	500
2631–37–0	Promecarb	d	1,000	500/10,000
106–96–7	Propargyl Bromide		10	10
57–57–8	Propiolactone, Beta-		10	500
107–12–0	Propionitrile		10	500
542–76–7	Propionitrile, 3-Chloro-		1,000	1,000
70–69–9	Propiophenone, 4-Amino-	c	100	100/10,000
109–61–5	Propyl Chloroformate		500	500
75–56–9	Propylene Oxide	f	100	10,000
75–55–8	Propyleneimine		1	10,000
2275–18–5	Prothoate		100	100/10,000
129–00–0	Pyrene	b	5,000	1,000/10,000
140–76–1	Pyridine, 2-Methyl-5-Vinyl-		500	500
504–24–5	Pyridine, 4-Amino-	d	1,000	500/10,000
1124–33–0	Pyridine, 4-Nitro-,l-Oxide		500	500/10,000
53558–25–1	Pyriminil	d	100	100/10,000
14167–18–1	Salcomine		500	500/10,000
107–44–8	Sarin	d	10	10
7783–00–8	Selenious Acid		10	1,000/10,000
7791–23–3	Selenium Oxychloride		500	500
563–41–7	Semicarbazide Hydrochloride		1,000	1,000/10,000
3037–72–7	Silane, (4-Aminobutyl)Diethoxymethyl-		1,000	1,000
7631–89–2	Sodium Arsenate	1	1,000/10,000	
7784–46–5	Sodium Arsenite		1	500/10,000
26628–22–8	Sodium Azide (Na(N_3))	a	1,000	500
124–65–2	Sodium Cacodylate		100	100/10,000
143–33–9	Sodium Cyanide (Na(CN))	a	10	100
62–74–8	Sodium Fluoroacetate		10	10/10,000
13410–01–0	Sodium Selenate		100	100/10,000
10102–18–8	Sodium Selenite	d	100	100/10,000
10102–20–2	Sodium Tellurite		500	500/10,000
900–95–8	Stannane, Acetoxytriphenyl-	c	500	500/10,000
57–24–9	Strychnine	b	10	100/10,000
60–41–3	Strychnine Sulfate		10	100/10,000
3689–24–5	Sulfotep		100	500
3569–57–1	Sulfoxide, 3-Chloropropyl Octyl		500	500

CAS No.	Chemical name	Notes	Reportable quantity* (pounds)	Threshold planning quantity (pounds)
7446–09–5	Sulfur Dioxide	f	500	500
7783–60–0	Sulfur Tetrafluoride		100	100
7446–11–9	Sulfur Trioxide	a	100	100
7664–93–9	Sulfuric Acid		1,000	1,000
77–81–6	Tabun	b, d	10	10
7783–80–4	Tellurium Hexafluoride	e	100	100
107–49–3	TEPP		10	100
13071–79–9	Terbufos	d	100	100
78–00–2	Tetraethyllead	b	10	100
597–64–8	Tetraethyltin	b	100	100
75–74–1	Tetramethyllead	b, f	100	100
509–14–8	Tetranitromethane		10	500
10031–59–1	Thallium Sulfate	d	100	100/10,000
6533–73–9	Thallous Carbonate	b, d	100	100/10,000
7791–12–0	Thallous Chloride	b, d	100	100/10,000
2757–18–8	Thallous Malonate	b, d	100	100/10,000
7446–18–6	Thallous Sulfate		100	100/10,000
2231–57–4	Thiocarbazide		1,000	1,000/10,000
39196–18–4	Thiofanox		100	100/10,000
297–97–2	Thionazin		100	500
108–98–5	Thiophenol		100	500
79–19–6	Thiosemicarbazide		100	100/10,000
5344–82–1	Thiourea, (2-Chlorophenyl)-		100	100/10,000
614–78–8	Thiourea, (2-Methylphenyl)-		500	500/10,000
7550–45–0	Titanium Tetrachloride		1,000	100
584–84–9	Toluene 2,4-Diisocyanate		100	500
91–08–7	Toluene 2,6-Diisocyanate		100	100
110–57–6	Trans-1,4-Dichlorobutene		500	500
1031–47–6	Triamiphos		500	500/10,000
24017–47–8	Triazofos		500	500
76–02–8	Trichloroacetyl Chloride		500	500
115–21–9	Trichloroethylsilane	d	500	500
327–98–0	Trichloronate	e	500	500
98–13–5	Trichlorophenylsilane	d	500	500
1558–25–4	Trichloro(Chloromethyl)Silane		100	100
27137–85–5	Trichloro(Dichlorophenyl) Silane		500	500
998–30–1	Triethoxysilane		500	500
75–77–4	Trimethylchlorosilane		1,000	1,000

CAS No.	Chemical name	Notes	Reportable quantity* (pounds)	Threshold planning quantity (pounds)
824–11–3	Trimethylolpropane Phosphite	d	100	100/10,000
1066–45–1	Trimethyltin Chloride	500	500/10,000	
639–58–7	Triphenyltin Chloride		500	500/10,000
555–77–1	Tris(2-Chloroethyl)Amine	d	100	100
2001–95–8	Valinomycin	b	1,000	1,000/10,000
1314–62–1	Vanadium Pentoxide		1,000	100/10,000
108–05–4	Vinyl Acetate Monomer	f	5,000	1,000
81–81–2	Warfarin		100	500/10,000
129–06–6	Warfarin Sodium	d	100	100/10,000
28347–13–9	Xylylene Dichloride		100	100/10,000
58270–08–9	Zinc, Dichloro(4,4-Dimethyl-5((((Methylamino)Carbonyl) Oxy)Imino)Pentanenitrile)-, (T-4)-		100	100/10,000
1314–84–7	Zinc Phosphide	a	100	500

*Only the statutory or final RQ is shown. For more information, see 40 CFR 355.61.

Notes:

a. This material is a reactive solid. The TPQ does not default to 10,000 pounds for non-powder, non-molten, non-solution form.

b. The calculated TPQ changed after technical review as described in a technical support document for the final rule, April 22, 1987.

c. Chemicals added by final rule, April 22, 1987.

d. Revised TPQ based on new or re-evaluated toxicity data, April 22, 1987.

e. The TPQ was revised due to calculation error, April 22, 1987.

f. Chemicals on the original list that do not meet toxicity criteria but because of their acute lethality, high production volume and known risk are considered chemicals of concern ("Other chemicals"), November 17, 1986 and February 15, 1990.

g. The TPQ was recalculated (September 8, 2003) since it was mistakenly calculated in the April 22, 1987 final rule under the wrong assumption that this chemical is a reactive solid, when in fact it is a liquid. RQ for this chemical was adjusted on September 11, 2006.

Appendix E - EPA Standards for Pesticide Chemical Residues in Food

The EPA regulation of chemical residue in food, including produce and animal products, is based on the listings of hazardous and extremely hazardous chemicals in 40 CFR 502.4 and 355, the latter of which is included in the previous *Appendix*.

40 CFR - Protection of Environment, Part 180 - Tolerances and Exemptions for Pesticide Residues in Food, includes the following four categories: chlorinated organic pesticides, arsenic-containing chemicals, metallic dithiocarbamates, and cholinesterase-inhibiting pesticides. It is important to note that the FDA action levels for food and feed contaminants involve a much more restrictive listing of ecotoxins than the 100 or more categories listed in 40 CFR 180.3. These in turn are only a small component of the thousands of ecotoxins listed as hazardous chemicals in 40 CFR 502.4, which also includes all of the extremely hazardous chemicals listed in 40 CFR 555.

The Food and Drug Administration (FDA) is responsible for enforcement of pesticide tolerances and food additive regulations that have been established by the EPA. This enforcement authority is derived from section 402(a)(2)(B) and of the FFDCA (Federal Food, Drug, and Cosmetic Act),

The FDA Action Levels include the following categories of pesticides, which represent chemical groups currently widely used in industrial and agricultural activities throughout the world. Most of the chemicals in the FDA Action Level listing have been barred from use as POPs (persistent organic pollutants), but their persistence facilitates their continued presence as ecotoxins circulating in food webs and biogeochemical cycles. Of particular note is the widespread production of closely-associated chlorinated hydrocarbons and other POPs, which, due to a slight change in their chemical configuration, can be manufactured and distributed despite their status as POPs.

FDA Action Levels for Unavoidable Pesticide Residues in Food and Feed Commodities:

Pesticides Covered:

> Aldrin & Dieldrin, Benzene Hexachloride (BHC), Chlordane, Chlordecone (Kepone), DDT, DDE, & TDE, Dicofol (Kelthane), Ethylene Dibromide (EDB), Heptachlor & Heptachlor, Epoxide, LINDANE, MIREX

Unless otherwise a specified, an action level listed for:

1. A raw agricultural commodity (other than grains) may also apply to the corresponding processed food intended for human consumption;
2. Grains may also apply to both raw and processed grains intended for human or animal consumption;
3. Fish may also apply to shellfish and processed fish intended for human consumption; and
4. Processed animal feed may include mixed feeds and feed ingredients.

Several important observations need to be made about the thousands of hazardous and extremely hazardous chemicals regulated by EPA Guidelines, only a few of which are listed in the FDA Action Levels as pesticides covered by their enforcement activities. All of the pesticides listed in the FDA's ORA (Office of Regulatory Affairs) CPG (Compliance Policy Guide) 575.100 have numerous cogeners, metabolites, derivatives, and combinations (e.g. DDT as DDE and TDE), many of which are specifically listed in the more comprehensive 40 CFR 502.4.

Neither the EPA nor the FDA has control over the fact that a particular food commodity may be contaminated by multiple different chemicals in its 502.4 listing, a broad selection of emerging chemicals not included within the 502.4 listings, and/or multiple pulses of pharmaceutical ecotoxins that are just now becoming environmental toxins of concern as illustrated by the following quotation.

> "Where residues from two or more chemicals in the same class are present in or on a raw agricultural commodity, the tolerance for the total of such residues shall be the same as that for the chemical having the lowest numerical tolerance in this class, unless a higher tolerance level is specifically provided for the combined residues by a regulation in this Part." (40 CFR Part 180)

Industrial agricultural activities often result in multiple examples of the following anthropogenic ecotoxins occurring as contaminant signals in the common, everyday foods sold in supermarkets throughout the world. Foreign export of American-made, or at least American-invented, pesticides and herbicides provides one of the most lucrative sources of income for agro-corporations and the Wall St. bankers who fund their operations and profit from their debts. Most pesticide residue contaminant levels may be well below observable health physics impact levels, but the omnipresence of an increasingly wide variety of ecotoxins circulating in the commercial exchange networks of global consumer society are the most significant 21st century threat to public health.

The following two tables have been removed from the text in order to save space. They may be found online at:
http://www.davistownmuseum.org/PDFs/EPA%20Standards%20for%20Pesticide%20Chemical%20Residues%20in%20Food.pdf

Chemicals with specific tolerances covered under the CFR, Title 40, Chapter I, part 180

Chemicals with exemptions from tolerances under the CFR, Title 40, Chapter I, part 180

Appendix F - EPA Drinking Water Standards

Drinking water standards were first established in 1976. The most interesting characteristic of the EPA Drinking Water Standards is the restricted number of ecotoxins chosen for regulation in contrast to the large number of hazardous and extremely hazardous chemicals listed in 40 CFR 502.4 and 555. See commentary on the EPA listings in *Appendices C, D,* and *E.* The EPA maintains and updates this list at: http://www.epa.gov/ogwdw/contaminants/index.html

CHEMICAL	MAXIMUM ALLOWED, MG/L:	CHEMICAL	MAXIMUM ALLOWED, MG/L:
BROMATE:	0.010	CHLORITE:	1.0
HALOACETIC ACIDS:	0.060	TRIHALOMETHANES:	0.080
CHLORAMINES:	4.0	CHLORINE:	4.0
CHLORINE DIOXIDE:	.8	ANTIMONY:	.006
ARSENIC:	0.010	ASBESTOS (IN FIBERS PER LITER):	7,000,000
BARIUM:	2	BERYLLIUM:	0.004
CADMIUM:	0.005	CHROMIUM:	0.1
COPPER:	1.3	CYANIDE:	0.2
FLUORINE:	4.0	LEAD:	0.015
MERCURY:	0.002	NITRATE:	10
NITRITE:	1	SELENIUM:	0.05
THALLIUM:	0.002	ACRYLAMIDE:	1
ALACHLOR:	0.002	ATRAZINE:	0.003
BENZENE:	0.005	BENZO(A)PYRENE (PAHs):	0.0002
CARBOFURAN:	0.04	CARBON TETRACHLORIDE:	0.005
CHLORDANE:	0.002	CHLOROBENZENE:	0.1
2,4-D:	0.07	DALAPON:	0.2
1,2-DIBROMO-3-CHLOROPROPANE (DBCP):	0.0002	O-DICHLOROBENZENE:	0.6
P-DICHLOROBENZENE:	0.075	1,2-DICHLOROETHANE:	0.005
1,1-DICHLOROETHYLENE:	0.007	CIS-1,2-DICHLOROETHYLENE:	0.07
TRANS-1,2-DICHLOROETHYLENE	0.1	DICHLOROMETHANE:	0.005
1,2-DICHLOROPROPANE:	0.005	DI(2-ETHYLHEXYL) ADIPATE:	0.4
DI(2-ETHYLHEXYL) PHTHALATE:	0.006	DINOSEB:	0.007
DIOXIN (2,3,7,8-TCDD):	0.00000003	DIQUAT:	0.02
ENDOTHALL:	0.1	ENDRIN:	0.002
EPICHLOROHYDRIN:	20	ETHYLBENZENE:	0.7
ETHYLENE DIBROMIDE:	0.00005	GLYPHOSATE:	0.7
HEPTACHLOR:	0.0004	HEPTACHLOR EPOXIDE:	0.0002
HEXACHLOROBENZENE:	0.001	HEXACHLOROCYCLOPENTADIENE:	0.05
LINDANE:	0.0002	METHOXYCHLOR:	0.04
OXAMYL (VYDATE):	0.2	POLYCHLORINATED BIPHENYLS:	0.0005
PENTACHLOROPHENOL:	0.001	PICLORAM:	0.5
SIMAZINE:	0.004	STYRENE:	0.1
TETRACHLOROETHYLENE:	0.005	TOLUENE:	1
TOXAPHENE:	0.003	2,4,5-TP (SILVEX):	0.05
1,2,4-TRICHLOROBENZENE:	0.07	1,1,1-TRICHLOROETHANE:	0.2
1,1,2-TRICHLOROETHANE:	0.005	TRICHLOROETHYLENE:	0.005
VINYL CHLORIDE:	0.002	XYLENES:	10

Appendix G – Oak Ridge National Laboratory (ORNL): U.S. Spent Nuclear Fuel and Radioactive Waste Inventories

Oak Ridge National Laboratory. (December 1997). *Integrated Data Base Report, 1996: U.S. Spent Nuclear Fuel and Radioactive Waste Inventories, projections, and characteristics (revision 13)*. Report No. DOE/RW-0006, Rev. 13. Oak Ridge National Laboratory, Oak Ridge, TN.

The Oak Ridge National Laboratory (ORNL) publication, *U.S. Spent Fuel and Radioactive Waste Inventories, Projections, and Characteristics*, is among the most important of all U.S. government publications pertaining to the evaluation of the public safety consequences of a nuclear accident or terrorist attack on a U.S. nuclear reactor, or for that matter, on any nuclear reactor. While no longer publicly available, this database provides "mass, radioactivity, and thermal power of nuclides in domestic commercial LWR spent fuel at the end of calendar year 1991." It should be noted that, in addition to the spent fuel inventories at an operational (or decommissioned) reactor, the isotopic inventories within the fuel rods of an operating reactor are approximately 10% of those in the spent fuel just prior to spent fuel discharge. The ORNL database thus does not include the radioactive inventory of the operating reactor core.

Two contemporary factors in particular make the ORNL spent fuel inventories document of exceptional importance. First, the rapidly increasing cost of non-renewable energy resources and the worldwide desire to curb CO_2 emissions provide a huge incentive for the re-licensing of aging nuclear reactors in the U.S. and elsewhere. Second, the prolonged operation of aging reactors in an age of energy shortages ensures the increasing chances for a loss of reactor coolant accident (LORCA) or other type of mishap at an operating nuclear reactor. The 1986 Chernobyl accident provides an example of the unexpected consequences of one variation of the many types of nuclear accidents that can occur at any operating nuclear facility. In the case of Chernobyl, the atmospheric distribution of the biologically significant radioisotope Cesium 137 was hemispheric; the extent of its distribution patterns and the extraordinarily high terrestrial deposition levels in distant locations were totally unexpected. Chernobyl thus serves as a warning and a lesson pertaining to possible accident emission inventories and their fallout distribution patterns for future nuclear mishaps.

The Oak Ridge document is also of particular importance in the modern era of global warfare and the increasing possibility of a terrorist attack on an operating nuclear reactor. It is highly unlikely that such a terrorist attack will be directed at an operational U.S. nuclear power plant. Unfortunately, in a world with numerous unstable governments (Pakistan, etc.) or vulnerable government nuclear facilities (India, many Eastern European nations), a terrorist attack on any operational nuclear power plant will have a worldwide safety impact. Fallout deposition of biologically significant radioisotopes in terrestrial environments universally results in the rapid transport of these radioactive ecotoxins in pathways to human consumption. The most dangerous of all scenarios is a nuclear attack by one nation, or a terrorist group, on an operational nuclear reactor that results in the vaporization of both the fuel inventory of the reactor itself and the inventory of its on site spent fuel pool. In the case of Chernobyl, had all its inventory of Cesium 137 been vaporized by a nuclear attack, the worldwide spread of radioactive Cesium would have been at least an order of magnitude greater than what occurred in 1986.

One of the most disconcerting phenomena of the cold war was the targeting of Russian nuclear reactors by U.S. strategic missiles, including multiple reentry nuclear warheads. It can be assumed that the Russians also targeted our nuclear reactors. In the case of an actual nuclear war, the vaporization of the radioactive fuel and waste inventories at only one or two operational reactors would have resulted in a worldwide disaster. Unfortunately, the U.S. still continues to target Russian as well as Chinese reactor facilities. We can assume that our governmental leaders would have enough common sense not to cause a worldwide catastrophe by attacking any of these facilities. We cannot assume that some rogue terrorist group in possession of a nuclear weapon would not vaporize an operating nuclear reactor in some distant, unstable political community. There are now 440 operational nuclear reactors generating electricity in a world with increasing costs and shortages of non-renewable energy resources. That any of these facilities might be the subject of an attack in the future cannot be discounted.

Table C-4 in the ORNL *Integrated Database for 1992* contains an inventory of accumulated radioactivity at U.S. nuclear reactors. This inventory includes only spent fuel radioisotopes, and does not document the variable inventory of radioisotopes in an operational reactor core, which can be near zero for new fuel rods and as much as 10% of the total spent fuel inventory just prior to fuel rod exchange. As of September 2008, out of 440 operating nuclear reactors, 104 were located in the US. To obtain a conservative but typical estimate of the inventory of radioactivity in a reactor undergoing decommissioning, divide the data in Table C-4 by 100. While knowledge of the accumulated reactor inventory of a facility undergoing decommissioning is essential for calculating decommissioning costs and waste disposal strategies and options, these inventories are also critical for the evaluation of the quantities of radioactivity discharged by any type of accident or terrorist attack. Such incidents can result in the discharge of only a tiny percentage of reactor inventories; for example, the Chernobyl accident resulted in the release and hemispheric transport of only about 10% of its inventory of Cesium 137. Nonetheless, this accident had catastrophic health effects in many areas of Belarus, Ukraine, and to a lesser extent in Turkey, Russia, and the northern sections of the U.K. It also should be noted that the ORNL database was compiled before its 1992 publication date; the data is a 1991 cumulative inventory estimate. Any estimate of current reactor inventory for a particular isotope would be extremely conservative given that the 2009 cumulative inventories are much greater than those in 1991. Unfortunately for anyone making a public safety evaluation of any such releases, these reactor inventories are now classified information. Nonetheless, dividing the data in Table C-4 by 100 still provides a handy, if conservative, guide to reactor inventories.

Readers of Table C-4 should be advised of the following:

- The entire Table C-4 listing of reactor isotopes is online at: www.davistownmuseum.org/cbm/; this Appendix only republishes a selection of the most biologically significant reactor wastes, particularly cesium-137, strontium-90, and the most important of the alpha radiation emitters, and plutonium-239.

- The ONRL publication is the only source of information on the exact inventory of spent fuel at US nuclear reactors.

- This component of the database is now classified information and no longer available to the general public.

- See the *Abbreviations Section* in *Volume 2* for explanations of reporting units used in this report and summary.

Atomic #	Element	Mass # of nuclide	Half-life	Mass, g		Radioactivity, Ci	
				Annual	Cumulative	Annual	Cumulative
27	Cobalt	60	5.2714 y	1.29E+04	7.00E+04	1.46E+07	7.92E+07
38	Strontium	90	28.90 y	9.93E+05	8.84E+06	1.35E+08	1.21E+09
53	Iodine	129	15.7×106 y	3.40E+05	3.53E+06	6.00E+01	6.23E+02
55	Cesium	137	30.07 y	2.24E+06	2.01E+07	1.95E+08	1.75E+09
94	Plutonium	238	88 y	2.65E+05	2.39E+06	4.53E+06	4.10E+07
94	Plutonium	239	2.41 × 104 y	9.52E+06	1.14E+08	5.92E+05	7.08E+06
94	Plutonium	241	14 y	2.32E+06	1.92E+07	2.39E+07	1.98E+09

Source: U.S. Department of Energy. (October 1992). *Integrated Data Base for 1992: U.S. Spent Fuel and Radioactive Waste Inventories, Projections, and Characteristics.* Oak Ridge National Laboratory, Oak Ridge, Tennessee.

Cesium-137 Cumulative Inventory Synopsis

The 1992 cumulative inventory of the biologically significant beta-emitting isotope Cesium 137 is 1.75E+09 or 1,750,000,000 curies for all U.S. light water reactors. Dividing this number by 100 gives a conservative cumulative inventory of 17,500,000 Ci. of radiocesium in the spent fuel of one U.S. nuclear reactor. As much as 10% of this quantity would be in the fuel rods in an operating reactor; any in situ reactor isotopic inventories are not listed in Table C-4.

Plutonium-239 Synopsis

The 1992 cumulative inventory of Plutonium 239 in all light water reactors (LWR) was 7.08E+06 or 7,080,000 curies. To estimate the inventory in one reactor, divide this figure by 100, which would give a conservative cumulative inventory of 70,800 Ci. Significant additional inventories of Plutonium 239 will have accumulated at all currently operating U.S. nuclear reactors since that date.

Fuel leakage and fuel cladding failure accidents can mean that a small but significant percentage of the inventory of radiocesium, or of any other long-lived radioisotope, can be discharged into the environment through reactor water systems, as filtered waste, or as loose inventory in the reactor containment vessel. Such leakage may amount to as much as one percent or more of the total radiocesium inventory in a worst case scenario, e.g. tens of thousands of curies of radiocesium at any reactor site at which a major fuel cladding failure accident or other significant discharge has occurred. Most fuel cladding failures, including those where one or two fuel rods are broken open, as have occurred at the Maine Yankee Atomic Power Plant in Wiscasset, Maine, result in much smaller releases, which are often contained within the reactor cooling system. Efficient vacuum systems can remove much of this contamination, which may amount to a curie or less of actual discharges. In the case of the Maine Yankee accident in 1973, the quantity of radiocesium released to the reactor cooling system is unknown. For a review of licensee and NRC (Nuclear Regulatory Commission) documentation of the first of the two fuel cladding failures at Wiscasset, see *Patterns of Noncompliance: The Nuclear Regulatory Commission and the Maine Yankee Atomic Power Company, Generic and Site-specific Deficiencies in Radiological Surveillance Programs* (Brack 1998). See in particular pages 51 through 55 for data compiled at decommissioning (Site Characterization Survey Report) by GTS Duratek on radiocesium 137 that eventually escaped the reactor water systems at the Wiscasset facility. The soil contamination documented by the GTS Duratek survey report occurred in 1983 during a reactor water storage system leak, well before a second incident of fuel cladding failure closed the plant in 1997.

The unfortunate significance of the ORNL Integrated Database for 1992, the last publicly available compilation of reactor inventories, is that in a world of aging nuclear reactors, prospective new reactor facilities, global warfare, and potential terrorist attacks, this database will be necessary for evaluating the environmental impact of any future accident or attack.

Appendix H - RISO National Laboratory Cumulative C_{137} Fallout Record

The database compiled by the RISO National Laboratory is the most comprehensive record of yearly fallout (Column 1 of the following table) and cumulative fallout (column 2) for radiocesium available in the public domain. The RISO Laboratory has also compiled an equally comprehensive record of Sr_{90} (Strontium 90), which is not reprinted in this publication. The US has also collected detailed data about the dietary intake of fallout nuclides, cited in Section 9 of Radnet: Nuclear Information on the Internet at: www.davistownmuseum.org/cbm/Rad5.html, but most unclassified U.S. environmental monitoring reports consist of summaries of composites, usually monthly or quarterly averages, which successfully mask real time localized pulses of radiation, especially in areas close to the weapons testing site in Nevada. The research and publications of the RISO National Laboratory set the standard for media specific analysis of environmental radiation which remained unchallenged until the Chernobyl accident, at which time Finland compiled the most comprehensive reports of the impact of the accident.

Denmark was considered to have received a minimal amount of Chernobyl-derived radiocesium and other radionuclides. The mean deposition of 1,210 Bq/m^2 recorded in Denmark in 1986 was, nonetheless, higher than the highest annual cesium deposition from weapons testing fallout during any preceding year, but still insignificant compared to the peak depositions of 60,000 to 100,000 Bq/m^2 recorded in sections of Norway, Sweden, Finland and England. These countries are among the few locations that have maintained adequate record keeping on the impact of the Chernobyl accident. The tradition of freedom of information (FOI) pertaining to radioactivity in the environment needs to be expanded to include a comprehensive survey of the areas adjacent to fuel reprocessing, weapons production, and nuclear power production facilities. There is an urgent need for comprehensive radiological surveys, including radiometric surveys of contamination deposition expressed as radioactivity per square meter, media specific pathway analyses for the population groups most impacted by anthropogenic source points of radioactive contamination, and real-time nuclide-specific monitoring of air concentrations of radioactive effluents. The research and record keeping of the RISO National Laboratory pertaining to accumulated fallout from weapons tests, as well as its comprehensive media-specific yearly reports, represent the first step in updating antiquated radiological surveillance paradigms. Unfortunately, radiological monitoring data compiled in Denmark since 1991 appears to be no longer available.

The RISO baseline is a handy guideline for evaluating Cs_{137} fallout in the US, which experienced a pulse of Chernobyl-derived fallout in 1986. The cumulative fallout record also serves these three purposes: first to provide an approximate history of the cumulative fallout from weapons testing-derived contamination as it probably occurred in Maine and other parts of the US; secondly, to provide an illustration of the hemispheric impact of an accident at a nuclear power installation (Chernobyl - see fallout data in Appendix I from 1986) that resulted in tropospheric contamination that was then washed out by rainfall events (Denmark did not experience any rainfall events during the passage of the Chernobyl plume, therefore, contamination was slight compared to other areas in Europe; nonetheless, in a few hours, Chernobyl fallout exceeded the maximum annual weapons testing fallout) and thirdly, to provide a point of comparison for evaluating fallout from any future nuclear accident or war utilizing the indicator nuclide Cs_{137}. The RISO index is extracted and reposted from the RADNET labyrinth of nuclear information for ease of visitor access.

	DENMARK		JUTLAND		ISLANDS	
FALLOUT RATES AND CUMULATIVE FALLOUT (BQ CS$_{137}$ M-2) IN DENMARK, JUTLAND, AND THE FAROE ISLANDS						
1950-1991 *(1)*						
DI = ANNUAL DEPOSITION, AI = CUMULATIVE RADIATION						
YEAR	**DI**	**AI(30.2)**	**DI**	**AI(30.2)**	**DI**	**AI(30.2)**
1950	1.243	1.215	1.302	1.273	1.184	1.157
1951	5.979	7.030	6.749	7.838	5.210	6.221
1952	11.722	18.323	13.261	20.618	10.182	16.029
1953	29.600	46.830	33.507	52.889	25.693	40.770
1954	112.539	155.731	127.398	176.173	97.680	135.290
1955	148.059	296.857	167.595	335.922	128.523	257.792
1956	183.579	469.471	207.792	531.304	159.366	407.637
1957	183.579	638.145	207.792	722.227	159.366	554.062
1958	254.678	872.445	288.245	987.409	221.053	757.424
1959	361.238	1205.526	408.954	1364.492	313.582	1046.561
1960	67.488	1243.959	76.427	1408.032	58.608	1079.940
1961	87.675	1301.241	99.219	1472.849	76.072	1129.632
1962	439.738	1701.242	472.179	1900.635	407.296	1501.849
1963	988.344	2628.199	1092.418	2924.739	884.270	1331.659
1964	616.390	3170.535	691.752	3533.949	541.029	2807.121
1965	234.077	3326.905	248.877	3696.486	219.277	2957.324
1966	126.984	3375.057	128.227	3737.418	125.741	3012.697
1967	61.982	3358.593	69.619	3720.145	54.346	2997.040
1968	83.058	3363.098	92.826	3725.944	73.230	3000.195
1969	61.272	3346.212	73.467	3712.693	49.077	2979.675
1970	97.502	3365.115	117.986	3743.247	77.019	2986.928
1971	89.155	3375.430	102.179	3757.659	76.131	2993.148
1972	25.752	3323.554	27.054	3698.331	24.450	2948.724
1973	11.366	3258.804	12.728	3626.358	9.946	2891.141
1974	42.032	3225.498	46.117	3588.654	38.066	2862.350
1975	24.509	3175.828	26.758	3532.894	22.259	2818.771
1976	6.098	3109.302	6.867	3458.970	5.328	2759.642
1977	22.733	3060.549	23.976	3403.451	21.430	2717.597
1978	27.410	3017.479	31.850	3356.893	22.970	2678.016
1979	9.827	2958.211	10.301	3290.341	9.235	2625.917
1980	5.606	2896.171	6.766	3221.854	4.591	2570.470
1981	17.059	2846.738	18.316	3166.216	15.948	2527.385
1982	2.706	2784.409	2.851	3096.736	2.561	2472.203
1983	2.151	2722.959	2.126	3028.134	2.175	2417.902
1984	1.751	2662.521	1.935	2960.911	1.567	2364.247

FALLOUT RATES AND CUMULATIVE FALLOUT (BQ Cs_{137} M-2) IN DENMARK, JUTLAND, AND THE FAROE ISLANDS 1950-1991 *(1)* DI = ANNUAL DEPOSITION, AI = CUMULATIVE RADIATION						
	DENMARK		JUTLAND		ISLANDS	
YEAR	DI	AI(30.2)	DI	AI(30.2)	DI	AI(30.2)
1985	1.290	2603.012	1.191	2894.495	1.388	2311.642
1986	1210.000	3725.984	1340.000	4137.847	1080.000	3314.232
1987	29.000	3669.280	32.000	4074.674	26.000	3263.994
1988	11.900	3597.161	13.400	3994.768	10.300	3199.562
1989	3.500	3518.480	4.510	3907.998	2.530	3129.007
1990	2.63	3440.744	3.85	3822.564	1.41	3058.968
1991	1.63	3363.805	1.92	3737.194	1.36	2990.480

(1) Aarkrog, A., Botter-Jensen, L., Jiang, Chen Quing, Dahlgaard, H., Hansen, H., Holm, E., Lauridsen, B., Nielsen, S. P., Strandberg, M. and Sogaard-Hansen, J. (1992). *Environmental radioactivity in Denmark in 1990 and 1991*. Roskilde, Denmark: Riso National Laboratory.

Appendix I - FDA Guidelines for Radioisotopes in Food and Water

United States Food and Drug Administration. (March 5, 1997). Draft: *Accidental radioactive contamination of human food and animal feeds: Recommendations for state and local agencies*. Center for Devices and Radiological Health, U.S. FDA, Washington, D.C.

- Following the Chernobyl nuclear accident in 1986, the FDA revised its earlier 1982 guidelines by finally issuing new guidelines in 1997. Their Derived Intervention Level (DIL) listing provides insight into what the FDA considers the most biologically significant isotopes that are likely to be in pathways to human consumption following a nuclear accident or attack. Any use of a dirty bomb would likely be restricted to either a cobalt 60 or CS137 source from medical devices or food irradiation facilities, or for smaller dirty bombs, Americium 241 from smoke detectors. The impact from most such devices would be primarily psychological rather than of radiological health significance.

- Derived intervention levels are far stricter (more conservative) than the 1982 regulations. Derived intervention levels for the radiocesium group (1,160 Bq/kg for 15 year old = 31,320 picocuries/kg) are far closer to the "levels of concern" that resulted in seizure of food containing 10,000 picocuries/kg of radiocesium following the Chernobyl accident.

- "DILs [Derived Intervention Levels] are limits on the concentrations permitted in human food distributed in commerce. Comparable limits were not provided in the 1982 FDA recommendations. DILs apply during the first year after an accident." (pg. 3).

Table D-5 (pg. D-13)						
DERIVED INTERVENTION LEVELS (Bq/kg)						
(individual radionuclides, by age group, most limiting of either PAG)						
Radionuclide	**3 months**	**1 year**	**5 years**	**10 years**	**15 years**	**Adult**
Sr-90	308	362	616	389	160	465
I-131	196	167	722	1200	1690	2420
Cs-134	1600	2190	1940	1530	958	930
Cs-137	2000	2990	2810	2180	1370	1360
Cs group[a]	1800	2590	2380	1880	1160	1150
Ru-103	6770	8410	12200	16400	25000	28400
Ru-106	449	621	935	1340	2080	2360
Pu-238	2.5	21	17	14	12	10
Pu-239	2.2	18	14	13	10	9.8
Am-241	2.0	17	13	11	9.1	8.8
Pu+Am group [b]	2.2	19	15	13	9.6	9.3
[a] Computed as: (DIL for C2-134 + DIL for Cs-137) /2						
[b] Computed as: (DIL for Pu-238 + DIL for Pu-239 + DIL for AM-241) /3						

- "The equation given below is the basic formula for computing DILs.

	intervention level of dose (Sv)
DIL (Bq/kg) =	———————————————————
	f x Food Intake (kg) x DC (Sv/Bq)

Where:

DC = Dose coefficient; the radiation dose received per unit of activity ingested (Sv/Bq).

f = Fraction of the food intake assumed to be contaminated.

Food Intake = Quantity of food consumed in an appropriate period of time (kg)." (pg. 8).

- "The food monitoring results from FDA and others following the Chernobyl accident support the conclusion that I-131, Cs-134 and Cs-137 are the principal radionuclides that contribute to radiation dose by ingestion following a nuclear reactor accident, but that Ru-103 and Ru-106 also should be included (see *Appendix C*)." (pg. 10). "DIL is equivalent to, and replaces the previous FDA term Level of Concern (LOC)." (pg. 12).

- "The types of accidents and the principal radionuclides for which the DILs were developed are:
    1. Nuclear reactors - (I-131; Cs-134 + Cs-137; Ru-103 + Ru-106),
    2. Nuclear fuel reprocessing plants - (Sr-90; Cs-137; Pu-239 + Am-241),
    3. Nuclear waste storage facilities - (Sr-90; Cs-137; Pu-239 + Am-241),
    4. Nuclear weapons - (i.e., dispersal of nuclear material without nuclear detonation) (Pu-239), and
    5. Radioisotope thermoelectric generators (rtgs) and radioisotope heater units (rhus) used in space vehicles (Pu-238)." (pg. 13).

- "For each radionuclide, DILs were calculated for six age groups using Protective Action Guides, dose coefficients, and dietary intakes relevant to each radionuclide and age group. The age groups included 3 months, 1 year, 5 years, 10 years, 15 years and adult (>17 years). The dose coefficients used were from ICRP Publication 56 (ICRP 1989)." (pg. 14).

Table 2 (pg. 16) Recommended Derived Intervention Levels (DILs) or Criterion for Each Radionuclide Group All Components of the Diet		
Radionuclide Group	(Bq/kg)	(pCi/kg)
Sr-90	160	4300
I-131	170	4600
Cs-134 + Cs-137	1200	32,000
Pu-238 + Pu-239 + Am-241	2	54
Ru-103 + Ru-106	$\frac{C_3}{6800} + \frac{C_6}{450} < 1$	$\frac{C_3}{180{,}000} + \frac{C_6}{12{,}000} < 1$

- "Typical precautionary actions include covering exposed products, moving animals to shelter, corralling livestock and providing protected feed and water." (pg. 20). "The blending of contaminated food with uncontaminated food is not permitted because this is a violation of the Federal Food, Drug and Cosmetic Act (FDA 1991)." (pg. 22).

- Neither the FDA nor any other U. S. Government agency has the capacity for the rapid and timely monitoring, analysis, and reporting of radioactive contamination in any significant quantity of foodstuffs during a nuclear accident of any type. 1950s era laboratory capacity remains unimproved in an era of funding shortages. In the event of a nuclear accident, neither the FDA nor any other agency would be able to determine whether intervention to prevent consumption of contaminated foodstuffs is justified or necessary. These revised DIL's are a step in the right direction but have no real world credibility in the event that extensive foodstuffs monitoring becomes necessary. (**Editor's note**: the above observation was made shortly after the FDA issued its new guidelines. It is highly unlikely, even after the 9/11/01 attack, that the Bush Administration made any effort to update its analytic capabilities for evaluating contamination in the public food supply. Additional information on this subject would be welcomed.)

- The peak pulse of Chernobyl derived radiocesium in imported foods was observed by the FDA in 1987, ten to sixteen months after the accident began. At no time since the Chernobyl accident has the full body of raw data been available to the general public; the Center for Biological Monitoring obtained this information via a Freedom of Information Request. The FDA report on Chernobyl contamination took nine years to prepare and the result was only a few pages of clever disinformation.

Appendix J - Polybrominated Diphenyl Ethers in Selected Biological Media

PBDE Concentration Data from University of Stockholm Data:

Table 3.5. Some published data on ΣPBDEs concentrations in non-biological matrices from around the world.

Compartment	Location	Mean/median	Min-Max	Ref
Air	Canadian Arctic	7.7 pg/m^3	0.4–47	(85)
	Baltic Sea	8.6	0.4–79	(86)
	Canada		3.0–30	(87)
Indoor dust	USA	1.9 µg/g	0.59–34	(88)
	Singapore	1.2	0.11–13	(89)
	USA (17 houses)		0.78–30	(90)
Soil	China	1.0 ng/g dw	0.1–3.8	(91)
	Spain		21	(92)
	Sweden (5 sites)		0.03–1.9	(93)
	China		305	(94)
Amended soil	Spain (6 sites)	ng/g dw	30–690	(92)
	Sweden (2 sites) low dose		0.58–1.2	(93)
	Sweden (2 sites) high dose		0.84–2.1	(93)
	Sweden (2 sites) sludge applied		0.063–3900	(93)
Sediment	China	ng/g dw	4434–16088	(95)
	Spain (3 sites)		30–14395	(96)
	USA (3 sites)		1.7–4	(97)
	Australia (35 sites)	0.30		(98)
Sewage sludge	Sweden (50 STPs)*	µg/kg dw	18–260	(46)
	Danmark	238± 23		(46)
	Spain (6 STPs)		844–18100	(46)
	Spain (5 STPs)		197–1185	(92)
	Germany (11 STPs)	108	13–288	(99)

*Sewage treatment plants (STPs)

PBDEs are ubiquitous contaminats in wildlife and humans with PBDEs substituted with 4-6 bromines being the most abundant congeners (100) and these have been reported in biota at high trophic levels in variable concentrations and congener patterns (45,56). BDE-209 and other highly brominated diphenyl ethers have been reported in wildlife (45). A few examples of wildlife PBDE concentration are summarized in Table 3.6.

Table 3.6. Some selected data on ΣPBDEs concentrations (ng/g fat) in wildlife from around the globe, including references.

Species	Location	Mean/median	Min-Max	Ref
Terrestrial				
Red Fox (muscle)	Belgium	3.4	1.0-44	(101)
Birds				
Kestrel (muscle)	China	12300 ± 5540	279-31700	(102)
Guillemot (egg)	Baltic Sea	77		(103)
Sparrowhawk (liver)	Belgium	4900	280-26000	(104)
Fish				
Brown trout (liver)	Switzerland		16-7400	(105)
Brown trout (fillet)	Norway		21-1215	(106)
Burbot (liver)	Norway		125-915	(106)
Lake trout (whole fish)	USA		1395 ± 56	(107)
Marine mammal				
Beluga whale (female)	Canadian arctic	540	300-1060	(108)
Pilot whale (male, young)	Faroe Islands		3160	(100)
Harbor seal (male)	USA	5100	1900-8300	(108)
Ringed seal (male)	Canadian arctic	4600	2900-6300	(108)
Bottlenose dolphins	USA	5860± 4285	429-22783	(109)
Arctic species				
Polar bear	E-Greenland	70	22-192	(110)
Polar bear (female)	Svalbard	50	27-114	(111)
Polar bear (female)	Alaska	6.7	4.6-11	(111)
Walrus (male)	Svalbard	15	9-27	(112)
Penguin (egg)	Antarctica	3.1		(113)

Humans are exposed to both lower and the higher brominated diphenyl ether congeners, via food, indoor air and possibly through dermal uptake (100). In Sweden, a human milk time-related study from 1972-1997 was made, showing that concentrations of PBDEs in human milk had increased over the past two decades (114). This trend peaked in 1997 possibly due to the voluntary ban on the production and use of the PentaBDE in Europe as early as the 1990's (114). Human milk from Sweden, Japan, Canada and USA have been compared, it demonstrated large differences between the concentrations from Sweden and Japan with median levels of 3.2 and 1.4 ng/g fat, respectively, compared to 25 and 41 ng/g fat in milk from Canada and the USA, respectively. The concentrations from Canada and the USA are 10 times higher then those from Sweden and Japan. This could be a result of a more abundant use of PentaBDE in North America than in Europe and Asia (44). However, also high concentration of PBDEs in human from Nicaragua and the Faroe Island have been reported (19,115). Human serum/plasma concentrations of PBDEs, in different part of the world, are presented in Table 3.7, including levels of BDE-47, 99, 153, 183 and 209.

Table 3.7. Human serum PBDE concentrations (ng/g fat) from different parts of the world are presented.

Location	Year	N	BDE-47 median	BDE-47 min-max	BDE-99 median	BDE-99 min-max	BDE-153 median	BDE-153 min-max	BDE-183 median	BDE-183 min-max	BDE-209 median	BDE-209 min-max	ref
Sweden*	2000	17 (M)[a]	1.2	<LOQ[b]-6.3	0.20	<LOQ-3.1	1.9	1.1-3.7	<LOQ	<LOQ	2.4	0.88-93	(58)
Sweden	2000	50 (F)[c]	0.91	0.27-81	0.20	<LOQ-31	1.1	0.29-4.7	<LOQ		0.46	<LOQ-3.3	(116)
Norway	1977-03	20 (M)	1.3[d]		0.3[d]		1.3[d]		0.18[d]		<10[d]		(117)
Norway	1977-03	20 (F)	1.6[d]		0.35[d]		0.8[d]		<0.1[d]		<10[d]		(117)
Faroe Islands	1994-95	57 (F)	1.3	<LOQ-11	0.33	<LOQ-4.7	1.0	0.26-7.1	0.30	<0.14-1.8	0.77	<LOQ-3.6	(118)
UK	2003	154[e]	0.82	<0.30-180	<0.16	<0.16-150	1.7	<0.26-87			<15	<15-240	(119)
Spain	2003-04	61 (F)	2.4	0.30-9.0	2.6	1.4-6.9	0.86	<LOQ-2.5	0.47	<LOQ-2.3	1.1	<LOQ-20	(120)
Spain	2003-04	51 (M)	2.3	0.34-7.3	2.3	1.4-5.3	0.81	<LOQ-3.2	0.60	<LOQ-2.6	1.1	<LOQ-59	(120)
Belgium	1999-04	11[f]	1.17	0.2-3.07	0.20	0.20-1.02	1.55	0.99-3.07	0.21	0.10-0.41	111	3.6-33.1	(121)
Netherlands	2001-02	114 (F)	0.8	0.04-6	0.2	ND[g]-2.1	1.6	0.3-20					(122)
New Zealand	2001	23[h]	3.17	0.76-12.7	0.88	0.32-2.34	1.02	0.43-2.31	0.23	0.06-0.97			(123)
Korea	2001	10 (M)	5.74[d]	2.20-12.12	2.68[d]	1.46-5.39	4.69[d]	2.78-7.74	2.04[d]	0.51-4.86			(124)
Korea	2001	12 (F)	4.49[d]	1.84-7.74	2.29[d]	1.11-4.92	2.99[d]	1.64-7.48	2.10[d]	0.59-5.85			(124)
China	2006	21 (F)	1.0	0.36-3.6	0.36	0.08-7.4	1.4	0.36-6.4	0.31	0.0-1.3			(125)
China	2007	20 (F)	11	0.5-3.6	0.4	0.1-7.4	1.3	0.46-4	0.3	ND-1.3	5.7	ND-63.2	(126)
USA	1999-01	24 (F)	11	2.5-205	2.9	0.5-54	1.5	0.4-35	0.1	0.1-39			(127)
USA	2001	12 (F)	28	9.2-310	5.7	2.4-68	2.9	1.0-83	0	0.0-2.7			(128)
Nicaragua	2002	11[i]	218[j]		119[j]		20[j]		1.7[j]		8.9[j]		(19)
Mexico	2004	5 (F)	9.0[d]	3.0-14.5	2.0[d]	0.6-3.6	3.9[d]	0.6-6.6			9.5[d]	4.8-14.6	(129)

[a] Male. [b] Limit of quantification. [c] Female. [d] Mean. [e] 50/154 M, 104/154 F. [f] Unknown gender. [g] Not Detected. [h] 10/23 M, 13/23 F. [i] Children [j] Pooled samples
*Referents, abattoir workers
N is number of sample

Source: Teclechiel, Daniel. (2008). *Synthesis and Characterization of Highly Polybrominated Diphenyl Ethers*. Department of Environmental Chemistry, Stockholm University, Stockholm, Sweden. www.diva-portal.org/diva/getDocument?urn_nbn_se_su_diva-7410-2__fulltext.pdf

PBDE Concentration data from Washington state plan of action:

Table 5. Levels of PBDEs in food from the U.S., Japan and Europe.

Location (date)	Type of sample	PBDE congeners	Food (Sample size)	Total PBDE Concentration, ppt wet weight, except where noted[a]	Ref.
Texas (2003)	Grocery stores	13 total including BDE-47, 99, 100, 153, 154, 209	Fish (9)	Median 1725; range 8.5– 3078	103
			Meat (9)	Median 283; range 0.9-679	
			Dairy products (9)	Median 31.5; range ND -1373	
			Soy formula (1)	16.9	
			Eggs (1)	73.7	
			Calf liver (1)	115	
California (2003 and 2004)	Grocery stores	25 total including BDE-47, 99, 100, 153, 154, 209	Fish, wild (8)	Range 255 – 4955	104
			Fish, farmed (5)	Range 506 – 3063	
			Meat (3)	Range 164 – 379	
			Fowl (6)	Range 196 – 2516	
Nine U.S. cities (2001)	Grocery stores	BDE-28, 47, 99, 153, 154, 183. 209 data given separately.	Bacon (11)	Mean 296; range ND – 7831; BDE-209 ND	105
			Chicken fat (17)	Mean 1593; range 86 – 8965; mean BDE-209 1845	
			Steak fat (11)	Mean 165; range ND – 586; BDE-209 ND	
			Pork fat (9)	Mean 1282; range 17 – 7831; mean BDE-209 1913	
Spain (2000)	Grocery stores	BDE-47, 99, 153, 154, 183	Fish & shellfish (8)	Mean 333.9[a]	106
			Meat (15)	Mean 109.2[a]	
			Eggs (2)	Mean 64.5[a]	
			Milk (2)	Mean 16.9[a]	
			Dairy products (2)	Mean 47.9[a]	
			Fats and oils (3)	Mean 587.7[a]	
			Fruits (6)	Mean 5.8[a]	
			Cereals (4)	Mean 35.7[a]	
			Pulses (2)	Mean 10.7[a]	
			Tubers (2)	Mean 7.4[a]	
			Vegetables (8)	Mean 7.9[a]	
Japan (2001)	Grocery stores	BDE-28, 47, 99, 100, 153, 154	Fish (16)	Median 1400; range 17.7-1720	107
			Shellfish (2)	Median 52; range 43 – 61	
			Meat (3)	Range 6.25 – 63.6	
			Vegetables (3)	Range 38.4 – 134	
U.S., U.K., Norway, and Canada (2001 and 2002)	Fish farms and fish suppliers	43 congeners including BDE-28, 47, 99, 153, 154, 183	Salmon, farmed (153)	Median 2500 (approx.); range 500 – 4000 (approx.)	108
			Salmon, wild (45)	Median 150 (approx.); range 100 – 4200 (approx.)	
Sweden (1999)	Grocery stores	BDE-47, 99, 100, 153, 154	Diary Products (sample size not provided)	Mean 360 ppt (lipid basis)	109
			Meat Products	Mean 360 ppt (lipid basis)	
			Eggs	Mean 420 ppt (lipid basis)	
Europe (inc. North Sea and Baltic Sea) (various years)	Not provided	BDE-47 only or various congeners	Herring (Sample size not provided)	Range 17,000 – 528,000 ppt (lipid basis) total PBDEs; Range 9,000 – 100,000 ppt (lipid basis) BDE-47 only.	110
Scotland and Belgium (1999 and 2001)	Fish markets	BDE-28, 47, 71, 75, 66, 99, 100, 153, 154	Salmon, farmed and wild (13)	Range 1,100 – 85,200 ppt (lipid basis)	111

ppt = parts per trillion; ND = non-detectible; Ref = Reference
[a] Mean values are assumed because actual method of calculation is unclear from the report.

Table 6. Estimates of PBDE daily human intake for different countries.

Daily PBDE intake (mg/kg-bw i/day)	Country	Age	Sources of exposure	PBDE congeners	Ref.
0.0000007	Sweden	adult	food	47, 99, 100, 153, 154	133
0.00001[a]	Sweden	infant (0-6 mo.)	breast milk	47, 99, 100, 153, 154	134
0.00000062[b] (0.044 µg/day)	Canada	adult	food	28, 47, 99, 100, 153, 154	135
0.00000019 – 0.000003[b] (0.013-0.213 µg/day)	The Netherlands	adult	food	28, 47, 99, 100, 153, 154	136
0.0000014 - .0000011[b] (0.097-0.082 µg/day)	Spain	adult	food	Sum of tetra- to octa-BDEs	137
0.00000059[b] (0.041 µg/day)	Sweden	adult	food	47, 99, 100, 153, 154	138
0.0000013[b] (0.091 µg/day)	U.K.	adult	diet, air, occupational	47, 99, 100, 153, 154	139
0.00000073[b] (0.051 µg/day)	Canada	infant	breast milk	Sum of tri-BDEs to hepta-BDEs	140
0.00000043[b] (0.030 µg/day)	Canada	adult	diet, air, occupational	Sum of tri- to hepta-BDEs	141
0.0002 -0.0026	Canada	0-6 mo., 0.5-4, 5-11, 12-19, 20-59, 60+ yrs.	air, water, food, breast milk, and dust	Sum of (tetra- to deca-BDEs)	142
0.000355 (U.S) 0.000011 (Germany)	U.S. and Germany	nursing infants	breast milk	17, 28, 47, 66, 77, 85, 99, 100, 138, 153, 154, 183, 209	143
max 0.000004 (child); max 0.000003 (adult)	U.S. (CA)	children (<18 yrs) and adults	food (fish, meat, fowl)	Sum of (mono- to deca-BDEs)	144
0.00004-0.0009	U.S.	<1 yr, 1-2 yrs, 3-5 yrs.	multiple pathways	Penta-BDE congeners	145, 146
0.000014 – 0.000054[c]	U.S.	adult women	back-calculated from tissue levels	Total; mostly 47, 99, 100, 153, 154	147

Notes: mg/kg, milligram per kilogram bodyweight per day; µg/day, microgram per day; Ref., reference
[a] Calculated from value in cited reference using an assumed 7.5 kg bodyweight for infant.
[b] Calculated from value in cited reference using an assumed 70 kg bodyweight for adult.
[c] Calculated from value in cited reference using an assumed 62 kg bodyweight for adult woman.

Table 8. Measured concentrations of PBDEs in North American biota

Organism	Location; year	Total PBDEs	Reference
Biota measured in Pacific Northwest			
Dungeness crab hepatopancreas	West coast, Canada; 1993 - 1995	4.2 – 480 µg/kg lipid	[211]
Bald eagle egg	Lower Columbia River, Washington and Oregon, 1994 - 1995	446 – 1,206 µg/kg ww	[212]
Heron egg	British Columbia; 1983 - 2000	1.308 – 288 µg/kg ww	[213]
Orca blubber	Northeastern Pacific Ocean; 1993 - 1996	87 – 1,620 µg/kg lipid	[214]
Mountain whitefish (muscle)	Columbia River, British Columbia; 1992 - 2000	0.726 – 131 µg/kg ww	[215]
Rainbow trout	Spokane River, Washington; 1999	297 µg/kg ww	[216]
Mountain whitefish		1250 µg/kg ww	
Largescale sucker		105 µg/kg ww	
Biota measured in other areas of North America			
Murre egg	Northern Canada; 1975 - 1998	0.442 – 2.93 µg/kg ww	[217]
Fulmar egg	Northern Canada; 1975 - 1998	0.212 – 2.37 µg/kg ww	
Herring gull egg	Great Lakes; 1981 - 2000	9.4 – 1544 µg/kg ww	[218]
Beluga whale blubber	Canadian Arctic	81.2 – 160 µg/kg lipid	[219]
Beluga whale blubber	St. Lawrence Estuary, Canada, 1988 - 1999	17.2 – 935 µg/kg lipid	[220]
Lake trout	Lake Ontario; 1997	95 µg/kg ww	[221]
	Lake Erie; 1997	27 µg/kg ww	
	Lake Superior; 1997	56 µg/kg ww	
	Lake Huron; 1997	50 µg/kg ww	
Carp	Virginia; 1998 - 1999	1140 µg/kg ww	[222]

ww = wet weight

Source: Department of Ecology. (2006). *Washington State Polybrominated Diphenyl Ether (PBDE) Chemical Action Plan: Final Plan*. Department of Ecology Publications Distributions Office, Olympia, WA, http://www.ecy.wa.gov/biblio/0507048.html.

Appendix K - Detection Frequencies and Median Concentrations for Selected Volatile Organic Compounds in Samples from Aquifer Studies.

USGS (WIP – Need full citation)

[µg/L, micrograms per liter; ND, compound not detected; <, less than; --, not applicable]

Compound name	Number of samples	Detection frequency at selected assessment levels[1] (percent)	No assessment level	Number of samples	Median concentration[3] (µg/L) at selected assessment levels[2]					All samples	Samples with detections
					0.02 µg/L	**0.2 µg/L**	1 µg/L	5 µg/L	10 µg/L		
Refrigerants											
Dichlorodifluoromethane	1,687	4.1	4.1	3,496	1.9	0.34	0.029	ND		< 0.20	0.28
Trichlorofluoromethane	1,686	2.4	2.1	3,495	1.1	.37	.057	.029		< .20	.20
Trichlorotrifluoroethane	1,686	1.0	.77	2,666	.26	.038	ND	ND		< .060	.092
Solvents											
Carbon tetrachloride	1,686	1.3	1.1	3,497	0.31	0.086	ND	ND		< 0.20	0.077
Chlorobenzene	1,687	1.3	.41	3,498	.17	.11	0.029	ND		< .20	.007
Chloroethane	1,686	.30	.30	3,113	.29	.064	ND	ND		< .12	.20
Chloromethane	1,676	7.9	6.0	2,988	1.1	.13	.033	0.033		< .20	.035
1,2-Dichlorobenzene	1,687	.53	.41	3,464	.12	.087	.029	.029		< .20	.041
1,3-Dichlorobenzene	1,687	.41	.12	2,347	ND	ND	ND	ND		< .054	.008
1,1-Dichloroethane	1,686	2.8	2.4	3,496	.86	.17	.029	ND		< .20	.085
1,2-Dichloroethane	1,687	.18	.18	3,438	.47	.15	ND	ND		< .20	.30
cis-1,2-Dichloroethene	1,686	2.3	1.7	2,847	.42	.070	ND	ND		< .050	.047
trans-1,2-Dichloroethene	1,686	.30	.24	3,200	.91	.38	.12	.062		< .050	.60
Hexachloroethane	1,683	ND	ND	1,759	ND	ND	ND	ND		< .19	--
Methylene chloride	1,685	5.0	3.6	3,487	.89	.37	.057	.057		< .20	.040
Perchloroethene	1,656	13.2	8.3	3,449	3.7	1.5	.70	.32		< .20	.090
n-Propylbenzene	1,687	.30	.24	2,461	.041	.041	ND	ND		< .042	.048
1,2,4-Trichlorobenzene	1,687	.059	.059	2,509	ND	ND	ND	ND		< .20	.020
1,1,1-Trichloroethane	1,687	8.2	4.4	3,498	1.7	.57	.17	.14		< .20	.043
1,1,2-Trichloroethane	1,687	.18	.12	3,119	ND	ND	ND	ND		< .10	.028
Trichloroethene	1,686	5.2	3.8	3,497	2.6	1.1	.46	.26		< .20	.20
Trihalomethanes											
Bromodichloromethane	1,686	4.9	3.7	3,497	1.1	0.46	0.20	0.11		< 0.20	.080
Bromoform	1,685	2.2	1.4	3,496	1.0	.31	.029	ND		< .20	.30
Chloroform	1,686	29.8	21.2	3,495	7.4	2.3	.69	.31		< .20	.079
Dibromochloromethane	1,686	2.0	1.7	3,497	.94	.34	.14	.11		< .20	.20
Total trihalomethanes[5]	1,686	30.5	21.6	3,497	8.1	2.5	.71	.34		< .20	.090

1. These detection frequencies are for the subset of samples that were analyzed with the U.S. Geological Survey's low-level method 0–4127–96. At this assessment level, detection frequencies are estimates. (19)

2. These detection frequencies are for all samples included in this assessment, regardless of the analytical method.

3. The analytical methods used for this assessment have varied sensitivity among compounds and comparison of the median concentrations between compounds is not appropriate. No assessment level was applied to determine the median.

4. Considered as 2 of the 55 compounds included in this assessment.

5. Not considered as 1 of the 55 compounds included in this assessment.

Appendix L - Maine Human Body Burdens Chemicals

From the *Body of Evidence* study by Alliance for a Clean and Healthy Maine
(http://www.cleanandhealthyme.org/)

The following survey of chemical contaminants in Maine's citizens is an important recent survey issued by the Alliance for a Clean and Healthy Maine. The organizers of the *Body of Evidence* report are associated with a much larger network of biomonitoring organizations, the links to some of which are listed below. Of special interest are other surveys of POPs and other ecotoxins in humans. Space restrictions in this publication prevent us from adding additional appendices, which would include their research. A number of recently issued surveys are listed below and include those by the Coming Clean Network as well as affiliated organizations. For a more detailed introduction to the Pandora's Box of anthropogenic ecotoxins in biotic media, please explore the websites and reports listed below.

Coming Clean Network
Chemical Body Burdens
http://www.chemicalbodyburden.org/home.htm

Chemical Trespass: Report on Pesticide Body Burden Data
http://www.panna.org/docsTrespass/chemicalTrespass2004.dv.html

Flame Retardant Study in Washington State

Biomonitoring Results in the U.K.

Chemicals in U.S. Population
http://www.chemicalbodyburden.org/rr_cheminus.htm

Fire Retardants (PBDEs) in Breast Milk
http://www.ewg.org/reports/mothersmilk/

On-line Body Burden/ Community Monitoring Handbook
http://www.ewg.org/reports/mothersmilk/

PCBs in People of St. Lawrence Island
http://www.ewg.org/reports/mothersmilk/

Phthalates in Cosmetics
http://www.ewg.org/reports/mothersmilk/

The following tables summarize some of the most important findings of the survey done by the Alliance for a Clean and Healthy Maine.

RESULTS FROM 13 MAINE PARTICIPANTS				RESULTS FROM OTHER STUDIES			
Phthalates units = ug/gCr-L (creatinine corrected)				from federal CDC 3rd National Exposure Report[91] n = 2,536 for MEP; n = 2,772 for all other phthalates[1]			
	Minimum	Maximum	Median – or 50th %tile	Median – or 50th %tile	75th %tile	90th %tile	95th %tile
MMP	< 1.16	46.5	8.19	1.33	2.62	5.00	7.97
MEP	10.6	205	54.7	147	388	975	1860
MBP	21.8	92.2	50.5	26.0	51.6	98.6	149
MBzP	6.29	68.8	29.1	13.5	26.6	55.1	90.4
MEHP	1.62	66.9	10.6	3.89	7.94	18.2	32.8
MEOHP	4.13	132	15.9	11.2	21.3	45.1	87.5
MEHHP	8.39	324	40.7	16.6	32.3	70.8	147
Sum TOTAL	105	793	223	219	530	1,268	2,375
PBDEs units = pg/g on a lipid weight basis				from McDonald 2005[92] n = 62 women from CA & IN		n = 10	n = 11
	Minimum	Maximum	Median – or 50th %tile	Median – or 50th %tile	95th %tile	Washington Median[93]	California Median[94]
BDE-15	60.6	603	144			275	·
BDE-17/25	36.7	506	84.4			61.7	·
BDE-28/33	350	2200	694			1128	·
BDE-35	< 27.7	119	*68.5			< 5.64	·
BDE-37	< 19.1	*55.3	< 27.7			10.0	·
BDE-47	2900	33500	8380	included below	included below	19950	14100
BDE-49	48	275	98.3			178	·
BDE-51	< 19.1	83.8	< 31.9			~ 12	·
BDE-66	*55.2	506	122			170	·
BDE-71	< 19.1	70.8	< 31.3			< 17.4	·
BDE-75	< 19.1	85.1	< 27.7			25.0	·
BDE-79	< 27.2	140	*38.8			*61.1	·
BDE-85	53.2	745	148			346	·
BDE-99	987	9280	1870	included below	included below	4255	3100
BDE-100	454	7230	1550	included below	included below	3115	2100
BDE-116	< 22.1	*51.7	< 36.9			< 22.4	·
BDE-119/120	< 19.1	*56.2	< 31.9			22.8	·
BDE-138/166	21.2	121	47.9			73.8	·
BDE-140	< 24.6	77.8	49.6			~ 44	·
BDE-153	1390	15300	4060	included below	included below	2725	3400
BDE-154	96.8	746	200	included below	included below	368	280
BDE-155	< 27.2	*78.6	*45.9			43.4	·
BDE-183	*147	1400	328			218	·
BDE-203	90.1	303	134			152	·
Sum TOTAL	6,918	59,869	19,971	40,700	305,000	47,500	22,980

As an introduction to *Body of Evidence – Maine Human Body Burdens Chemicals Survey of Chemicals*, the authors comment on some of the possible community action strategies that individuals can take to mitigate their intake of ecotoxins in the food web.

In the 'What is Body Burden' section we learned that each of us carries an estimated 700 chemicals in our bodies. So now that you are informed, what can you do?

- Get Organized
- Learn More
- Take Action against phthalates

There are some lifestyle choices that can minimize your chemical load. When you have enough information, you can choose products made with safer materials. You can also:

- Choose certified organic produce, meat and dairy products to limit short term exposure
- Seek out the least toxic household products (e.g., environmentally friendly cleaning agents, body care products, cosmetics)
- Eliminate indoor and outdoor pesticide use in the home by choosing safer alternatives to conventional tick and flea collars for pets and avoiding weed killers, insect sprays and termite treatments
- Reduce intake of fat because many chemicals are stored in fat

In the long term, however, the best way to reduce the load of chemicals we all carry is to stop using them. This means creating public policies that encourage the production of safer products produced without dangerous chemicals. It means relying less and less on chemical pesticides to grow our food. We need policies that are truly protective of human health, so that future generations are not born with a chemical body burden that grows throughout their lifetimes. Moving toward a cleaner economy and reducing our chemical body burden means changing policies, challenging chemical companies, changing consumer behavior and supporting cleaner industrial and agricultural production. This can only happen with widespread involvement of concerned individuals in communities across the country and around the world. In this section, you will find groups working on various issues who share the goal of cleaning up toxic chemicals and moving toward safer alternatives. Your involvement will make a difference - contact one or more of these groups to join the effort (http://www.cleanandhealthyme.org/).

Appendix M - Biodiversity Research Institute Report: *Preliminary Findings of Contaminant Screening of Maine Bird Eggs 2007 Field Season*

(Goodale 2008)

The BioDiversity Research Institute is located in Gorham, Maine, and works in conjunction with U. S. Fish and Wildlife Service, the Canadian Wildlife Service, and Environment Canada as well as the State of Maine Department of Environmental Protection. During the period from 2001 to 2005, BRI participated in the compilation of a comprehensive database on mercury sources and impacts in northeast freshwater environments. Some of the data from this research are cited in *Appendix O*. The BioDiversity Research Institute is among the most important environmental research organizations currently collecting data on the proliferation of anthropogenic ecotoxins in the environment. The Executive Summary and Overall Conclusions of their preliminary screening of Maine bird eggs during the 2007 field season are reprinted below in their entirety. Of particular interest is not only the wide variety of synthetic chemicals detected in Maine bird species, but also the fact that all species subject to sampling analysis had been contaminated with detectable levels of multiple anthropogenic ecotoxins.

1. Executive Summary and Primary Findings

Starting in May 2007, BioDiversity Research Institute (BRI) and collaborators initiated a broadbased contaminant study on Maine birds, measuring both historical and emerging chemicals. This comprehensive project measured 192 synthetic contaminants in 23 species across Maine to determine in which species, habitats, and locations these anthropogenic compounds are concentrating. The compounds we analyzed in 60 egg composites were mercury (Hg), polychlorinated biphenyls (PCB), polybrominated diphenyl ethers (PBDE), perfluorinated compounds (PFCs), and organochlorine pesticides (OCs). Our preliminary findings are:

- Hg, PCBs, PBDEs, PFCs, and OCs are found in all species sampled across marine, estuarine, riverine, lacustrine (lake), and terrestrial ecosystems; these are the first records of PFCs in Maine birds.

- Hg, PCBs, PFCs are all found at levels that may cause adverse effects—there are currently no established adverse effects thresholds established for PBDEs in bird eggs. OCs are all significantly below adverse effects thresholds.

- Our Hg, PCB, and OC levels were generally consistent with levels recorded around the country. Certain species had PBDEs higher than other locations, while other species had lower levels. PFOS have not been widely studied in eggs; therefore, we could not directly compare our results to other areas.

- The total PCBs levels we recorded are lower than those in the past, indicating a continued decline in PCBs.

- Bald eagles have the highest overall contaminant load of the 23 species measured.

- We found all of the compounds across the entire state, but overall contaminant loading tends to be highest in southern coastal Maine. This geographic pattern suggests that these compounds are entering the environment both through atmospheric deposition, because they are found across the entire state, and through local point sources, because we detected higher levels in urban and industrial areas.

- PCBs, PBDEs, PFCs, and OCs levels are positively correlated, indicting that birds with high levels of one compound tend to have higher levels of the others. PBDEs and PCB have the strongest relationship.

- Birds that feed on terrestrial prey accumulated higher brominated PBDEs; DecaBDE is found in eight species with gulls and peregrine falcon having the highest levels.

- Of the samples we analyzed, birds feeding in estuaries have the lowest contaminant levels.

- The mouth of the Kennebec and Isles of Shoals tended to have high concentrations of contaminants.

2. INTRODUCTION

2.1 Project overview

Starting in May 2007, BioDiversity Research Institute (BRI) and collaborators initiated a broad-based contaminant study on Maine birds, measuring both historical and emerging chemicals. This comprehensive project measured 192 synthetic contaminants in 23 species across Maine to determine in which species, habitats, and locations these anthropogenic compounds are concentrating. The chemicals we analyzed in 60 egg composites were mercury, polychlorinated biphenyl (PCB) congeners, polybrominated diphenyl ether (PBDE) congeners, perfluorinated compounds (PFCs; e.g., PFOS, PFOSA, PFHxS, PFOA, PFNA, PFDA, PFDoDA, PFUnDA, PFHxA, PFHpA), and organochlorine pesticides (OCs) (DDTs, HCHs, chlordanes, HCB).

The project had two components. The first was evaluating geographic differences by analyzing eggs of seven marine species from six sites near the outflows of Maine's largest rivers (Figure 1). Since studies indicate that levels of PCBs and other organics in eagles are higher along the coast than inland (Matz 1998), and contaminants bioaccumulate[1] in coastal cormorants (Mower 2006), terns, and plovers (Mierzykowski and Carr 2004), we focused geographic contaminant screening along the coast. We selected sites near the largest river outflows and areas of high population density. The sites were: 1) Isles of Shoals (Piscataqua River, Kittery); 2) Casco Bay (Portland); 3) Popham Beach and Sheepscot Bay (Androscoggin, Kennebec, and Sheepscot rivers, Phippsburg); 4) Penobscot Bay (Penobscot River, Islesboro); and 5) Cobscook Bay (St. Croix River, Eastport) (Figure 1, Table 2).

We evaluated geographic variation in freshwater ecosystems with common loon and bald eagle eggs (Figure 1). The species we selected have a broad range of foraging strategies and represent most of Maine's primary ecosystems. The second component evaluated exposure in major habitat types through analyzing eggs from multiple species in the same area. In the Portland, Maine area we collected eggs from marine, estuarine, riverine, lake, and terrestrial habitats, focusing on high trophic[2] level predators (Figure 2). Species include insectivores, piscivores, and bird and mammal predators. Additionally, to ensure direct comparison among habitats, we collected eggs from tree swallows—a low trophic level insectivore. Collectively, this sampling effort provided a baseline and initial screening of contaminant levels, and helped determine if contaminants are concentrating in certain areas.

1 Increase in an organism over time because they take in more than they can expel.

2 How high in the food web a bird eats (i.e. eagles are high trophic level and eiders

4.9 Overall conclusions

We found both established (Hg, PCBs, chlordane, HCB, DEE) and emerging (PBDEs, PFCs) bioaccumaltive toxic pollutants of concern in all the bird eggs we analyzed. Our results are the first records of PFCs in Maine birds. Since the birds we selected act as bioindicators of multiple ecosystems across the state, our results indicate that the compounds we measured are present in the offshore marine, coastal marine, estuarine, riverine, lacustrine, and terrestrial ecosystems. Although we found most of the compounds across the entire state, there tended to be higher levels in coastal southern Maine. This geographic pattern suggests that these compounds are entering the environment both through atmospheric deposition, because they are found across the entire state, and through local point sources, because we detected higher levels in urban and industrial areas. In particular, several areas consistently have higher levels than the rest of the state: Isles of Shoals, Portland, and the mouth of the Kennebec.

As expected, a number of loon eggs have Hg levels above known effects thresholds. One eagle egg has PCB levels within a range of known effects, although the congener pattern is not dominated by the most toxic PCBs. Since no effects threshold have been established for PBDEs in bird eggs, the residues we detected may or may not have negative effects. Twenty-three of our samples have PFOS levels above effects threshold established for chicken eggs—the species we studied may be more or less sensitive than the chickens. OCs are all substantially below effects thresholds.

Our Hg results are consistent with other studies conducted in the region. Our PCB results are also consistent with those across the United States, and when compared to earlier studies in Maine, herring gull, common eider, and bald eagle all have lower levels than in the past. Overall our PBDE result are not consistently higher or lower than other areas across the globe, but some species had higher or lower PBDEs than in other areas. Like recent studies on terrestrial birds we detected higher brominated PBDEs, including decaBDE, in terrestrial predators: American kestrel and peregrine falcon. Only one study has analyzed PFOS in bird eggs of two species in the Great Lakes region. Our results are similar to these. Moreover, although a direct comparison is not possible, our results are consistent with PFOS in the liver of multiple species around the world. Our OC results are generally within the range of other studies.

Our results show that many of the compounds we measured increase in concert with each other. The strongest relationship we found is between PCBs and PBDEs, indicating that species and areas with high PCB levels may also have high PBDE levels. These relationships suggest that some species may have higher levels simultaneously of multiple compounds, which together may have greater negative impact on reproductive success, the neurological system, endocrine function, and overall physiology. Consequently, high trophic level predators may have a combined negative effect of these compounds despite having individual contaminants below known effects thresholds.

In general our results followed the expected pattern- with high trophic level predators having the highest overall contaminant levels. Bald eagle in particular have PCB, PBDEs, PFCs, chlordane, and DDE multiple times higher than other species. Two species did not follow the expected pattern: belted-kingfisher and piping plover. The kingfisher eggs were collected from an urbanized river that may have overall higher pollution levels than other sites where we collected samples; if we were able to collect samples from other species at this site, their levels also may have been higher. The reason for the higher contaminant levels in plovers in not clear.
Estuaries consistently have lower levels of all the compounds. However, these results may be confounded by the lower trophic level of species we collected samples from in estuaries. As expected, lakes have higher levels of Hg than other habitats, and PCBs have higher levels along the coast.

In summary, our results indicate that both historical and emerging chemicals of concern are accumulating in birds that forage in diverse ecosystems across the entire state of Maine.

Appendix N - Maine Fish Consumption Advisories

The Maine fish and game guidelines may be found at: http://www.maine.gov/dhhs/eohp/fish/index.htm.

Maine Center for Disease Control and Prevention. (June 3, 2009). *Warning about eating saltwater fish and lobster tomalley*. Environmental and Occupational Health Programs, Division of Environmental Health, Augusta, ME. http://www.maine.gov/dhhs/eohp/fish/saltwater.htm.

WARNING About Eating Saltwater Fish and Lobster Tomalley

Warning: Chemicals in some Maine saltwater fish and lobster tomalley may harm people who eat them. Women who are or may become pregnant and children should carefully follow the Safe Eating Guidelines.

It's hard to believe that fish that looks, smells, and tastes fine may not be safe to eat. But the truth is that some saltwater fish have mercury, PCBs and Dioxins in them.

All these chemicals settle into the ocean from the air. PCBs and Dioxins also flow into the ocean through our rivers. These chemicals then build up in fish.

Small amounts of mercury can damage a brain starting to form or grow. That's why babies in the womb, nursing babies, and young children are at most risk. Mercury can also harm older children and adults, but it takes larger amounts.

PCBs and Dioxins can cause cancer and other health problems if too much builds up in your body. Since some saltwater fish contain several chemicals, we ask that all consumers of the following saltwater species follow the safe eating guidelines.

Revised June 3, 2009

SAFE EATING GUIDELINES

Striped Bass and Bluefish: Pregnant and nursing women, women who may get pregnant, nursing mothers and children under 8 years should not eat any striped bass or bluefish. **All other individuals** should eat no more than **4 meals per year.**

Shark, Swordfish, King Mackerel, and Tilefish: Pregnant and nursing women, women who may get pregnant and children under 8 years of age are advised to not eat any swordfish or shark. **All other individuals** should eat no more than **2 meals per month.**

Canned Tuna: Pregnant and nursing women, women who may get pregnant and children under 8 years of age can eat no more than **1 can** of "white" tuna or **2 cans** of "light" tuna **per week.**

All other ocean fish and shellfish, including canned fish and shellfish: Pregnant and nursing women, women who may get pregnant and children under 8 years of age can eat no more than **2 meals per week.**

Lobster Tomalley: No Consumption. While there is no known safety considerations when it comes to eating lobster meat, consumers are advised to refrain from eating the tomalley. The tomalley is the soft, green substance found in the body cavity of the lobster. It functions as the liver and pancreas, and test results have shown the tomalley can accumulate contaminants found in the environment.

For more information, including warnings on freshwater fish call (866)-292-3474 or visit our web site http://www.maine.gov/dhhs/eohp.

Maine Bureau of Health. (August 29, 2000). *Warning about eating freshwater fish.* Environmental Toxicology Program, Maine Bureau of Health, Augusta, ME. http://www.maine.gov/dhhs/eohp/fish/documents/2KFCA.pdf

WARNING About Eating Freshwater Fish

Warning: **Mercury in Maine freshwater fish may harm the babies of pregnant and nursing mothers, and young children.**

SAFE EATING GUIDELINES

Pregnant and nursing women, women who may get pregnant, and children under age 8 SHOULD NOT EAT any freshwater fish from Maine's inland waters. Except, for brook trout and landlocked salmon, 1 meal per month is safe.

All other adults and children older than 8 CAN EAT 2 freshwater fish meals per month. For brook trout and landlocked salmon, the limit is 1 meal per week.

It's hard to believe that fish that looks, smells, and tastes fine may not be safe to eat. But the truth is that fish in Maine lakes, ponds, and rivers have mercury in them. Other states have this problem too. Mercury in the air settles into the waters. It then builds up in fish. For this reason, older fish have higher levels of mercury than younger fish. Fish (like pickerel and bass) that eat other fish have the highest mercury levels.

Small amounts of mercury can harm a brain starting to form or grow. That is why unborn and nursing babies, and young children are most at risk. Too much mercury can affect behavior and learning. Mercury can harm older children and adults, but it takes larger amounts. It may cause numbness in hands and feet or changes in vision. The Safe Eating Guidelines identify limits to protect everyone.

Warning: Some Maine waters are polluted, requiring additional limits to eating fish.

Fish caught in some Maine waters have high levels of PCBs, Dioxins or DDT in them. These chemicals can cause cancer and other health effects. The Bureau of Health recommends additional fish consumption limits on the waters listed below. Remember to check the mercury guidelines. If the water you are fishing is listed below, check the mercury guideline above and follow the most limiting guidelines.

SAFE EATING GUIDELINES

Androscoggin River Gilead to Merrymeeting Bay:	6-12 fish meals a year.
Dennys River Meddybemps Lake to Dead Stream:	1-2 fish meals a month.
Green Pond, Chapman Pit, & Greenlaw Brook(Limestone):	Do not eat any fish from these waters.
Little Madawaska River & tributaries(Madwaska Dam to Grimes Mill Road):	Do not eat any fish from these waters.
Kennebec River Augusta to the Chops:	Do not eat any fish from these waters.
Shawmut Dam in Fairfield to Augusta:	5 trout meals a year, 1-2 bass meals a month.
Madison to Fairfield:	1-2 fish meals a month.
Meduxnekeag River:	2 fish meals a month.
North Branch Presque Isle River	2 fish meals a month.
Penobscot River below Lincoln:	1-2 fish meals a month
Prestile Stream:	1 fish meal a month.
Red Brook in Scarborough:	6 fish meals a year.
Salmon Falls River below Berwick:	6-12 fish meals a year.
Sebasticook River (East Branch, West Branch & Main Stem)(Corinna/Hartland to Winslow):	2 fish meals a month.

Appendix O - Methylmercury in Biotic Media

A review of the literature pertaining to the proliferation of ecotoxins throughout the food webs of the biosphere reveals that a small group of highly toxic persistent organic pollutants banned by the Stockholm Convention show some signs of declining. Other anthropogenic chemicals such as PCBs, one of the chemicals banned by the Stockholm Convention agreement, continue to be important ecotoxins of concern in most biotic media. Among the most alarming developments in the spread of chemical fallout throughout the biosphere is the continued increase in methylmercury in most biotic media. The following observation by Barry Mower, the Maine DEP's research analyst for ecotoxins in biotic media in Maine pinpoints the fundamental challenge of evaluating the impact of mercury discharges to the environment from human activities. "Through the process of biomagnification, the tissues of predatory freshwater fish near the top of the food chain may contain levels of methyl mercury that are 100,000 to 1,000,000 times higher than the concentration in the water." (Casco Bay Estuary Partnership 2007, 54). This Appendix consists of citations from a number of research projects on the proliferation of methylmercury in the biosphere as a result of human activity. The following quotation summarizes the basic parameters of methylmercury as an anthropogenic ecotoxin, recapitulating many of the observations made in the previous sections of this publication. "The heavy metal mercury can enter the environment through industrial processes, such as chlorine manufacturing (Evers 2005), and through combustion of coal, oil, wood, natural gas and mercury-containing trash. Over the past century, anthropogenic inputs of mercury into the environment have significantly increased (Evers 2004). Once in the environment, elemental mercury can be transformed by bacteria into a highly toxic compound (methyl mercury) which is readily absorbed into living tissues." (Casco Bay Estuary Partnership 2007, 51).

The Casco Bay Estuary Partnership publication contains important information about the biomagnification of mercury in the environment, a summary of which follows (Casco Bay Estuary Partnership 2007, 53).

Media	Typical Concentration Levels (ng/kg)
Water	1 ng/kg
Bacteria and Phytoplankton	10 ng/kg
Protozoa and Zooplankton	100 ng/kg
Insect larvae	1,000 ng/kg
Fish fry	10,000 ng/kg
Minnows	100,000 ng/kg
Medium size fish	1,000,000 ng/kg
Predators (Humans and predatory birds)	10,000,000 ng/kg

This summary on the biomagnification of mercury in the environment helps place the findings of other researchers in the context of the rapid biomagnification of methylmercury in pathways to human consumption. An ongoing program to monitor mercury in the environment is sponsored by the USGS, the Seabird Tissue Archival and Monitoring Project (STAMP).

Centers for Disease Control and Prevention. (2005). *Third national report on human exposure to environmental chemicals.* pg. 45. Table 19. "Mercury in Blood".

Geometric mean and selected percentiles of blood concentrations (in μg/L) in the U.S. population.

Age Group	Survey years	Geometric mean (95% conf. interval)	75th percentile (95% conf. interval)	95th percentile (95% conf. interval)	Sample size
1-5 years (females and males)	99-00	.343 (.297-.395)	.500 (.500-.600)	2.30 (1.20-3.50)	705
	01-02	.318 (.255-.377)	.700 (.500-.800)	1.90 (1.40-2.90)	872
Females	99-00	.377 (.299-.475)	.800 (.500-1.10)	2.70 (1.30-5.50)	318
	01-02	.329 (.265-.407)	.700 (.500-.800)	2.60 (1.30-4.90)	432
Males	99-00	.317 (.269-.374)	.500 (.500-.600)	2.10 (1.10-3.50)	387
	01-02	.307 (.256-.369)	.600 (.400-.700)	1.70 (1.40-2.00)	440
16-49 years (females only)	99-00	1.02 (.825-1.27)	2.00 (1.50-3.00)	7.10 (5.30-11.3)	1709
	01-02	.833 (.738-.940)	1.70 (1.40-1.90)	4.60 (3.70-5.90)	1928
Race/ethnicity (females, 16-49 years)					
Mexican Americans	99-00	.820 (.664-1.01)	1.40 (1.20-2.00)	4.00 (2.70-5.50)	579
	01-02	.667 (.541-.824)	1.10 (1.00-1.40)	3.50 (2.30-4.40)	527
Non-Hispanic blacks	99-00	1.35 (1.06-1.73)	2.60 (1.80-3.40)	5.90 (4.20-11.7)	370
	01-02	1.06 (.871-1.29)	1.80 (1.50-2.20)	4.10 (3.30-6.00)	436
Non-Hispanic whites	99-00	.944 (.726-1.23)	1.90 (1.30-3.30)	6.90 (4.50-12.0)	588
	01-02	.800 (.697-.919)	1.50 (1.30-2.00)	4.60 (3.30-6.80)	806

The data above "are similar or slightly lower than levels found in other population studies. In Germany, for example, the geometric mean for blood mercury was 0.58 μg/L for 4.645 adults aged 18 to 69 years participating in a 1998 representative population survey (Becker et al., 2002). During the years 1996 through 1998, Benes et al. (2000) studied 1,216 blood donors in the Czech Republic (896 men and 320 women; average age 33 years) and 758 children (average age 9.9 years). The median concentration of blood mercury for adults was 0.78 μg/L and 0.46 μg/L for the juvenile population." (Centers for Disease Control and Prevention 2005, 48).

Evers, D. C., et al. (2007). Biological mercury hotspots in the northeastern U.S. and southeastern Canada. *BioScience*. 57(1). pg. 31.

Category/species	Sample size	Hg concentration Range (ppm)	Hg level of concern (tissue type)	% of samples with concentrations > level of concern
Human health				
Yellow perch	4089	< 0.05-5.24	0.30 (fillet)	50
Largemouth bass	934	< 0.05-2.66	0.30 (fillet)	75
Ecological health				

Category/species	Sample size	Hg concentration Range (ppm)	Hg level of concern (tissue type)	% of samples with concentrations > level of concern
Brook trout	319	<0.05-2.07	0.16 (whole fish)	75
Yellow perch	(841)	<0.05-3.18	0.16 (whole fish)	48
Common loon	1546	0.11-14.20	3.0 (blood)	11
Bald eagle	217	0.08-1.27	1.0 (blood)	6
Mink	126	2.80-68.50	30.0 (fur)	11
River otter	80	1.14-37.80	30.0 (fur)	15

Appendix P - Evidence of Mercury Toxicity in Autism

The following interpretation of the literature review of methylmercury by Bernard (2001) provides one of the many takes on the possible relationship between methylmercury as a neurotoxin and autism as a brain developmental disorder. Many experts content autism has a predominantly genetic basis (Abrahams 2008). Up to 15% of autism cases have been attributed to chromosome abnormalities or other genetic conditions (Folstein 2001); it is also correlated with mental retardation (Dawson 2008), anxiety disorder (White 2009), and metabolic defects. The fact that the incidence of autism has greatly increased since the 1970s has been attributed by most experts to increased surveillance and ASD (autism spectrum disorder) syndrome awareness. Only a minority of experts suggest that autism may have a link to environmental chemicals. The widespread prevalence of methylmercury in biotic media, especially as a biomagnified anthropogenic ecotoxin in higher trophic levels, raises the compelling question of its relationship with the increasing incidence of autism, conservatively reported as 6 per 1,000 people with over four times as many cases in boys as in girls (Newschaffer 2007). Diagnosed cases of autism have grown dramatically, as measured by those served under part B of the Individuals with Disabilities Education Act (IDEA); the 1996 report of 21,669 cases rose to 64,094 in 2001 and to 110,529 in 2005 (Montes 2007). In one Minnesota county the cumulative incidence of autism increased as much as 22-fold during the same time period (Volkmar 2009).

Recent linkage of autism to Thermisol in childhood vaccines has been widely discounted due to "lack of any convincing scientific evidence." (Rutter 2005). Such is not the case with its possible links to Hg contamination of the biosphere. A variety of information sources on Hg as an environmental toxin follow the Bernard synopsis. Numerous other information sources, including the CDC, Environmental Working Group, and other governmental and NGO organizations are listed in the links section.

Evidence of Mercury Toxicity in Autism

A literature review by Bernard, et al. (2001) reported that all of the criteria necessary for a diagnosis of autism were also observed in cases of mercury toxicity. For example, the major effect of toxic mercury compounds is on the central nervous system, although immune and gastrointestinal systems are sometimes affected. The same abnormalities in the same systems have been found in children with autism. Mercury binds with sulfur in the body and is reflected in widespread dysfunction of enzymes, transport mechanisms, and structural proteins, causing, among other conditions, speech and hearing problems. One of the primary features of autism is language delay.

Susceptibility to mercury appears to have a genetic component as boys are more sensitive than girls. Autism occurs about four times more frequently in boys than in girls.

As with autism, sensory disturbances, including numbness in the mouth, hands, and feet, sensitivity to loud noises, aversion to touch, and over or under response to pain are symptoms of mercury toxicity. On another parallel, autism and mercury exposure can both cause cognitive impairment and difficulty with complex thinking and acting processes, social withdrawal, anxiety, and obsessive-compulsive behaviors.

Mercury disrupts serotonin, dopamine, glutamate, and acetylcholine neurotransmitters, some of the same abnormalities found in autistic children; and the same brain function pathology is seen in those affected by mercury as in those affected by autism.

Mercury toxicity causes damage to the immune system and triggers autoimmune responses, as also occur in autism. A subset of autistic children has been found to have evidence of chronic viral infections, including the measles virus, another symptom shared by those with mercury exposure, which can increase susceptibility to certain virus strains. Mercury poisoning can cause gastrointestinal disturbances and inhibit digestive enzymes and peptides. Many children with autism develop gastrointestinal problems and have difficulty digesting dairy and wheat products.

In summary, according to Bernard, all the symptoms reported in the literature for autism have also been reported for mercury toxicity, and vice versa. It seems very likely that some children who are actually suffering from mercury toxicity could be given the diagnosis of "autism," which is simply a label indicating they have a communication/behavior/social disorder of unknown cause. Overall, infants have limited ability to excrete mercury, and children with autism have an unusually low ability to excrete mercury due to low glutathione and excessive oral antibiotics (for frequent viruses). Furthermore, antibiotics increase the toxicity of mercury, further complicating the picture.

Citations:

Abrahams, B. S. and Geschwind, D. H. (2008). Advances in autism genetics: On the threshold of a new neurobiology. *Nat. Rev. Genet.* 9(5). pg. 341-55.

Dawson, M., Mottron, L. and Gernsbacher, M. A. (2008). Learning in autism. In: Byrne, J. H., Ed. *Learning and memory: A comprehensive reference*. Vol. 2. Academic Press, Burlington, MA.

Folstein, S. E. and Rosen-Sheidley, B. (2001). Genetics of autism: Complex aetiology for a heterogeneous disorder. *Nat. Rev. Genet.* 2(12). pg. 943-55.

Montes, G. and Halterman, J. S. (2007). Psychological functioning and coping among mothers of children with autism: A population-based study. *Pediatrics.* 119(5).

Newschaffer, C. J., Croen, L. A., Daniels, J., et al. (2007). The epidemiology of autism spectrum disorders. *Annu. Rev. Public Health.* 28. pg. 235-58.

Rutter, M. (2005). Incidence of autism spectrum disorders: Changes over time and their meaning. *Acta Paediatr.* 94(1). pg. 2-15.

Volkmar, F. R., State, M. and Klin, A. (2009). Autism and autism spectrum disorders: Diagnostic issues for the coming decade. *J. Child Psychol. Psychiatry.* 50(1-2). pg. 108-15.

White, S. W., Oswald, D., Ollendick, T. and Scahill, L. (2009). Anxiety in children and adolescents with autism spectrum disorders. *Clin. Psychol. Rev.* 29(3). pg. 216-29.

Links

The Sustainable Hospitals Project: http://www.sustainablehospitals.org/cgi-bin/DB_Index.cgi

ATSDR Mercury: http://www.atsdr.cdc.gov/substances/toxsubstance.asp?toxid=24

Women's guide to eating fish safely:
http://www.ct.gov/dph/lib/dph/environmental_health/eoha/pdf/womens_guide_05.pdf

EPA national advice on mercury in fish: http://www.epa.gov/ost/fishadvice/factsheet.html

Center for Food Safety and Applied Nutrition: http://vm.cfsan.fda.gov/~dms/admehg.html

Health Care Without Harm mercury issues: http://www.noharm.org/us/mercury/issue

Mercury Policy Project: http://www.mercurypolicy.org

Appendix Q - Water Quality Data for Pharmaceuticals, Hormones, and Other Organic Wastewater Contaminants in U.S. Streams, 1999-2000

Barnes, Kimberlee K., Kolpin, Dana W., Meyer, Michael T., Thurman, E. Michael, Furlong, Edward T., Zaugg, Steven D. and Barber, Larry B. (2002). United States Geological Survey, Iowa City, IA. http://toxics.usgs.gov/pubs/OFR-02-94/

The following graph depicting the frequency of detection of a small selection of pharmaceutical and other anthropogenic ecotoxins in U. S. streams illustrates one component of a much larger pattern of contamination of the global atmospheric water cycle by human activities. All of the contaminants in this survey, including, surprisingly, caffeine and cholesterol, are the effluents of modern global consumer culture and were not present in streams before 1945 in any significant quantities, if at all. Even more ominous is the fact that most surface waters as well as most rainfall events contain the contaminant signals of thousands of other environmental chemicals (anthropogenic ecotoxins,) often at levels near or below their LOD (limit of detection). This USGS survey raises an important question: to what extent are these and other pharmaceutical ecotoxins biomagnified in the food webs that they inevitably enter?

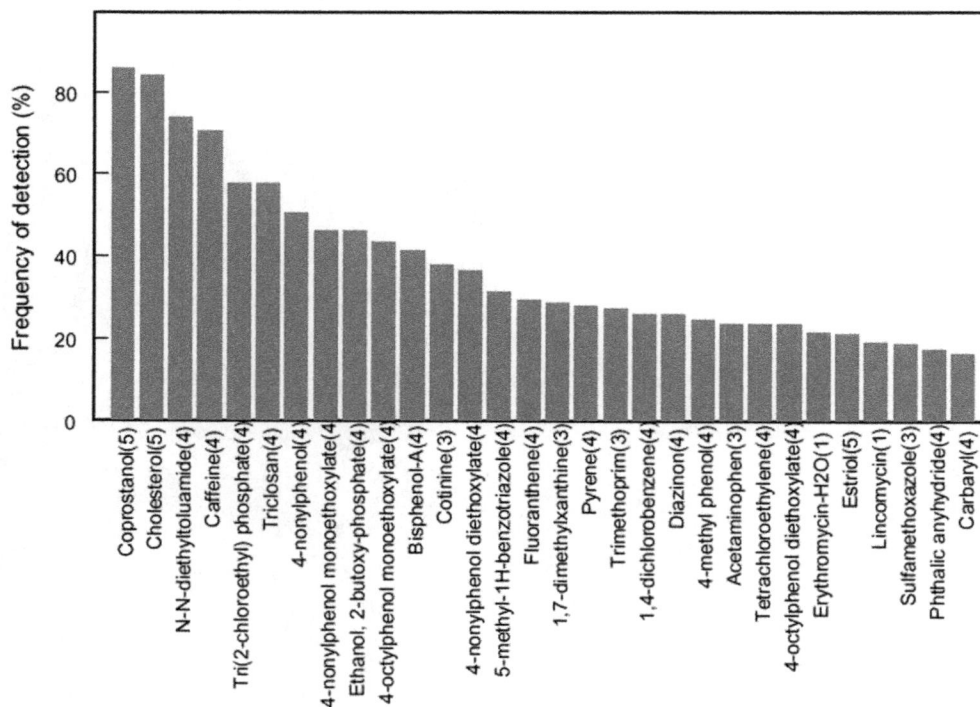

Figure 2. Most frequently detected compounds. The analytical method number is provided (in parentheses) at the end of each compound name.

Appendix R - Growth Hormones

The proliferation of the use of growth hormones, such as rBGH, in cows and cattle in the United States (but not in Europe) has been a very profitable bioengineering innovation for American dairy and cattle industries. The use of this growth hormone causes the well documented and well publicized increase of the naturally occurring hormone IGF-1 in the milk of treated cows. The health impact of elevated levels of IGF-1 is well documented; its presence has resulted in the banning of the sale of American dairy products in Europe. There exists no better example of the proliferation of a "toxic asset" in human food supplies, and thus, within the infrastructure of a large component of global consumer culture than this compelling example of the imposition of human technology on non-human organisms.

IGF-1 Levels in Milk (ng/mL, mean)

Treatment Cycle	rBGH Treated Cows	Untreated Cows	Difference
Week 1	3.80	3.22	18%
Week 2	5.39	2.62	106%
Week 3	4.98	3.78	31.7%

Source: Torkelson, A.R., Lanza, G.M., Birmingham, B.K., Vicini, J.L., White, T.C., Dyer, S.E., Madsen, K.S., and Collier, R.J. 1988. Concentrations of insulin-like growth factor 1 (IGF-1) in bovine milk: Effect of herd, stage of lactations, and sometribove, *Journal of Dairy Science,* 71: 52.

IGF-1 Levels in Milk (ng/mL, mean)

Treatment Cycle	rBGH Treated Cows	Untreated Cows	Difference
Week 1	3.50	3.17	10%
Week 2	5.33	3.34	60%
Week 3	4.68	3.35	40%

Source: Miller, M.A., Hildebrandt, J.R., White, T.C., Hammond, B.G., Madsen, K.S. and R.J. Collier. 1989. Determinations of IGF-I concentrations in raw, pasteurized, and heat-treated milk. Unpublished report MSL 8673 dated January 3, 1989, Monsanto Agricultural Company, St. Louis, MO.

Appendix S – IPPC's Carbon Dioxide Equivalency for Common Chemical Emissions (Global Warming Potential)

Industrial Designation or Common Name (years)	Chemical Formula	Lifetime (years)	Radiative Efficiency (W m^{-2} ppb^{-1})	Global Warming Potential for Given Time Horizon			
				SAR‡ (100-yr)	20-yr	100-yr	500-yr
Carbon dioxide	CO_2	See below[a]	[b]1.4×10^{-5}	1	1	1	1
Methane[c]	CH_4	12[c]	3.7×10^{-4}	21	72	25	7.6
Nitrous oxide	N_2O	114	3.03×10^{-3}	310	289	298	153
Substances controlled by the Montreal Protocol							
CFC-11	CCl_3F	45	0.25	3,800	6,730	4,750	1,620
CFC-12	CCl_2F_2	100	0.32	8,100	11,000	10,900	5,200
CFC-13	$CClF_3$	640	0.25		10,800	14,400	16,400
CFC-113	CCl_2FCClF_2	85	0.3	4,800	6,540	6,130	2,700
CFC-114	$CClF_2CClF_2$	300	0.31		8,040	10,000	8,730
CFC-115	$CClF_2CF_3$	1,700	0.18		5,310	7,370	9,990
Halon-1301	$CBrF_3$	65	0.32	5,400	8,480	7,140	2,760
Halon-1211	$CBrClF_2$	16	0.3		4,750	1,890	575
Halon-2402	$CBrF_2CBrF_2$	20	0.33		3,680	1,640	503
Carbon tetrachloride	CCl_4	26	0.13	1,400	2,700	1,400	435
Methyl bromide	CH_3Br	0.7	0.01		17	5	1
Methyl chloroform	CH_3CCl_3	5	0.06		506	146	45
HCFC-22	$CHClF_2$	12	0.2	1,500	5,160	1,810	549
HCFC-123	$CHCl_2CF_3$	1.3	0.14	90	273	77	24
HCFC-124	$CHClFCF_3$	5.8	0.22	470	2,070	609	185
HCFC-141b	CH_3CCl_2F	9.3	0.14		2,250	725	220
HCFC-142b	CH_3CClF_2	17.9	0.2	1,800	5,490	2,310	705
HCFC-123	$CHCl_2CF_3$	1.3	0.14	90	273	77	24
HCFC-124	$CHClFCF_3$	5.8	0.22	470	2,070	609	185
HCFC-141b	CH_3CCl_2F	9.3	0.14		2,250	725	220
HCFC-142b	CH_3CClF_2	17.9	0.2	1,800	5,490	2,310	705
HCFC-225ca	$CHCl_2CF_2CF_3$	1.9	0.2		429	122	37
HCFC-225cb	$CHClFCF_2CClF_2$	5.8	0.32		2,030	595	181
Hydrofluorocarbons							
HFC-23	CHF_3	270	0.19	11,700	12,000	14,800	12,200
HFC-32	CH_2F_2	4.9	0.11	650	2,330	675	205
HFC-125	CHF_2CF_3	29	0.23	2,800	6,350	3,500	1,100
HFC-134a	CH_2FCF_3	14	0.16	1,300	3,830	1,430	435
HFC-143a	CH_3CF_3	52	0.13	3,800	5,890	4,470	1,590
HFC-152a	CH_3CHF_2	1.4	0.09	140	437	124	38
HFC-227ea	CF_3CHFCF_3	34.2	0.26	2,900	5,310	3,220	1,040
HFC-236fa	$CF_3CH_2CF_3$	240	0.28	6,300	8,100	9,810	7,660
HFC-245fa	$CHF_2CH_2CF_3$	7.6	0.28		3,380	1030	314
HFC-365mfc	$CH_3CF_2CH_2CF_3$	8.6	0.21		2,520	794	241
HFC-43-10mee	$CF_3CHFCHFCF_2CF_3$	15.9	0.4	1,300	4,140	1,640	500
Perfluorinated compounds							
Sulphur hexafluoride	SF_6	3,200	0.52	23,900	16,300	22,800	32,600
Nitrogen trifluoride	NF_3	740	0.21		12,300	17,200	20,700
PFC-14	CF_4	50,000	0.10	6,500	5,210	7,390	11,200
PFC-116	C_2F_6	10,000	0.26	9,200	8,630	12,200	18,200

Industrial Designation or Common Name (years)	Chemical Formula	Lifetime (years)	Radiative Efficiency (W m⁻² ppb⁻¹)	Global Warming Potential for Given Time Horizon			
				SAR‡ (100-yr)	20-yr	100-yr	500-yr
Perfluorinated compounds (continued)							
PFC-218	C_3F_8	2,600	0.26	7,000	6,310	8,830	12,500
PFC-318	$c\text{-}C_4F_8$	3,200	0.32	8,700	7,310	10,300	14,700
PFC-3-1-10	C_4F_{10}	2,600	0.33	7,000	6,330	8,860	12,500
PFC-4-1-12	C_5F_{12}	4,100	0.41		6,510	9,160	13,300
PFC-5-1-14	C_6F_{14}	3,200	0.49	7,400	6,600	9,300	13,300
PFC-9-1-18	$C_{10}F_{18}$	>1,000ᵈ	0.56		>5,500	>7,500	>9,500
trifluoromethyl sulphur pentafluoride	SF_5CF_3	800	0.57		13,200	17,700	21,200
Fluorinated ethers							
HFE-125	CHF_2OCF_3	136	0.44		13,800	14,900	8,490
HFE-134	CHF_2OCHF_2	26	0.45		12,200	6,320	1,960
HFE-143a	CH_3OCF_3	4.3	0.27		2,630	756	230
HCFE-235da2	$CHF_2OCHClCF_3$	2.6	0.38		1,230	350	106
HFE-245cb2	$CH_3OCF_2CHF_2$	5.1	0.32		2,440	708	215
HFE-245fa2	$CHF_2OCH_2CF_3$	4.9	0.31		2,280	659	200
HFE-254cb2	$CH_3OCF_2CHF_2$	2.6	0.28		1,260	359	109
HFE-347mcc3	$CH_3OCF_2CF_2CF_3$	5.2	0.34		1,980	575	175
HFE-347pcf2	$CHF_2CF_2OCH_2CF_3$	7.1	0.25		1,900	580	175
HFE-356pcc3	$CH_3OCF_2CF_2CHF_2$	0.33	0.93		386	110	33
HFE-449sl (HFE-7100)	$C_4F_9OCH_3$	3.8	0.31		1,040	297	90
HFE-569sf2 (HFE-7200)	$C_4F_9OC_2H_5$	0.77	0.3		207	59	18
HFE-43-10pccc124 (H-Galden 1040x)	$CHF_2OCF_2OC_2F_4OCHF_2$	6.3	1.37		6,320	1,870	569
HFE-236ca12 (HG-10)	$CHF_2OCF_2OCHF_2$	12.1	0.66		8,000	2,800	860
HFE-338pcc13 (HG-01)	$CHF_2OCF_2CF_2OCHF_2$	6.2	0.87		5,100	1,500	460
Perfluoropolyethers							
PFPMIE	$CF_3OCF(CF_3)CF_2OCF_2OCF_3$	800	0.65		7,620	10,300	12,400
Hydrocarbons and other compounds – Direct Effects							
Dimethylether	CH_3OCH_3	0.015	0.02		1	1	<<1
Methylene chloride	CH_2Cl_2	0.38	0.03		31	8.7	2.7
Methyl chloride	CH_3Cl	1.0	0.01		45	13	4

Notes:

a The CO_2 response function used in this report is based on the revised version of the Bern Carbon cycle model used in Chapter 10 of this report (Bern2.5CC; Joos et al. 2001) using a background CO_2 concentration value of 378 ppm. The decay of a pulse of CO_2 with time t is given by

$$a_0 + \sum_{i=1}^{3} a_i \cdot e^{-t/\tau_i}$$

Where $a_0 = 0.217$, $a_1 = 0.259$, $a_2 = 0.338$, $a_3 = 0.186$, $\tau_1 = 172.9$ years, $\tau_2 = 18.51$ years, and $\tau_3 = 1.186$ years.

b The radiative efficiency of CO_2 is calculated using the IPCC (1990) simplified expression as revised in the TAR, with an updated background concentration value of 378 ppm and a perturbation of +1 ppm (see Section 2.10.2).

c The perturbation lifetime for methane is 12 years as in the TAR (see also Section 7.4). The GWP for methane includes indirect effects from enhancements of ozone and stratospheric water vapour (see Section 2.10.3.1).

d Shine et al. (2005c), updated by the revised AGWP for CO_2. The assumed lifetime of 1,000 years is a lower limit.

Source: IPCC Fourth Assessment Report by Working Group 1

Appendix T - Hydrofracking (Hydraulic Fracturing) for Natural Gas Recovery

The following excerpts from the New York State Water Resources Institute website, TEDX (The Endocrine Disruption Exchange), and the Pennsylvania Department of Environmental Protection provide an overview of the numerous ecotoxins that are used during the high pressure water injection process to fracture gas-containing shale rock formations. Hydrofracking has been used in western states for natural gas recovery for decades, but the recent discovery of significant natural gas deposits in the Marcellus shale deposits in New York, Pennsylvania, West Virginia, Maryland, and elsewhere brings the threat of both ground water and aquifer contamination, a frequent occurrence during hydrofracking, to the water resources of the city of New York, to mention only the largest population group impacted by the growing threat of this rapidly expanding technology. Of critical importance is the fact that oil and gas companies that use the hydraulic fracturing process are exempt from the regulations under the Safe Drinking Water Act that require disclosure of the chemicals used in the fracturing process. An act is pending in Congress that would require the disclosure of the identity of these chemicals: The Fracturing Responsibility and Awareness of Chemicals (FRAC) Act of 2009 introduced by Representative Diana DeGette (Democrat – CO) and Senator Bob Casey (Democrat – PA) (http://frwebgate.access.gpo.gov/cgi-bin/getdoc.cgi?dbname=111_cong_bills&docid=f:h2766ih.txt.pdf.)

Two important reports on hydrofracking have been presented by Amy Goodman on Democracy Now! The War and Peace Report on Alternative Public Radio: *Fracking and the Environment: Natural Gas Drilling, Hydraulic Fracturing and Water Contamination* (Goodman 2009) and *Congress to Investigate Safety of Natural Gas Drilling Practice Known as Hydraulic Fracturing* (Goodman 2010). In the February 2010 presentation Goodman interviewed Josh Fox director of *Gasland*, which recently won the Special Jury Prize for a Documentary at the Sundance Film Festival. This film presents a compelling narration of the widespread contamination that results from hydrofracking, including the ability of some households to turn their water faucets on, light a match, and burn off the resulting gas fumes, which have contaminated surface water supplies.

The New York State Water Resources Institute states on their website:
(http://wri.eas.cornell.edu/gas_wells_waste.html)

Hydrofracking Fluids

Hydrofracking fluids are injected into wells under pressure in order to create cracks or fractures in the rock formation. These cracks accelerate gas flow out of the rock and into the well. Hydrofracking fluids are created by adding a proppant (commonly sand)
204

to water. The role of the proppant is to keep the cracks from resealing once the hydrofracking fluid is withdrawn from the well. In addition to the proppant, several types of chemicals are added to the hydrofracking fluid to serve a number of purposes.

- A friction reducer is added to reduce the friction pressure during pumping operations.
- A surfactant is used to increase the recovery of injected water into a well.
- A biocide is used to inhibit the growth of organisms that could produce gases (particularly hydrogen sulfide) that could be dangerous as well as contaminate the methane gas.
- Scale inhibitors are used to control the precipitation of carbonates and sulfates.

TEDX

The Endocrine Disruption Exchange
P.O. Box 1407, Paonia, CO 81428
970-527-4082
www.endocrinedisruption.org
tedx@tds.net
http://www.endocrinedisruption.com/files/ProductsandChemicalsUsedinFracturing2-16-09.pdf

Products and Chemicals Used in Fracturing

February 2009

TEDX (The Endocrine Disruption Exchange) has compiled a list of the names of products and their chemical constituents reportedly used during the fracturing of natural gas wells. Nalco1, World Oil2 and J.D. Arthur 3 list the functional categories of these chemicals as follows:

• acids	• defoamers	• iron control	• resins
• biocides	• emulsifiers	• non-emulsifiers	• sand
• breakers	• fluid loss control	• pH control	• scale control
• clay stabilizers	• foamers	• polymers	• solvents
• corrosion inhibitors	• friction reducers	• pseudo-polymers	• surfactants
• crosslinkers	• gellants	• proppants	• viscosifiers

TEDX's list includes the names of 435 fracturing products containing 344 chemicals as of February 2009. Information about the products and the chemicals they contain came from several states and a variety of sources including Material Data Safety Sheets (MSDSs), state Emergency Planning and Community Right-to-Know Act (EPCRA) Tier II reports, Environmental Impact Statement and Environmental Assessment Statement disclosures, rule-making documents, and from accident and spill reports, etc. The quantity and quality of information varied among these data sources. TEDX makes no claim that the following information is complete either in the scope of the products used during fracturing operations, or in the chemical composition of the products.

Posted by Jeff (December 9, 2009) on www.planputnam.org
http://www.planputnam.org/ntm/2009/12/hydro-fracking-whats-in-it/

Thanks to Dr. Peter Rostenberg for passing this along: The Pennsylvania Department of Environmental Protection provides a list of chemicals used in fracking for Marcellus gas.

Frac Water Chemicals

Chemical Components (From MSDS)
2.2-Dibromo-3-Nitrilopropionamide Glycol Ethers
2-butoxyethanol Guar gum
2-methyl-4-isothiazolin-3-one Hemicellulase Enzyme
5-chloro-2-methyl-4-isothiazotin-3-one Hydrochloric Acid
Acetic Acid Hydrotreated light distillate
Acetic Anhydride Hydrotreated Light Distilled
Alphatic Acid Isopropanol
Alphatic Alcohol Polyglycol Ether Isopropyl Alcohol
Ammonia Persulfate Magnesium Nitrate
Aromatic Hydrocarbon Mesh Sand (Crystalline Silica)
Aromatic Ketones Methanol
Boric Acid Mineral Spirits
Boric Oxide Monoethanolamine
Butan-1-01 Petroleum Distallate Blend
Citric Acid Petroleum Distillates
Crystalline Silica: Cristobalite Polyethoxylated Alkanol (1)
Crystalline Silica: Quartz Polyethoxylated Alkanol (2)
Dazomet Polyethylene Glycol Mixture
Diatomaceus Earth Polysaccharide
Diesel (use discontinued) Potassium Carbonate
Ethane-1,2-diol Potassium Hydroxide
Ethoxlated Alcohol Prop-2-yn-1-01
Ethoxylated Alcohol Propan-2-01
Ethoxylated Octylphenol Propargyl Alcohol
Ethylene Glycol Propylene

206

Ethylhexanol Sodium Bicarbonate
Ferrous Sulfate Heptahydrate Sodium Chloride
Formaldehyde Sodium Hydroxide
Glutaraldehyde Sucrose
Tetramethylammonium Chloride

Frac Stage #1

Hydrochloric Acid
Propargyl Alcohol
Methanol
Acetic Acid
Acetic Anhydride

Frac Stage #2

Methanol
Boric Oxide
Petroleum Distallate Blend
Polysaccharide
Potassium Carbonate
Sodium Chloride
Potassium Hydroxide
Ethylene Glycol
Boric Acid
Sodium Bicarbonate
Monoethanolamine

Frac Stage #3

Hydrotreated light distillate
Ethoxylated Alcohol
Glutaraldehyde
Dazomet
Sodium Hydroxide
Methanol
Diesel (use discontinued)
2,2-Dibromo-3-Nitrilopropionamide
Polyethylene Glycol Mixture
Mesh Sand (Crystalline Silica)

Appendix U – Endocrine Disrupting Chemicals

The following information is excerpted from *Hormonal Chaos* by Sheldon Krimsky (2000).

Wingspread I consensus position

We are certain of the following.

— A large number of man-made chemicals that have been released into the environment, as well as a few natural ones, have the potential to disrupt the endocrine system of animals, including humans. Among these are the persistent, bioaccumulative, organohalogen compounds that include some pesticides (fungicides, herbicides, and insecticides) and industrial chemicals, other synthetic products, and some metals.

— Many wildlife populations are already affected by these compounds. The impacts include thyroid dysfunction in birds and fish; decreased fertility in birds, fish, shellfish, and mammals; decreased hatching success in birds, fish, and turtles; gross birth deformities in birds, fish, and turtles; metabolic abnormalities in birds, fish, and mammals; behavioral abnormalities in birds; demasculinization and feminization of male fish, birds, and mammals; defeminization and masculinization of female fish and birds; and compromised immune systems in birds and mammals.

— The patterns of effects vary among species and among compounds. Four general points can nonetheless be made: (1) the chemicals of concern may have entirely different effects on the embryo, fetus, or perinatal organism than on the adult; (2) the effects are most often manifested in offspring, not in the exposed parent; (3) the timing of exposure in the developing organism is crucial in determining its character and future potential; and (4) although critical exposure occurs during embryonic development, obvious manifestations may not occur until maturity.

— …The effects seen in utero DES-exposed humans parallel those found in contaminated wildlife and laboratory animals, suggesting that humans may be at risk to the same environmental hazards in wildlife.

(Krimsky 2000, 28-9).

The Experimental Domain. The environmental endocrine hypothesis posits and association between a class of chemicals defined by their functional effects and a variety of health consequences mediated through the human endocrine system. These

208

linkages are postulated and supported through one of more experimental systems, including:

— Epidemiological studies of humans in the general population

— Occupational epidemiological studies of targeted populations

— Animal bioassays

— Wildlife studies

— Clinical observations

— Exposure studies of unique populations

— In vitro cell culture

(Krimsky 2000, 122).

Preliminary List of Chemicals Associated with Endocrine System Effects in Animals and Humans or in Vitro

Known	Probable	Suspect
Atrazine	Alachlor	Aldicarb
Chlordanes	Aldrin	Butyl benzyl phthalate
Chlordecone (Kepone)*	Amitrole (Aminotriazole)	Tert-Butylhydroxy-anisole[†]
_____ (DDD)	Benomyl	p-sec-Butylphenol[†]
1,I-Dichloro-2,2-bis ethylene (DDE)	Bisphenol-A[†]	p-tert-Butylphenol[†]
Dichloro-diphenyl trichloroethane (DDT)	Cadmium*	Carbaryl
1,2-Dibromo-3-chloro-propane*	2,4-Dichlorophenoxy-acetic acid (2,4-D)	Cypermethrin
Dicofol (Kelthane)	Di(2-ethylhexyl)phthalate	2,4-Dichlorophenol[†]
Dieldrin	Endrin	Dicyclohexyl phthalate
Diethylstilbestrol (DES)*	Heptachlor	Di(2-ethylhexyl)adipate[†]
Dioxins (2,3,7,8-)	Heptachlor epoxide	Di-n-butyl phthalate[†]
Endosulfans	Hexachlorobenzene	Di-n-hexyl phthalate
Furans (2,3,7,8-)	ß-Hexachlorocyclo-hexane	Di-n-pentyl phthalate
Lindane	Lead*	Di-n-propyl phthalate

Known	Probable	Suspect
Methoxychlor	Maneb	Esfenvalerate
p-Nonylphenol	Mancozeb	Fenvalerate
Polychlorinated biphenyls (PCBs)	Mercury*	Malathion
Toxaphene	Methyl parathion	Methomyl
Tributyl tin	Metiram	Metribuzin
	Mirex	Nitrofen
	p-Octylphenol	Octachlorostyrene
	Parathion	p-iso-Pentylphenol[†]
	Pentachlorphenol	p-tert-Pentylphenol[†]
	Polybrominated biphenyls (PBBs)	Permethrin
	Styrene[*,†]	Polynuclear aromatic hydrocarbons (PAHs)
	2,4,5-Trichlorophenoxy-acetic acid (2,4,5-T)	Ziram
	Trifluralin	
	Vinclozolin	
	Zineb	

Source: Illinois Environmental Protection Agency (1997:3). (Krimsky 2000, 121)

*In animals and humans.

†In vitro.

Postulated Effects of Exposure to Environmental Endocrine Disrupters in Humans and Wildlife

1.	Reduction in sperm count/infertility	15. Wasting syndrome
2.	Testicular cancer	16. Ovarian cancer
3.	Prostate cancer	17. Diminished cognitive function
4.	Cryptorchidism (undescended testes in baby boys)	18. Intersex (feminized or demasculinized populations)
5.	Abnormally short penises	19. Breast cancer
6.	Endometriosis (occurrence of endometrial tissue in the pelvic cavity)	20. Thyroid deficiency

210

7. Impaired immune system/autoimmune disease

8. Abnormal testicular development

9. Premature breast development

10. Precocious puberty

11. Vitellogenin (egg yolk protein) in male fish

12. Lowered biosynthesis of steroids

13. Eggshell thinning

14. Altered sex ratio–diminished males

21. Assorted reproductive and developmental abnormalities

22. Goiter

23. Gynecomastia (excessive male breast development)

24. Decreased testosterone

25. Higher embryo mortality

26. Hypospadias (abnormal male urinary canal)

27. Behavioral abnormalities, such as ADHD

(Krimsky 2000, 125)

Appendix V - What Causes Autism Spectrum Disorders?

The rapidly accelerating emergence of a wide range of autism spectrum disorders and other related developmental defects and health issues (diabetes, obesity, asthma, immunological disorders, etc.) is one of the most unfortunate legacies of the age of chemical fallout. The following excerpts and citations from the growing literature on autism and endocrine disrupting chemicals summarizes the contemporary take on these issues by two of America's most noted analysts of the health impact of chemical fallout. The summary of Landrigan's article is followed by a summary of the health effects of endocrine disrupting chemicals, compiled by Theo Colborn and available via her comprehensive website on endocrine disrupting chemicals (TEDX) (http://www.ourstolenfuture.org/New/recentimportant.htm).

"What causes autism? Exploring the environmental contribution"
by Philip J. Landrigan (*Current Opinion in Pediatrics*, January 16, 2010)

Abstract

Purpose of review: Autism is a biologically based disorder of brain development. Genetic factors - mutations, deletions, and copy number variants - are clearly implicated in causation of autism. However, they account for only a small fraction of cases, and do not easily explain key clinical and epidemiological features. This suggests that early environmental exposures also contribute. This review explores this hypothesis.

Recent findings: Indirect evidence for an environmental contribution to autism comes from studies demonstrating the sensitivity of the developing brain to external exposures such as lead, ethyl alcohol and methyl mercury. But the most powerful proof-of-concept evidence derives from studies specifically linking autism to exposures in early pregnancy - thalidomide, misoprostol, and valproic acid; maternal rubella infection; and the organophosphate insecticide, chlorpyrifos. There is no credible evidence that vaccines cause autism.

Summary: Expanded research is needed into environmental causation of autism. Children today are surrounded by thousands of synthetic chemicals. Two hundred of them are neurotoxic in adult humans, and 1000 more in laboratory models. Yet fewer than 20% of high-volume chemicals have been tested for neurodevelopmental toxicity. I propose a targeted discovery strategy focused on suspect chemicals, which combines expanded toxicological screening, neurobiological research and prospective epidemiological studies.

Appendix W - The Health Physics Impact of Endocrine Disrupting Chemicals

The following list of the health effects of endocrine disrupting chemicals is reprinted from the website of *Our Stolen Future* by Theo Colborn, Dianne Dumanoski, and John Peterson Myers (http://www.ourstolenfuture.org/New/recentimportant.htm). Additional information about each potential health effect may be obtained by clicking on the link in this website, which is the most comprehensive survey of the impact of endocrine disrupting chemicals on human health in the medical literature.

Cancer

- Early DDT exposure linked to breast cancer later in life
- *Bisphenol A* interferes with treatment for prostate cancer
- *Bisphenol A* causes breast cancer in rats
- *Bisphenol A* increases sensitivity to carcinogen that causes breast cancer
- Epigenetic inheritance of adult chronic diseases, including tumors
- *Bisphenol A* tied to prostate cancer in rats
- *Bisphenol A* alters mammary gland development in mice
- Genotype increases breast cancer risk associated with PCBs
- Testicular cancer linked to environmental exposures early in life
- Are current declines in non-Hodgkin's lymphoma rates tied to chemical bans in the '70s?
- Mother's organochlorine contamination linked to son's testicular cancer
- Cancer and EDCs: are we asking the wrong question in human studies?
- Miscalculation revealed: we are not winning the war on cancer after all
- *Bisphenol A* stimulates proliferation of prostate cancer cells
- Seveso, Italy: breast cancer risk rises with dioxin exposure
- Childhood leukemia linked to household pesticides
- A 2001 review of breast cancer epidemiology
- Soy phytoestrogen causes uterine cancer, like diethylstilbestrol (DES)
- Low-level arsenic interferes with glucocorticoid's role in tumor suppression
- PCBs interact synergistically with virus to elevate risk of non-Hodgkin's lymphoma
- Dieldrin associated with increased risk of breast cancer
- Dieldrin linked to higher mortality rates in breast cancer victims

Behavior and intelligence

- Low doses of BPA alter sexual dimorphism in brain structure and behavior in mice
- Low doses of *bisphenol A* alter maternal behavior in mice
- A single low exposure to PBDE in the womb causes ADHD in pubertal rats
- Brain development patterns different for autistic children, starting in the womb
- High PBDE levels in California and Indiana
- 10-fold higher rate of autism in Atlanta found by CDC study

- Evidence builds that soaring autism rate in California is not an artifact
- ADHD associated with smaller brain size in children
- Dioxin and PCBs alter sex-specific behavior in children
- Dioxin at extremely low levels reduces motivation in rats
- Low level carbaryl disrupts frog behavior
- Brominated flame retardant disrupts "brain growth spurt" in mice
- Polybrominated flame retardants are powerful thyroid disrupters
- Nonylphenol alters dopamine uptake in hypothalamic cells in nonmontonic fashion
- Brominated flame retardants act as developmental neurotoxicants in mice

Endometriosis

- Strong links apparent in animal studies of dioxin and endometriosis
- Immune and endocrine disorders much more likely in women with endometriosis

Sperm count and other testicular maladies

- Phthalate linked to reduced sperm quality in men
- Phthalate levels linked to genital abnormalities in boys
- A single low exposure to PBDE in the womb reduces adult sperm count
- Dramatic link between sperm quality and pesticide exposure in mid-west men
- Phthalate levels linked to deterioration in semen quality
- PCB congener levels linked to deterioration in semen quality
- Phthalate exposure associated with DNA damage in men's sperm
- Big geographic differences in sperm quality among US men
- Testicular dysgenesis syndrome: linking testicular cancer, hypospadias, cryptorchidism and lower sperm quality to a common origin
- Are sperm maladies contributing to fertility declines?
- Cryptorchidism linked to estrogenic organochlorine exposure
- Genetic basis of endocrine disruption leading to cryptorchidism
- Expanded data set corroborates broad-based decline in sperm count
- Dramatically low levels of sperm count prevalent in young Danish men. This study is the best attempt at a representative sample of a population yet obtained in studies of human sperm count.
- PCBs exposure in the womb leads to lower sperm quality

Reproduction and fertility

- New CDC data confirm ubiquitous exposure to *bisphenol A*
- Neonatal BPA exposure causes uterine abnormalities in middle-aged mice, like DES
- BPA scrambles the chromosomes of grandchildren in mice
- Testosterone is declining in Massachusetts men
- Epigenetic inheritance of adult chronic diseases, including testicular dysfunction

- *Bisphenol A* linked to recurrent miscarriage in people
- *Bisphenol A* alters mammary gland development in mice
- Low dose effects on prostate development confirmed and extended
- Several xenoestrogens stimulate prolactin secretion at extremely low levels
- Risks of infertility higher in women using herbicides and fungicides
- Phthalate linked to preterm birth
- Research links herbicides used on wheat to birth defects in the Great Plains
- DDT in mother's serum linked to more difficulty for daughter to become pregnant, 30 yrs later
- Plastic compound causes chromosomal error that in people leads to miscarriage, Down Syndrome
- Common lawn herbicide mixture causes fetal loss in mice at very low exposure level
- Frog reproductive development is disrupted by extremely low levels of atrazine
- DDT linked to preterm birth. May have caused 15% of infant mortality in US during the 1960s
- A compilation of impacts of contamination on the timing of puberty in animals
- Pregnant women living within a few miles of fields where agricultural pesticides are used are at increased risk of losing their fetus
- DES daughters are more likely to have bad pregnancy outcomes, including increased rates of spontaneous abortion
- Eating contaminated fish impairs conception in women living around the Great Lakes
- Polybrominated biphenyls *in utero* associated with early menstruation

Immune system

- Phthalate tied to asthma in human study
- Phthalate increases allergic reaction in mice at low exposure levels
- Epigenetic inheritance of adult chronic diseases, including immune dysfunction
- Pesticide mixture suppresses tadpole immune system, causes infection
- Do phthalates cause asthma via endocrine disruption?
- Immune system impacts of dioxin linger for decades
- Low-level arsenic interferes with glucocorticoid and may undermine immune system function
- Children with higher (but still low) background exposure to PCBs suffer increased rates of childhood diseases
- PCBs interact synergistically with virus to elevate risk of non-Hodgkin's lymphoma

Wildlife and ecosystem

- Pesticide mixture suppresses tadpole immune system, causes infection
- Low doses of nonylphenol have profoundly adverse effects on oysters
- Atrazine induces gonadal abnormalities in wild frogs at very low levels
- Pesticides render frog immune system less able to resist deformity-causing parasite

- Frog malformities linked to parasites in American west
- Frog reproductive development is disrupted by extremely low levels of atrazine.
- Nitrogen fixation of plants disrupted by EDCs
- Salamander behavior disrupted by low level endosulfan
- Frog behavior disturbed by low level carbaryl
- Widespread feminization of wild Chinook salmon in the Columbia River
- *Orca* (killer whales) in British Columbia, Canada, heavily contaminated with PCBs
- Endocrine disruption by nonylphenol interferes with osmoregulatory change in Atlantic salmon, contributing to their endangerment.
- Salmon transport persistent bioaccumulative contaminants up-river during their migrations
- A single dose of DDT during development causes complete sex reversal in medaka, a species of fish. Chromosomal males become fertile females.

Low dose effects

- New CDC data confirm ubiquitous exposure to *bisphenol A*
- Scientists warn policymakers about adverse effects of *bisphenol A*
- Does the 'dose make the poison?'
- BPA scrambles the chromosomes of grandchildren in mice
- DEHP alters a brain enzyme important to male masculinization
- DEHP heightens allergic response in mice in non-monotonic fashion at low doses
- *Bisphenol A* increases sensitivity to carcinogen that causes breast cancer
- *Bisphenol A* tied to prostate cancer in rats
- Pesticide mixture suppresses tadpole immune system, causes infection
- Industry science on low doses effects of *bisphenol A* profoundly flawed
- Low doses of *Bisphenol A* cause insulin resistance in mice
- Scientists challenge irresponsible use of hormesis to loosen health standards
- Extremely low doses of *Bisphenol A* alters mammary gland development in mice
- Low dose effects on prostate development confirmed and extended
- Low doses of BPA alter sexual dimorphism in brain structure and behavior in mice
- Low doses of *bisphenol A* alter maternal behavior in mice
- Several xenoestrogens are as powerful as estradiol via cell membrane receptor
- Cadmium provokes estrogenic responses at extremely low levels of exposure
- *Bisphenol A* equipotent with estradiol in activating key gene transcription factor at <1ppb
- Analysis concludes high level testing cannot be used to detect low level effects
- Atrazine induces gonadal abnormalities in wild frogs at very low levels
- Common lawn herbicide mixture causes fetal loss in mice at very low exposure level
- *Bisphenol A* at nanomolar levels stimulates prostate cancer cell proliferation
- Frog reproductive development is disrupted by extremely low levels of atrazine
- Extremely low level dioxin undermines rat behavior
- Controversial low-dose impacts of *bisphenol A* on prostate confirmed independently

216

- Lab experiments find low level DDT and HCH exposure cause adverse effects in mice
- DDE and HCH found in amniotic fluid of 1/3rd of Los Angeles women measured
- Strains of mice differ dramatically in their genetic susceptibility to endocrine disruptors. One of the most widely used strains in tox testing is surprisingly insensitive, and thus inappropriate to use in establishing health risks.

Results important for regulatory science

- Scientists warn policymakers about adverse effects of *bisphenol A*
- With a non-monotonic dose-response curve, DEHP alters a brain enzyme important to male masculinization at environmental levels
- DEHP heightens allergic response in mice in non-monotonic fashion at low doses
- Epigenetic inheritance of adult chronic diseases across multiple generations
- Pesticide mixture suppresses tadpole immune system, causes infection
- Analysis concludes high level testing cannot be used to detect low level effects
- Lab contamination from polycarbonate plastic likely to have misled research
- Atrazine induces gonadal abnormalities in wild frogs at very low levels

Mixtures of chemicals

- Pesticide mixture suppresses tadpole immune system, causes infection
- Common lawn herbicide mixture causes fetal loss in mice at very low exposure level
- Weak xenoestrogens combine to have strong effects
- Even weak xenoestrogens can have additive effects in the presence of 17ß-estradiol
- Nitrate interacts with pesticides to produce unpredictable results in mice at levels often found in US drinking supplies, including increases in aggression. Strong synergy detected between PCBs and dioxin
- PCBs interact synergistically with virus to elevate risk of non-Hodgkin's lymphoma

Ubiquity of exposure
- New CDC data confirm ubiquitous exposure to *bisphenol A*
- Scientists warn policymakers about adverse effects of *bisphenol A*
- *Bisphenol A* ubiquitous in Americans
- Teflon-related chemicals found in people on 5 continents
- High levels of polybrominated flame retardants in American mothers and fetuses
- Nonylphenols ubiquitous in consumer food
- Brominated flame retardants common in sewage sludge used on farmland
- Phthalate exposure widespread in American public
- PCB exposure in remote Arctic is up to 70x higher than in areas closer to source of contaminant.
- POPs carried by transpacific winds to the west coast of North America
- Dioxin in Canadian arctic linked to specific sources to the south

Appendix X - Cord Blood Contaminants in Minority Newborns

The following table is from *Pollution in people: Cord blood contaminants in minority newborns* (Environmental Working Group 2009, 6-8). This survey is the most recent in the Environmental Working Group's (EWG) documentation of ecotoxins in humans. EWGs choice of chemicals of concern, such as Polychlorinated naphthalenes (PCNs), Polychlorinated dibenzodioxins and furans (PCDD/F), and Bisphenol A (BPA) provide a glimpse into what is, in fact, a very complicated labyrinth of anthropogenic ecotoxins circulating in a very vulnerable round-world biosphere. Many of the chemicals surveyed by EWG are now ubiquitous contaminants of our atmospheric water cycle; their crossplacental transfer to newborns of all races and social standings are their final stop as they move up the food chain and/or are ingested, inhaled, or absorbed directly by their mothers. The second component of this appendix is EGW's synopsis of the health problems associated with this cord blood survey.

This research demonstrates that industrial chemicals cross the placenta in large numbers to contaminate a baby before the moment of birth. Test results are shown below.

Chemicals Detected in Umbilical Cord Blood from 10 Minority Newborns

Chemical or chemical family	Geometric Mean (of the detections)	Range	Number of newborn umbilical blood samples with detections	Number of chemicals detected within chemical family
Metals [µg/dL (wet weight) in whole blood] - 3 of 3 found				
Lead [pollutant from lead-based paint in older homes, household dust, vinyl products; harms brain development and function]	0.348 µg/dL	(0.222 - 0.549)	10 of 10	NA
Mercury [pollutant from coal-fired power plants, mercury-containing products, and certain industrial processes; accumulates in seafood; harms brain development and function]	0.64 µg/dL	(0.09 - 3.91)	10 of 10	NA
Methylmercury [organic form of mercury typically found in contaminated fish and seafood]	0.513 µg/dL	(0.08 - 3.28)	10 of 10	NA

Chemical or chemical family	Geometric Mean (of the detections)	Range	Number of new-born umbilical blood samples with detections	Number of chemicals detected within chemical family
Polybrominated dibenzodioxins and furans (PBDD/F) [pg/g (lipid weight) in blood serum] – contaminants in brominated flame retardants; pollutants and byproducts from plastic production and incineration; accumulate in food chain; toxic to developing endocrine (hormone) system				
Brominated dioxin	34.57 pg/g	(0 - 41.8)	2 of 10	Tested for: 6 Found: 1
Brominated furan	292 pg/g	(0 - 1440)	4 of 10	Tested for: 6 Found: 5
Perfluorinated chemicals (PFCs) [ng/g (wet weight) in whole blood] - active ingredients or breakdown products of Teflon, Scotchgard, fabric and carpet protectors, food wrap coatings; global contaminants; accumulate in the environment and the food chain; linked to cancer, birth defects and more				
Perfluorochemicals (PFCs)	2.38 ng/g	(0.736 - 7.08)	10 of 10	Tested for: 13 Found: 6
Polybrominated diphenyl ethers (PBDEs) [ng/g (lipid weight) in blood serum] - flame retardant in furniture foam, computers and televisions; accumulates in the food chain and human tissues; adversely affects brain development and the thyroid				
Polybrominated diphenyl ethers (PBDEs)	7.28 ng/g	(3.05 - 15.1)	10 of 10	Tested for: 46 Found: 26 to 29[1]
Polychlorinated naphthalenes (PCNs) [ng/g (lipid weight) in blood serum] - wood preservatives, varnishes, machine lubricating oils, waste incineration; common PCB contaminant; contaminate the food chain; cause liver and kidney damage				
Polychlorinated naphthalenes (PCNs)	0.64 ng/g	(0.0743 - 3.43)	10 of 10	Tested for: 70 Found: 17 to 24[2]
Polychlorinated biphenyls (PCBs) [ng/g (lipid weight) in blood serum] - industrial insulators and lubricants; banned in the U.S. in 1976; persist for decades in the environment; accumulate up the food chain to humans; cause cancer and nervous system problems				
Polychlorinated biphenyls (PCBs)	22.1 ng/g	(9.68 - 39.7)	10 of 10	Tested for: 209 Found: 98 to 144[3]
Polychlorinated dibenzodioxins and furans (PCDD/F) [pg/g (lipid weight) in blood serum] - pollutants, by-products of PVC production, industrial bleaching and incineration; cause cancer in humans; persist for decades in the environment; very toxic to developing endocrine (hormone) system				
Chlorinated dioxin	52.6 pg/g	(5 - 383)	10 of 10	Tested for: 7 Found: 6
Chlorinated furan	16.3 pg/g	(0 - 278)	6 of 10	Tested for: 10 Found: 9
Bisphenol A (BPA) [ng/mL (wet weight) in blood serum] - building block of polycarbonate plastics and epoxy resins for thousands of consumer products, including baby bottles, drinking water containers, metal food and beverage can liners and dental sealants; linked to hormone disruption, birth defects and cancer				
Bisphenol A	2.8 ng/mL	(0 – 8.61)	9 of 10	NA
Brominated fire retardant [ng/g (lipid weight) in blood serum] - 1 of 1 found				
Brominated Fire Retardant	2986 ng/g	(0 - 3210)	3 of 10	NA

Chemical or chemical family	Geometric Mean (of the detections)	Range	Number of new-born umbilical blood samples with detections	Number of chemicals detected within chemical family
Perchlorate [µg/L (wet weight) in whole blood] - 1 of 1 found				
Perchlorate	0.231 µg/L	(0 - 0.6)	9 of 10	NA
Polycylic musks [ng/g (wet weight) in whole blood] – heavily used synthetic fragrances that mimic natural musk.				
Polycylic musks	0.835 ng/g	(0 - 2.74)	7 of 10	Tested for: 6 Found: 2

Notes:

1.) Numbers are expressed as a range because several PBDEs are tested for in pairs; a positive result may mean one or both are present. The range reflects the minimum and maximum number of possible positive results.

2.) Numbers are expressed as a range because many PCNs are tested for in groups of two or three chemicals; a positive result may mean that one, some, or all are present. The range reflects the minimum and maximum number of possible positive results.

3.) Numbers are expressed as a range because many PCBs are tested for in groups up to six at a time: a positive result may mean that one, some, or all are present. The range reflects the minimum and maximum number of possible positive results.

Chemical class and subclass	Concentrations of chemicals in umbilical cord blood from 10 newborns (average and range among individual umbilical cord blood samples)		Number of newborn umbilical blood samples with detections		
Metals [parts per billion wet weight]					
Methyl Mercury	0.947	(0.07 - 2.3)	10	of	10
Polybrominated dioxins and furans [parts per trillion lipid weight]					
Brominated dioxins	5.33	(0 - 53.3)	1	of	10
Brominated furans	50.5	(0 - 246)	7	of	10
Tetrabrominated dioxin	0	(0 - 0)	0	of	10
Pentabrominated dioxin	0	(0 - 0)	0	of	10
Hexabrominated dioxin	5.33	(0 - 53.3)	1	of	10
Octabrominated dioxin	0	(0 - 0)	0	of	10
Tetrabrominated furan	1.65	(0 - 11.1)	2	of	10
Pentabrominated furan	10.7	(0 - 48.5)	6	of	10
Hexabrominated furan	12.6	(0 - 73.3)	3	of	10
Heptabrominated furan	25.6	(0 - 118)	6	of	10
Octabrominated furan	0	(0 - 0)	0	of	10
Polychlorinated dioxins and furans [parts per trillion lipid weight]					
Chlorinated dioxins	53.4	(37 - 79.6)	10	of	10
Chlorinated furans	6.04	(0.758 - 35)	10	of	10
Tetrachlorinated dioxin	0	(0 - 0)	0	of	10
Pentachlorinated dioxin	0.291	(0 - 2.910)	1	of	10
Hexachlorinated dioxin	7.1	(3.79 - 12)	10	of	10
Heptachlorinated dioxin	8.92	(5.3 - 12.6)	10	of	10
Octachlorinated dioxin	37.1	(19.9 - 55)	10	of	10
Tetrachlorinated furan	0	(0 - 0)	0	of	10
Pentachlorinated furan	1.62	(0 - 8.660)	4	of	10
Hexachlorinated furan	2.31	(0.379 - 15.4)	10	of	10
Heptachlorinated furan	2.12	(0.379 - 11)	10	of	10
Octachlorinated furan	0	(0 - 0)	0	of	10
Organochlorine Pesticide (OC) [parts per trillion lipid weight]					
Organochlorine Pesticides (OCs)	18600	(8720 - 35400)	10	of	10
Perfluorochemical (PFCs) [parts per billion wet weight]					
Perfluorochemicals (PFCs)	6.17	(3.37 - 10.6)	10	of	10
Perfluorinated sulfonate	4.25	(2.26 - 7.760)	10	of	10
Perfluorinated carboxylic acid	1.92	(1.1 - 2.870)	10	of	10
Polyaromatic hydrocarbon (PAHs) [parts per trillion lipid weight]					
Polyaromatic hydrocarbon (PAHs)	285	(217 - 384)	5	of	5
Polybrominated diphenyl ether (PBDEs) [parts per trillion lipid weight]					
Polybrominated diphenyl ether (PBDEs)	6420	(1110 - 14200)	10	of	10
Dibrominated diphenyl ether	40.4	(0 - 82.7)	7	of	10
Tribrominated diphenyl ether	160	(75.6 - 303)	10	of	10
Tetrabrominated diphenyl ether	1660	(16.6 - 3950)	10	of	10
Pentabrominated diphenyl ether	574	(0 - 1750)	9	of	10
Hexabrominated diphenyl ether	1310	(272 - 7590)	10	of	10
Heptabrominated diphenyl ether	46.6	(12.2 - 117)	10	of	10
Octabrominated diphenyl ether	74.3	(41.2 - 134)	10	of	10
Nonabrominated diphenyl ether	859	(0 - 3250)	7	of	10
Decabrominated diphenyl ether	1700	(0 - 9630)	3	of	10
Polychlorinated biphenyl (PCBs) [parts per trillion lipid weight]					
Polychlorinated biphenyls (PCBs)	7880	(2990 - 19700)	10	of	10
Mono-PCB	95.3	(44.1 - 210)	10	of	10
Di-PCB	154	(0 - 304)	9	of	10
Tri-PCB	275	(41.3 - 540)	10	of	10
Tetra-PCB	366	(140 - 873)	10	of	10
Penta-PCB	671	(304 - 1300)	10	of	10
Hexa-PCB	2760	(766 - 6890)	10	of	10
Hepta-PCB	2400	(435 - 6870)	10	of	10
Octa-PCB	889	(172 - 2740)	10	of	10
Nona-PCB	191	(10.2 - 617)	10	of	10
Deca-PCB	75.4	(6.55 - 211)	10	of	10
Polychlorinated naphthalene (PCNs) [parts per trillion lipid weight]					
Polychlorinated naphthalenes (PCNs)	617	(295 - 964)	10	of	10
Monochlorinated naphthalene	65.7	(3.2 - 216)	10	of	10
Dichlorinated naphthalene	27.8	(1.1 - 79.3)	10	of	10
Trichlorinated naphthalene	164	(104 - 315)	10	of	10
Tetrachlorinated naphthalene	292	(127 - 409)	10	of	10
Pentachlorinated naphthalene	30	(2.2 - 64.5)	10	of	10
Hexachlorinated naphthalene	22.9	(2.2 - 111)	10	of	10
Heptachlorinated naphthalene	12.4	(0 - 68.4)	3	of	10
Octachlorinated naphthalene	2.81	(0 - 19.3)	2	of	10

Source: Chemical analyses of 10 umbilical cord blood samples conducted by AXYS Analytical Services (Sydney, BC) and Flett Research Ltd. (Winnipeg, MB).

41

221

Potential health problems associated with chemicals found in newborns

* Some chemicals are associated with multiple health impacts, and appear in multiple categories in this table.

Health Effect or Body System Affected	Number of chemicals found in 10 newborns tested that are linked to the listed health impact		
	Average number found in 10 newborns	Total found in all 10 newborns	Range (lowest and highest number found in individual newborns)
Cancer [1]	133	180 [2]	92 to 155
Birth Defects / Developmental Delays	151	208 [3]	101 to 176
Vision	1	1 [4]	0 to 1
Hormone System	153	211 [5]	104 to 179
Stomach Or Intestines	194	275 [6]	147 to 227
Kidney	128	174 [7]	84 to 149
Brain, Nervous System	157	217 [8]	108 to 183
Reproductive System	185	263 [9]	136 to 219
Lungs/breathing	144	200 [10]	93 to 170
Skin	159	226 [11]	115 to 187
Liver	40	46 [12]	30 to 45
Cardiovascular System Or Blood	162	226 [13]	117 to 190
Hearing	135	187 [14]	85 to 161
Immune System	130	177 [15]	89 to 151
Male Reproductive System	172	245 [16]	122 to 207
Female Reproductive System	142	196 [17]	92 to 168

222

Appendix Y – EWG US Cord Blood Contaminant Studies Summary 2009

The following appendix is another in the Environmental Working Group's (EWG) pathbreaking surveys of ecotoxins in humans. EWG has surveyed and summarized all known blood contaminant studies in the US as of 2009. The full report is available at: www.ewg.org/files/2009-Minority-Cord-Blood-Report.pdf. This appendix is a summary of their data, including date, chemicals, number of newborns tested, number of chemicals found, and mean total levels found. "To date, EWG has found 414 industrial chemicals, pollutants and pesticides in 186 people, from newborns to grandparents."

The following abbreviations will assist the reader in navigating the data reported by EWG and other researchers and in interpreting the results they discovered. This appendix cites the mean total levels in the studies cited by EWG, rather than including all quartile data, as reported in Gump (2008).

EWG (Environmental Working Group)
CBSL (Cord Blood Serum Lipids)
pg/g (picograms/gram) = trillionths of a gram per gram
ng/g (nanograms/gram) = billionths of gram per gram of CBSL
Mean: the arithmetic *mean* of a list of numbers is the sum of the list numbers divided by the number of items in the list
PAHs (Polyaromatic hydrocarbons)
PBDE (Polybrominated diphenyl ether)
PCB (Polychlorinated biphenyl)
PCDD (Polychlorinated dibenzodioxins)
PCDD/F (Polychlorinated dibenzodioxins and furans)
PCN (Polychlorinated naphthalene)
PV = Peak value
dL (deciliter) = 1/10th of a liter

1 liter of water weighs 1 kg (2.205lbs) at 4° C; this is approximately the same weight as a liter of cord blood.

Some of the following surveys have reporting units of µg/dL which approximately equals 1ug per 0.22 lbs of blood. Since there are 454 (453.592) gm in a lb, 0.22 lbs of blood contains 99.8 gms. Therefore, 1 deciliter of blood has a weight of 99.8. To express µg/dL as units per gram multiply the first reporting unit by 100.

Reporting Date	Chemical Subclass (Number of Chemicals Tested)	Number of Newborns Tested	Number of Chemicals Found	Mean Total Levels Found
EWG 2005	Brominated dioxin (12)	10	6-7	12 pg/g CBSL
EWG 2009	Brominated dioxin (12)	10	6	10.7 pg/g CBSL
EWG 2005	Chlorinated dioxin: Total PCDD (1 Group)	5	1	15 pg/g
EWG 2005	Chlorinated furan (17)	10	11	56.3 pg/g
EWG 2009	Chlorinated furan (17)	10	15	59.7 pg/g
EWG 2009	Brominated fire retardant (TBBPA) (1)	10	1	11 ng/g
Harvard 1984 (Rabinowith)	Cadmium (1)	94	1	0.045 µg /dl PV = 0.210 µg/dl
Gump 2008	Lead (1)	154	1	< 1.0 (1st quartile; n = 37); 1.1–1.4 µg/dL (2nd quartile; n = 39); 1.5–1.9 µg/dL (3rd quartile; n = 36); 2.0–6.3 µg/dL (4th quartile; n = 42)
Sagiv 2008	Lead (1)	527	1	1.45 µg/dL
Lederman 2008	Mercury (1)	289	1	7.82 µg/L
EWG 2009	Musk (2)	10	2	Galoxolide in 6 samples, 0.483 ng/g; Tonalide in 4 samples, 0.147 ng/g
Perera 2005	PAHs (1)	203	1	Not given

Reporting Date	Chemical Subclass (Number of Chemicals Tested)	Number of Newborns Tested	Number of Chemicals Found	Mean Total Levels Found
EWG 2005	PAHs (18)	5	9	279 ng/g
Herbstman 2007	PBDE (8)	297	7	94% contained at least one tested congener
Mazdai 2003	PBDE (6)	12	6	20-to-106 fold higher than levels reported in a similar study from Sweden
EWG 2005	PBDE (46)	10	27-32	4.53 ng/g
EWG 2009	PBDE (46)	10	26-29	72.9 ng/g
Herbstman 2008	PBDE (3)	288	3	All found in all subjects
Qiu 2009	PBDE	20	10	Not given
Herbstman 2007	PCB (35)	297	18	>99% (of samples) had at least one detectable PCB congener
Stewart 2003	PCB	10	98-147	Not given
EWG 2005	PCB (209)	10	98-147	6.2 ng/g
Sagiv 2008	PCB (51)	542	>4	Mean sum of PCB congeners 118,138,153, and 180 = 0.25 ng/g; TEF-weighted sum of monoortho PCB congeners 105, 118, 156, 167, and 189 = 6.75 pg/g
EWG 2005	PCN (70)	10	31-50	0.574 ng/g
EWG 2009	PCN (70)	10	17-24	0.637 ng/g

Reporting Date	Chemical Subclass (Number of Chemicals Tested)	Number of Newborns Tested	Number of Chemicals Found	Mean Total Levels Found
Whyatt 2003	Carbamate (3)	211	5	48% had exposure to 2-Isopropoxyphenol, 45% to carbofuran, and 36% to bendiocarb
Whyatt 2003	Fungicide (2)	211	4	83% of the babies had exposure to dicloran, 70% to phthalimide
Whyatt 2003	Herbacide (2)	211	5	38% had exposure to chlorthal-dimethyl and 20% had exposure to Alachor; all had exposure to at least one herbicide
Whyatt 2003	Imide (2)	211	1	83% had exposure to dicloran and 70% had exposure to phthalimide; all had exposure to at least one fungicide
Whyatt 2003	Mosquito Repellent (1)	211	1	33% of the babies had exposure to diethyltoluamide
Sagiv 2008	Organochlorine (2)	542	1	0.48 ng/g
EWG 2005	Organochlorine (28)	10	21	at least one in all 10 subjects
Whyatt 2003	Organophosphate Pesticides	211	8	49% had exposure to diazinon (mean 1.2 pg/g); 71% had exposure to chlorpyrifos (mean 4.7 pg/g); all babies had exposure to at least one of the organophosphates
Whyatt 2003	Pyrethroid (2)	211	2	7% had exposure to trans-permethrin and 13% had exposure to cis-permethrin

Reporting Date	Chemical Subclass (Number of Chemicals Tested)	Number of Newborns Tested	Number of Chemicals Found	Mean Total Levels Found
Apelberg 2007	PFC (10)	299	9	PFOS in 99% and PFOA in 100% of samples; eight other PFCs were detected at lesser frequency
EWG 2005	PFC (12)	10	9	5.86 ng/g
EWG 2009	PFC (13)	10	6	2.38 ng/g in whole blood
EWG 2009	Bisphenol A/BADGE (1)	10	1	2.18 ng/L
EWG 2009	Perchlorate (1)	10	1	0.209 ug/L

Appendix Z – Perchlorates in Breast Milk

One of many ecotoxins in breast milk, the presence of perchlorates in almost all samples of breast milk tested during the last ten years is a paradigm for the presence of numerous other environmental contaminants, both tested and untested, in the breast milk of humans as well as other species. Perchlorate is a biologically significant ecotoxin because it may inhibit iodide uptake both in the thyroids of women and in the thyroids of unborn and newborn children. Lactational and gestational exposure in newborns can also result in reduction in brain development and a wide variety of other reduced iodide uptake-related health issues. "Perchlorate can temporarily inhibit the thyroid gland's ability to absorb iodine from the bloodstream ('iodide uptake inhibition', thus perchlorate is a known goitrogen)." (Greer 2002). The Pearce study reprinted below is anomalous in failing to correlate perchlorate body burdens with iodide uptake. As noted, it is otherwise a most significant research report.

Small quantities of perchlorates occur naturally, but the majority of perchlorates in public drinking water supplies, breast milk, human blood, and other biotic media derive from anthropogenic activity. The most significant anthropogenic source is the use of lithium perchlorate in solid rocket fuel; the oxygen produced by the decomposition or combustion of lithium perchlorate is an important source of energy for spacecraft. As a result of rockets launched for space exploration and due to the worldwide proliferation of the military use of rockets of all kinds, as well as the use of perchlorates in fireworks and explosives, contamination of the atmospheric water supply by the hemispheric-wide distribution of perchlorates is now universal.

The first major study pertaining to perchlorate in pathways to human consumption was an EPA, 2002, perchlorate risk assessment that produced a reference dose (RfD) based on toxicology data. The EPAs original provisional reference dose for perchlorates in water was 1 part per billion. Shortly after this guideline was formulated by the EPA, California set a public health goal guideline of 6 parts per billion perchlorate in tap water. Massachusetts followed with a toxicity level of 2 parts per billion in milk. The Environmental Working Group has proposed a maximum safe level of perchlorate consumption in milk as 0.5 parts per billion.

In 2005, the National Academy of Science (NAS) reviewed the data on perchlorate consumption in the United States and its possible toxicological impact. The NAS issued guidelines that were much less stringent than those initially issued by the EPA in 2002, increasing the EPAs initial provisional RfD from 1 part per billion to 20 parts per billion. The NAS also delineated a safe perchlorate intake limit for babies of 4 µg/liter. As a result of the NAS report, the EPA issued a final risk assessment guideline of 24.5 µg/liter (ppb) for drinking water. Both the NAS analysis of data and suggested standards and the revised EPA guidelines have been the subject of extensive critical analysis by a number of researchers (Ginsberg 2005; EWG 2005).

Unlike many other anthropogenic ecotoxins, perchlorates, possibly due to their ubiquity and ease of measurement, have been the subject of extensive surveys. Among the most comprehensive analyses was the FDA (2004/2005) *Exploratory Survey Data on Perchlorate in Food*. This survey, which is available online, is summarized in 12 tables listing perchlorate values in lettuce, bottled water, milk, tomatoes, carrots, spinach, melon, fruit, fruit juice, vegetables, grains, and fish and seafood. California was a particular focus for lettuce and other vegetable samples due to the extensive known releases of percholates from US military installations in California and Nevada, which are now known to have contaminated the entire public water supply system in the state of California. The only good news in the FDA report is the lack of significant contamination in bottled water given a limit of detection of 0.50 ppb. In contrast, the FDA reported contamination of almost all raw 1% and whole milk samples tested, with peak values of contamination reported in California, Arizona, and Missouri, in the 9.0 – 10.4 range μg/liter (ppb), with a mean level of detection of about 5.75 μg/liter in samples ranging from New Jersey, Pennsylvania, and South Carolina west to Iowa, Kansas, Arizona, and California. The FDA did not test milk samples in every state. Among the most highly contaminated vegetables were organic carrots grown in Arvin, CA, coming in at 87.6 μg/liter. Samples of baby spinach from Santa Maria, CA, and two from Mexico had contamination levels of 111, 163, and 252 respectively. The highest levels reported were in Asian spinach from Riverside, CA at 680 μg/liter and Brawley, CA at 927 μg/liter, almost 1 ppm. In contrast, the highest reported values in catalopes (Goodyear, AZ) was 66.6 μg/liter. The good news in the FDA report was the nondetectable levels in most apples and low levels with a few exceptions in oranges, strawberries, watermelons, onions, and apple juice. Perchlorates were also nondetectable in most samples of corn meal, brown rice, and white rice, but appeared frequently in oatmeal and whole wheat flour. The FDA only measured a few specimens of fish or seafood, finding nondetectable levels in salmon from Bucks Harbor and Eastport, ME, and Canadian aquaculture facilities. Unfortunately shrimp from Thailand and Vietnam came in at 23.5 μg/liter and 50.5 μg/liter. The pervasive contamination of the world food supply with perchlorates is reflected in the extensive levels of perchlorate contamination documented in breast and dairy milk in the Texas survey summarized below, and in the controversial Boston study.

A survey sponsored by Texas Tech University at Lubbock (Kirk 2005) analyzed 47 dairy milk samples from grocery stores in 11 states and 36 breast milk samples from women in 18 states. Only one sample of dairy milk contained no detectable level of perchlorate. The average perchlorate concentration in breast milk was 10.5 μg/liter and in dairy milk was 2.0 μg/liter. This contrasts with the intial USEPA provisional suggested limit of 1.0 μg/liter in drinking water (2002). Both the FDA and Texas Tech studies have reporting levels in the context of a limit of detection for perchlorate in milk of 0.5 μg/liter in 2010.

Several years after the data collection by the Texas Tech team (Kirk 2005), Elizabeth N. Pearce and other researchers associated with the Boston University School of Medicine and the National Center for Environmental Health issued a survey of perchlorate concentrations in lactating Boston area women that graphically illustrates the tendency for anthropogenic ecotoxins to bioaccumulate in pathways to human consumption. Pearce and her team measured extraordinarily high peak value levels of perchlorates in breast milk (411 μg/liter), urine samples (127 μg/liter), with significantly lower peak value contamination levels in infant formula (4.1 μg/liter). The complete summary of Pearce report as published in the Journal of Clinical Endocrinology and Metabolism is reprinted below.

> **Context:** Breastfed infants rely on adequate maternal dietary iodine intake.
>
> **Objective:** Our objective was to measure breast milk iodine and perchlorate, an inhibitor of iodide transport into the thyroid and potentially into breast milk, in Boston-area women.
>
> **Participants:** The study included 57 lactating healthy volunteers in the Boston area.
>
> **Measurements:** Breast milk iodine and perchlorate concentrations and urine iodine, perchlorate, and cotinine concentrations were measured. For comparison, iodine and perchlorate levels in infant formulae were also measured.
>
> **Results:** Median breast milk iodine content in 57 samples was 155 μg/liter (range, 2.7–1968 μg/liter). Median urine iodine was 114 μg/liter (range, 25–920 μg/liter). Perchlorate was detectable in all 49 breast milk samples (range, 1.3–411 μg/liter), all 56 urine samples (range, 0.37–127 μg/liter), and all 17 infant formula samples (range, 0.22–4.1 μg/liter) measured. Breast milk iodine content was significantly correlated with urinary iodine per gram creatinine and urinary cotinine but was not significantly correlated with breast milk or urinary perchlorate.
>
> **Conclusions:** Perchlorate exposure was not significantly correlated with breast milk iodine concentrations. Perchlorate was detectable in infant formula but at lower levels than in breast milk. Forty-seven percent of women sampled may have been providing breast milk with insufficient iodine to meet infants' requirements. (Pearce 2007)

The Pearce study raises the question of what other ecotoxins can be documented in breast milk. The high levels of perchlorates documented by the Pearce study function as the canary in a coal mine; they are an indicator that other bioaccumulative ecotoxins will probably be documented in breast milk, maternal cord blood, and amniotic fluids in the fetal sack if we take the time and the expense to execute further biomonitoring studies. The Pearce study, along with the biomonitoring of CDC, Greenpeace, and EWG, prompts other questions. What are contemporary levels of methylmercury in breast milk, and are they increasing as methylmercury contamination of the biosphere becomes more pervasive? What endocrine disrupting chemicals can be identified in breast milk? Are the levels of these chemicals increasing over time, as appears to be the case with perchlorates? Will the CDC or an NGO monitoring group such as EWG execute a systematic survey of all known biologically significant ecotoxins in breast milk? The same question can be asked about ecotoxins in the cord blood (see *Appendices X* and *Y*).

Maximum concentration levels of environmental chemicals are found in breast milk, amniotic fluid, maternal cord blood, and thus the born and unborn children who are the dependent on what is essentially a highly contaminated atmospheric water cycle. With respect to human health, breast milk and maternal cord blood are the most biologically vulnerable components of the highest trophic (feeding) levels of the food webs of the biosphere. We may ignore high concentrations of ecotoxins in predators such as bald eagles or bluefin tuna, but the Pearce study suggests we should extend our surveillance of anthropogenic ecotoxins in humans to include many more environmental contaminants than just perchlorates.

Wikipedia is also a source of additional information on perchlorates and provides links to governmental information sources. "The US Environmental Protection Agency has issued substantial guidance and analysis concerning the impacts of perchlorate on the environment as well as drinking water. (http://www.epa.gov/safewater/contaminants/unregulated/perchlorate.html). California has also issued guidance regarding perchlorate use. (http://www.cdph.ca.gov/certlic/drinkingwater/Pages/Perchlorate.aspx)." (http://en.wikipedia.org/wiki/Perchlorate).

The following table pertaining to environmental chemicals in human breast milk samples is from the Agency for Toxic Substances and Disease Registry (2004) and illustrates the fact that the CDC is not the only source for biomonitoring information. Unfortunately, the data in this survey was collected in the years before 2001; this illustrates the urgent need to update the biomonitoring of ecotoxins in human breast milk.

Levels of Chemicals of Concern in Human Breast Milk Samples from General Populations

Chemical	Range of mean or median concentrations (ng/g lipid)	Newborn intake via breast milk[a] (µg/ kg/day)	Region	Reference
CDDs and CDFs	0.013–0.028[b]	0.00009–0.00057[c]	United States, Canada, Germany, New Zealand, Japan, Russia	Pohl and Hibbs 1996
CDDs and CDFs	0.162–0.485[b]	0.00115–0.00344[a]	South Vietnam (1970–1973)	Pohl and Hibbs 1996
Mercury (total)	130–793[c]	0.922–5.625	Japan, Germany, Sweden	Abadin et al. 1997
Hexachloro-benzene	5–63	0.035–0.447	New Zealand, Brazil, Arkansas, Australia, Canada, Mexico, Quebec Caucasians	Pohl and Tylenda 2000
Hexachloro-benzene	100–>1,000	0.709–>7.094	France, Spain, Quebec Inuits, Slovak Republic, Czech Republic	Pohl and Tylenda 2000
p,p'-DDE	500 1,200 2,000	3.547 8.513 14.188	Sweden 1989 Sweden 1979 Sweden 1967	Pohl and Tylenda 2000
p,p'-DDE	300–>3,000	2.128–21.281	New Zealand, Brazil, France, Australia, Quebec Caucasians and Inuits, Arkansas, Canada, Slovak Republic, Czech Republic, Germany, North Carolina	Pohl and Tylenda 2000; Rogan et al.1986a
PCBs	167–1,770	1.185–12.556	Japan, Quebec Caucasians and Inuits, New York, Michigan, Netherlands, Poland, Finland, Croatia, North Carolina	DeKoning and Karmaus 2000

[a]Converted from 0.6–3.6 µg Hg/dL, using a conversion factor of 45.4 g lipid/10 dL milk (DeKoning and Karmaus 2000). Organic forms accounted for about 7–50% of total mercury (Abadin et al. 1997).
[b]Measured in 2,3,7,8-TCDD toxic equivalents (TEQs).
[c]Calculated, based on assumptions of 3.2 kg body weight, 45.4 g fat/L milk and 0.5 milk/day (DeKoning and Karmaus 2000), as follows: 5 ng/g fat x 45.4 g fat/L x 0.5 L/day x 1/3.2 kg x 1 µg/1,000 ng=0.035 µg/kg/day.

VII. Bibliography

Bibliography -- Table of Contents

Ecology

Adams, Mary Kelly. (1999). *Toxics B. C.: A journey of imagination.* Self-published, St. George, ME.

Alverson, Dayton L., Freebert, Mark H., Murawski, Steven A. and Pope, J. G. (1994). *A global assessment of fisheries bycatch and discards.* FAO Technical Paper 339. Food and Agriculture Organization of the United Nations, NY.

American Chemistry Council. (2010). *The history of plastic.* http://www.americanchemistry.com/s_plastics/doc.asp?CID=1102&DID=4665.

Ames, Ted. (2004). Atlantic cod stock structure in the Gulf of Maine. *Fisheries.* 29(1). pg. 10-28. http://www.nefsc.noaa.gov/GARM-Public/2.%20Models%20Meeting/background/Atlantic%20Cod%20Structure%20in%20the%20Gulf%20of%20Maine%20_Ted%20Ames_Fisheries%20_Jan%202004.pdf.

Ames, Ted. (2010). Multispecies coastal shelf recovery plan: A collaborative, ecosystem-based approach. *Marine and Coastal Fisheries: Dynamics, Management, and Ecosystem Science.* 2. pg.217-31. http://www.penobscoteast.org/documents/C09-052.1GOMplan_001.pdf.

Anderson, John M., Whoriskey, Frederick G. and Goode, Andrew. (January/February 2000). Atlantic salmon on the brink. *Endangered Species Update.* 17(1). pg. 15-21. http://www.umich.edu/~esupdate/janfeb2000/anderson.htm.

Ashley, Clifford. (1938). *The Yankee whaler.* Houghton Mifflin, Boston, MA.

Associated Press, The. (August 18, 2008). City issues warning on abandoned boats. *The State.* http://www.thestate.com/local/story/493652.html.

Ausubel, Jesse H., Crist, Darlene Trew and Waggoner, Paul E. Eds. (2010). *First census of marine life 2010: Highlights of a decade of discovery.* Census of Marine Life International Secretariat, Washington, DC. http://www.coml.org/pressreleases/census2010/PDF/Highlights-2010-Report-Low-Res.pdf.

Baird, Joel Banner. (December 17, 2009). Bat death toll hits 90% from syndrome. *Burlington Free Press*. pg. B-1.

Barrett, G. W. and Rosenberg, Rutger, Eds. (1981). *Stress effects on natural ecosystems*. Wiley, NY.

Barringer, Felicity. (August 15, 2010). Fly fishers serving as transports for noxious little invaders. *The New York Times*. http://www.nytimes.com/2010/08/16/science/earth/16felt.html?scp=1&sq=fly%20fishers%20serving%20as%20transports&st=cse.

Baum, J. K. et al. (2003). Collapse and conservation of shark populations in the Northwest Atlantic. *Science*. 299. pg. 389-392.

Beder, Sharon. (2002). *Global spin: The corporate assault on environmentalism*. Green Books and Chelsea Green Publishing, White River Jct., VT.

Bender, F. L. (2003). *The culture of extinction: Toward a philosophy of deep ecology*. Humanity Books, Amherst, NY.

Berger, Tina L., Ed. (2009). *67th annual report of the Atlantic States Marine Fisheries Commission to the Congress of the United States and to the governors and legislators of the fifteen compacting states: 2008*. Atlantic States Marine Fisheries Commission, Washington, DC.

Berry, Wendell. (1978). *The unsettling of America: Culture & agriculture*. Avon, NY.

Blatt, Harvey. (2008). *America's food: What you don't know about what you eat*. MIT Press, Cambridge, MA.

Boesch, D. F. and Rabalais, Nancy N., Eds. (1990). *Long-term environmental effects of offshore oil and gas development*. Spon Press, London.

Borden, Richard J. Ed. (1985). *Human ecology, a gathering of perspectives: Selected papers from the first international conference of the Society for Human Ecology*. University of Maryland Press, College Park, Maryland.

Botkin, Daniel B. (1990). *Discordant harmonies: A new ecology for the twenty-first century*. Oxford University Press, NY.

Bowman, Leslie. (February/March 2011). Lobster growth trials among the work at Downeast Institute. *Working Waterfront*. 24(1). pg. 5.

Brand, Stewart. (2010). *Whole earth discipline: An ecopragmatist manifesto*. Viking Adult, NY.

Bright, Chris. (1998). *Life out of bounds: Bioinvasion in a borderless world*. W. W. Norton & Company, NY.

Butchart, S. H. M. et al. (2010). Global biodiversity: Indicators of recent declines. *Science*. 328. pg. 1164-8. http://www.sciencemag.org/content/328/5982/1164.abstract.

Cadbury, Deborah. (1999). *Altering Eden: The feminization of nature*. St. Martins Press, NY.

Callenbach, Ernest. (1975). *Ecotopia*. Banyan Tree Books, Berkeley, CA.

Callenbach, Ernest. (2008). *Ecology: A pocket guide*. University of California Press, Berkeley, CA.

Cameron, J. and Aboucher, J. (1991). The precautionary principle: A fundamental principle of law and policy for the protection of the global environment. *Boston College International Law Review*. 14. pg. 1-27.

Carson, Rachel. (1962). *Silent spring*. Houghton Mifflin Company, Boston, MA.

Casey, J. M. and Myers, R. A. (1998). Near extinction of a large, widely distributed fish. *Science*. 281. pg. 690-2.

Chiras, Daniel D. (2006). *Environmental science: A systems approach to sustainable development*. Jones & Bartlett Publishers, Boston, MA.

Chivian, Eric and Bernstein, Aaron, Eds. (2008). *Sustaining life: How human health depends on biodiversity*. Oxford University Press, NY.

Clover, Charles. (2006). *The end of the line: How overfishing is changing the world and what we eat*. New Press, NY.

Commoner, Barry. (1971). *The closing circle: Nature, man & technology*. Alfred A. Knopf, Inc., NY.

Collette, B. B. and Klein-MacPhee, G. (2002). *Fishes of the Gulf of Maine*. 3rd edition. Smithsonian Institute Press, Washington, DC.

Cottrell, D. (1971). The price of beef. *Environment*. 13. pg. 44-51.

Cronon, William (1983). *Changes in the land: Indians, colonists, and the ecology of New England*. Hill and Wang, NY.

Dantas, Gautam, Sommer, Morten O. A., Oluwasegun, Rantimi D. and Church, George M. (2008). Bacteria subsisting on antibiotics. *Science*. 302. pg. 100-103. www.sciencemag.org/cgi/content/abstract/320/5872/100.

Darling, Lois and Darling, Louis. (1968). *A place in the sun: Ecology and the living world*. William Morrow and Company, NY.

Darling, Frasier and Milton, John P. (1966). *Future environments of North America: Transformation of a continent*. Natural History Press, Garden City, NY.

Daskalov, Georgi M., et al. (June 19, 2007). Trophic cascades triggered by overfishing reveal possible mechanisms of ecosystem regime shifts. *Proceedings of the National Academy of Sciences*. pg. 10518-23.

Davis, Ronald B. (1966). Spruce-fir forests of the coast of Maine. *Ecological Monographs*. 36. pg. 79-94.

Dean, Cornelia. (November 2, 2006). Report warns of global collapse of fishing. *The New York Times*. http://www.nytimes.com/2006/11/02/science/03fishcnd.html?scp=1&sq=2006%20report%20warns%20of%20global%20collapse%20of%20fishing&st=cse

de Steiguer, J. E. (2006). *The origins of modern environmental thought*. The University of Arizona Press, Tucson, AZ.

Deese, Heather and Schmitt, Catherine. (June 2010). Fathoming: Tiny plankton, big problems. *Working Waterfront*. http://www.workingwaterfront.com/articles/Fathoming-Tiny-plankton-big-problems/13860/.

Deese, Heather and Schmitt, Catherine. (September 2010). Fathoming: Maine fish and birds in hot water? *Working Waterfront*. http://www.workingwaterfront.com/articles/Fathoming-Ancient-Fish-Modern-Methods/14034/.

Deese, Heather and Schmitt, Catherine. (December 2010-January 2011). Fathoming: An acid test for fisheries. *Working Waterfront*. 23 (10). pg. 7.

Devall, Bill and Sessions, George. (1985). *Deep ecology: Living as if nature mattered*. Gibbs M. Smith Inc., Salt Lake City, UT.

Dilworth, Craig. (2010). *Too smart for our own good: The ecological predicament of humankind*. Cambridge University Press, Cambridge, UK.

Doney, S. C. (2006). The dangers of ocean acidification. *Scientific American*. 294. pg. 58-65.

Dowie, M. (1998). What's wrong with *The New York Times* science reporting? *Nation*. 13/14. pg. 16-9.

Drengson, Alan. (1995). *Deep ecology movement: An introductory anthology*. North Atlantic Books, Berkeley, CA.

Duhigg, Charles and Roberts, Janet. (February 28, 2010). Toxic Waters: Rulings restrict Clean Water Act, foiling E.P.A. *The New York Times*. http://www.nytimes.com/2010/03/01/us/01water.html.

Durrell, Lee. (1986). *State of the ark*. GAIA Books, Bodeley Head, London.

ETC Group. (2007). *Extreme genetic engineering: An introduction to synthetic biology*. http://www.etcgroup.org/upload/publication/602/01/synbioreportweb.pdf.

Earle, Sylvia A. (1995). *Sea change: A message of the oceans*. G. P. Putnam's Sons, NY.

Earle, Sylvia A. (2009). *The world is blue: How our fate and the ocean's are one*. National Geographic, Washington, DC.

Economy, Elizabeth and Lieberthal, Kenneth. (March 3, 2009). Scorched earth: Will environmental risks in China overwhelm its opportunities? *Harvard Business Review*. http://hbr.org/2007/06/scorched-earth/ar/1.

Ehrlich, Paul R. (1968). *Population bomb*. Ballantine Books, NY.

Ehrlich, Paul R. (1986). *The machinery of nature*. Simon and Schuster, NY.

Ehrlich, Paul R. (1997). *A world of wounds: Ecologists and the human dilemma*. Ecology Institute, Oldendorf/Luhe, Germany.

Ehrlich, Paul R. and Ehrlich, Anne H. (1990). *The population explosion*. Simon and Schuster, NY.

Ehrlich, Paul R. and Ehrlich, Anne H. (1996). *Betrayal of science and reason*. Island Press, Washington, D.C.

Ehrlich, Paul R. and Ehrlich, Anne H. (2008). *The dominant animal: Human evolution and the environment*. Island Press, Washington, DC.

Ellstrand, N. C. (2003). *Dangerous liaisons? When cultivated plants mate with their wild relatives*. The Johns Hopkins University Press, Baltimore, MD.

Engineering and Public Policy Committee on Science, National Academy of Sciences, National Academy of Engineering, and Institute of Medicine. (2012). *Rising above the gathering storm:*

Developing regional innovation environments: A workshop summary. National Academies Press, Washington, DC.

Environmental Protection Agency. (1997). *Terms of environment: Glossary, abbreviations and acronyms*. United States Environmental Protection Agency, Washington, DC. www.epa.gov/OCEPAterms.

Farigone, Joseph, et al. (2008). Land clearing and the biofuel carbon debt. *Science*. 319. pg. 1235-38.

Field, C. B., Behrenfeld, M. J., Randerson, J. T. and Falkowsji, P. (1998). Primary production of the biosphere: Integrating terrestrial and oceanic components. *Science*. 281. pg. 237-40.

Fisher, Douglas. (March 10, 2005). The great experiment. *Inside Bay Area*. http://www.insidebayarea.com.

Flannery, T. F. (1994). *The future eaters*. Reed Books, Chatswood, NSW, Australia.

Flannery, Tim. (2002). *The eternal frontier: An ecological history of North America and its peoples*. Atlantic Monthly Press, Boston, MA.

Flatow, Ira. (March 12, 2010). Can biotech crops feed the developing world? *Talk of the Nation*. National Public Radio. http://www.npr.org/templates/story/story.php?storyId=124618560.

Ford, J. S. and Myers, R.A. (2008). A global assessment of salmon aquaculture impacts on wild salmonids. *PLoS Biology*. 6(2). http://www.lenfestocean.org/publications/FordMyersLOPRS2.08.pdf.

Franklin, H. Bruce. (2007). *The most important fish in the sea*. Island Press, Washington, DC.

Galuszka, Peter. (November 13, 2012). With China and India ravenous for energy, coal's future seems assured. *The New York Times*. http://www.nytimes.com/2012/11/13/business/energy-environment/china-leads-the-way-as-demand-for-coal-surges-worldwide.html.

Gleick, James. (1988). *Chaos: Making a new science*. Penguin Books, NY.

Goleman, Daniel. (2009). *Ecological intelligence: How knowing the hidden impacts of what we buy can change everything*. Broadway Business, NY.

Golley, F. B. (1993). *A history of the ecosystem concept in ecology*. Yale University Press, New Haven, CT.

Goodnough, Abby. (November 10, 2011). Panel votes to reduce a forage fish catch. *The New York Times*. pg. A17.

Greenberg, Paul. (2010). *Four fish: The future of the last wild food*. Penguin Press HC, NY.

Gross-Sorokin, Melanie Y., Roast, Stephen D. and Brighty, Geoffrey C. (2006). Assessment of feminization of male fish in English rivers by the Environment Agency of England and Wales. *Environmental Health Perspectives*. 114(S-1). pg. 147-51.

Gulf of Maine Area (GoMA) Census of Marine Life. (2009). *Atlantic cod (Gadus morhua) – distribution and abundance in relation to bottom temperature: Fall surveys, 1963-2004*. http://www.gulfofmaine-census.org/data-mapping/visualizations/atlantic-cod-and-temperature/.

Hagen, J. B. (1992). *An entangled bank: The origins of ecosystem ecology*. Rutgers University Press, Piscataway, NJ.

Hardin, G. (1993). *Living within limits*. Oxford University, UK.

Hardin, Garrett. (1968). The tragedy of the commons. *Science*. 162. pg. 1243-48.

Harmon, Amy. (July 28, 2013). A race to save the orange by altering its DNA: Contagion raging, Florida industry tries to build a better tree. *The New York Times*. pg. 1, 18-9. http://www.nytimes.com/2013/07/28/science/a-race-to-save-the-orange-by-altering-its-dna.html?pagewanted=all&_r=0.

Harris, Gardiner. (November 7, 2012). As dengue fever sweeps India, a lackluster response stirs experts' fears. *The New York Times*. pg. A8.

Harris, Michael. (1998). *Lament for an ocean: The collapse of the Atlantic cod fishery*. McClelland & Stewart, Inc. Toronto, Canada.

Hart Research Associates. (2009). *Nanotechnology, synthetic biology, & public opinion: A report of findings: Based on a national survey among adults*. Project on Emerging Nanotechnologies, The Woodrow Wilson International Center for Scholars, Washington, DC. http://www.nanotechproject.org/process/assets/files/8286/nano_synbio.pdf.

Hearne, Shelley A. (1984). *Harvest of unknowns*. Natural Resources Defense Council, Inc., San Francisco, CA. www.nrdc.org.

Heinberg, Richard. (2004). *Power down: Options and actions for a post-carbon world*. New Society Publishers, Gabriola Island, BC, Canada.

Heinrich, Bernd. (2010). *The nesting season: Cuckoos, cuckolds, and the invention of monogamy*. Belknap Press of Harvard University Press, Cambridge, MA.

Hoare, Philip. (2010). *The whale: In search of the giants of the sea*. Harper Collins, NY.

Hutchings, J. A. and Meyers, R. A. (1995). *The biological collapse of Atlantic Cod off Newfoundland and Labrador*. Vol. 3. An Island Living Series. Institute of Island Studies, Charlotte, P.E.I., Canada.

Humes, Edward. (2009). *Eco barons: The dreamers, schemers, and millionaires who are saving our planet*. Harper Collins, NY.

Innis, Harold A. (1940). *The cod fisheries: The history of an international economy*. Yale University Press, New Haven, CT.

International Water Management Institute. (July 10, 2008). Water: The forgotten crisis. http://www.iwmi.cgiar.org/News_Room/pdf/Water_The_forgotten_crisis.pdf.

Iudicello, Suzanne, Weber, Michael L. and Wieland, Robert. (1999). *Fish, markets, and fishermen: The economics of overfishing*. Island Press, Washington, DC.

Jackson, Jeremy B. C. et al. (July 2001). Historical overfishing and the recent collapse of coastal ecosystems. *Science*. 293. pg. 629-37.

Jacobson, G. L., Fernandez, I. J., Mayewski, P. A. and Schmitt, C. V. eds. (2009). *Maine's climate future: An initial assessment*. University of Maine, Orono, ME. http://www.climatechange.umaine.edu/mainesclimatefuture/.

Johnson Foundation at Wingspread. (September 1-3, 2009). *The Johnson Foundation Environmental Forum: Examining U.S. freshwater systems and services: Agriculture and food production*.

http://www.johnsonfdn.org/sites/default/files/conferences/whitepapers/10/03/30/TJF_AgricultureConference_0.pdf.

Jowit, Juliette. (September 29, 2010). One in five plant species face extinction. *Guardian News*. http://www.guardian.co.uk/environment/2010/sep/29/plant-species-face-extinction.

Kanner, Allen D., et al. (1995). *Ecopsychology: Restoring the Earth, healing the mind.*

Kareiva, P., Watts, S., McDonald, R. and Boucher, T. (2007). Domesticated nature: Shaping landscapes and ecosystems for human welfare. *Science*. 316. pg. 1866-9.

Katz, Eric, Light, Andrew and Rothenberg, David, Eds. (2000). *Beneath the surface: Critical essays in the philosophy of deep ecology*. MIT Press, Cambridge, MA.

Keefe, Jennifer. (July 18, 2010). Debate over wastewater, nitrogen rules continues. *Foster's Sunday Citizen*. http://www.fosters.com/apps/pbcs.dll/article?AID=/20100718/GJNEWS_01/707189857/-1/fosnews1404.

Keim, Brandon. (2009). 7 tipping points that could transform earth. *Wired Science*. http://www.wired.com/wiredscience/2009/12/tipping-elements/.

Kirgan, Harlan. (June 30, 2010). Researchers find evidence of oil spill in Gulf's food chain. *Gulflive.com*. http://blog.gulflive.com/mississippi-press-news/2010/06/research_discovers_oil_droplet.html.

Kletz, Trevor. (1993). *Lessons from disaster: How organizations have no memory and accidents recur*. Institution of Chemical Engineers, Great Britain.

Koerner, Brendan I. (February 22, 2010). Red menace: Stop the Ug99 fungus before its spores bring starvation. *Wired Magazine*. http://www.wired.com/magazine/2010/02/ff_ug99_fungus/.

Kongtorp, R. T., Kjerstad, A., Taksdal, T., Guttvik, A., and Falk, K. (2004). Heart and skeletal muscle inflammation in Atlantic salmon, Salmo salar L: A new infectious disease. *J Fish Dis*. 27. pg. 351-8.

Kormondy, Edward J. (1969). *Concepts of ecology*. Prentice Hall Inc., Englewood Cliffs, NJ.

Kunstler, James Howard. (2006). *The long emergency: Surviving the end of oil, climate change, and other converging catastrophes of the twenty-first century*. Grove Press, NY.

Lack, Larry. (August 2009). New Brunswick sea lice pesticide treatment generates opposition. *Fundy Tides*.

Lash, Jonathan. (2006). *Fisheries exhausted in a single generation*. World Resources Institute. http://www.wri.org/stories/2006/10/fisheries-exhausted-single-generation.

Leakey, Richard E. and Lewin, Roger. (1996). *The sixth extinction: Patterns of life and the future of humankind.* Anchor, NY.

Leonard, Annie. (2010). *The story of stuff: How our obsession with stuff is trashing the planet, our communities, and our health – and a vision for change*. Free Press, NY.

Leopold, Aldo. (2001). *A sand county almanac*. Oxford University Press, NY.

Liniger, Hanspeter and Critchley, William, Eds. (2007). *Where the land is greener: Case studies and analysis of soil and water conservation initiatives worldwide*. United Nations Environment Programme.

Loh, Jonathan and Wackernagel, Mathis, Eds. (2004). *Living planet report 2004*. World Wide Fund for Nature, Gland, Switzerland. assets.panda.org/downloads/lpr2004.pdf.

Lyne, James W. and Barak, Phillip. (2000). *Are depleted soils causing a reduction in the mineral content of food crops?* ASA/CSSA/SSSA Annual Meetings, Nov. 5 - 9, Minneapolis, MN. http://www.soils.wisc.edu/~barak/poster_gallery/minneapolis2000a/index.html.

Mackay, Richard. (2008). *The atlas of endangered species*. University of California Press, Berkeley, CA.

Maine People's Resource Center. (2004). *The persistent polluters of Maine: The routine release of persistent, bioaccumulative and toxic chemicals (PBTs) to air, water and land by major industries in Maine*. Maine People's Resource Center, Environmental Health Strategy Center, and the Alliance for a Clean and Healthy Maine, Bangor, ME.

Marchuk, G. I., Kagan, B. A. (1989). *Dynamics of ocean tides*. Kluwer Academic Publishers. pg. 225.

Marks, William E. (2001). *The holy order of water: Healing the Earth's waters and ourselves*. Bell Pond Books, Herndon, VA.

Mayer, A-M. (1997). Historical changes in the mineral content of fruits and vegetables: A cause for concern? In: *Agriculture production and nutrition: Proc. Sept. 1997*. Lockeretz, W. Ed. Tufts University, Medford, MA.

McDonough, William and Braungart, Michael. (2002). *Cradle to cradle: Remaking the way we make things*. North Point Press, NY.

McKibben, Bill. (1989). *The end of nature*. Random House, NY.

McKibben, Bill. (2010). *Eaarth: Making a life on a tough new planet*. Times Books, NY.

McNaughton, S. J. and Wolf, L. L. (1979). *General Ecology*. Holt, Rinehart and Winston, NY.

Meadows, Donella H., Meadows, Dennis L., Randers, Jorgen and Behrens, William W., III. (1972). *The limits to growth, a report to the Club of Rome*. Club of Rome, Italy.

Mesarovic, M. and Pestel, E. (1974). *Mankind at the turning point*. Club of Rome, Italy.

Millennium Ecosystem Assessment. (2003). *Ecosystems and human well-being: A framework for assessment*. Island Press, Chicago, IL.

Millennium Ecosystem Assessment. (March 30, 2005). *Living beyond our means: Natural assets and human well-being*. United Nations Environment Programme. http://www.millenniumassessment.org/en/index.aspx.

Mitchell, Alanna. (2009). *Seasick: Ocean change and the extinction of life on earth*. The University of Chicago Press, Chicago.

Mlot, Christine. (February 1985). Multimedia maneuvers: Shifting tactics for controlling shifting pollutants. *Science News*. 127. pg. 124-127.

Montaigue, Fen. (2010). *Fraser's penguin: A journey to the future in Antarctica*. John Mcrae Books, NY.

Montgomery, David R. (2007). *Dirt: The erosion of civilizations*. University of California Press, Berkeley, CA.

Muir, Diana. (2000). *Reflections in Bullough's Pond: Economy and ecosystem in New England.* University Press of New England, Hanover, NH.

Myers, R. A., Hutchings, J. A. and Barrowman, N. J. (1997). Why do fish stocks collapse? The example of cod in Atlantic Canada. *Ecological Applications.* 7. pg. 91-106.

Myers, Ransom A. and Worm, Boris. (May 15, 2003). Rapid worldwide depletion of predatory fish communities. *Nature.* 423. pg. 280-3.

Naess, Arne. (1989). *Ecology, community and lifestyle: Outline of an ecosophy.* Rothenberg, David trans. Cambridge University Press, UK.

Naess, Arne. (2002). *Life's philosophy: Reason and feeling in a deeper world.* University of Georgia Press, Athens, GA.

Naess, Arne. (2008). *The ecology of wisdom: Writings by Arne Naess, Alan Drengson, and Bill Devall.* Counterpoint, Berkeley, CA.

National Research Council. (1990). *Sea-level change.* Geophysics Study Committee, Commission on Physical Sciences, Mathematics, and Resources, National Research Council. National Academy Press, Washington, DC.

Novacer, Michael J. (2001). *The biodiversity crisis, losing what counts.* American Museum of Science, NY.

O'Brien, Mary. (2000). *Making better environmental decisions.* MIT Press, Cambridge, MA.

Odum, Eugene P. (1953). *Fundamentals of ecology.* Saunders, Philadelphia, PA.

Odum, Eugene P. (1975). *Ecology.* Holt, Rinehart and Winston, Inc, NY.

Odum, Eugene P. (1983). *Basic ecology.* Saunders College Publishing, NY.

Odum, Eugene P. (1993). *Ecology and our endangered life support systems.* Sunderland, MA.

Old, R. W. and Primrose, S. B. (1980). *Principles of gene manipulation: An introduction to genetic engineering.* University of California Press, Berkeley, CA.

Organization of Economic Co-operation and Development. (2003). *Voluntary approaches for environmental policy.* Paris. www.oecd.org/document/58/0,2340,fr_2649_34375_238437_1-1-1-1,00.html

Osborn, F. (1948). *Our plundered planet.* Little, Brown, Boston, MA.

PR Web. (August 7, 2007). How to solve the growing national problem of abandoned boats. *PR Web Press Release Newswire.* http://www.prweb.com/releases/2007/08/prweb544507.htm.

Pandolfi, J. M. et al. (2003). Global trajectories of the long-term decline of coral reef ecosystems. *Science.* 301. pg. 955-8.

Palacios, Gustavo, Lovoll, Marie, Tengs, Torstein, Hornig, Mady, Hutchison, Stephen, Hui, Jeffrey, Kongtorp, Ruth-Torill, Savji, Nazir, Bussetti, Ana V., Solovyov, Alexander, Kristoffersen, Anja B., Celone, Christopher, Street, Craig, Trifonov, Vladimir, Hirschberg, David L., Rabadan, Raul, Egholm, Michael, Rimstad, Espen and Lipkin, W. Ian. (July 9, 2010). Heart and skeletal muscle inflammation of farmed salmon is associated with infection with a novel reovirus. *PLoS ONE.* 5(7). http://www.plosone.org/article/info%3Adoi%2F10.1371%2Fjournal.pone.0011487.

Parr, Jeffrey F. and Sullivan, L. A. (2005). Soil carbon sequestration in phytoliths. *Soil Biology and Biochemistry*. 37. pg. 117-124. http://www.scu.edu.au/schools/esm/palaeo/Parr&Sullivan.pdf

Passmore, J. (1974). *Man's responsibility for nature*. Duckworth, London.

Pauley, D. and Christensen, V. (1995). Primary production required to sustain global fisheries. *Nature*. 374. pg. 255-7.

Pauley, D., Christensen, V. Dalsgard, J. Froese, R. and Torres, F. Jr. (1998). Fishing down marine food webs. *Science*. 279. pg. 860-3.

Paulson, Tom. (January 22, 2008). The lowdown on topsoil: It's disappearing: Disappearing dirt rivals global warming as an environmental threat. *Seattlepi*. http://www.seattlepi.com/local/348200_dirt22.html.

Pender, Geoff. (July 1, 2010). Oil found in Gulf crabs raises new food chain fears. *Biloxi Sun Herald*. http://www.mcclatchydc.com/2010/07/01/96909/oil-found-in-gulf-crabs-raising.html#storylink=misearch.

Pilson, D. and Prendeville, H. R. (2004). Ecological effects of transgenic crops and the escape of transgenes into wild populations. *Annual Review of Ecology, Evolution, and Systematics*. 35. pg. 149-74).

Pimm, Stuart L. (1982). *Food webs*. Chapman and Hall, NY.

Raffensberger, C. and Tickner, J., Eds. (1999). *Protecting public health & the environment: Implementing the precautionary principle*. Island Press, Washington, DC.

Ramsey, Paul. (1970). *Fabricated man: The ethics of genetic control*. Yale University Press, New Haven, CT.

Real, Natalia. (July 12, 2010). Salmon virus potential threat to wild stocks. *Fish Information & Services*. http://www.fis.com/fis/worldnews/worldnews.asp?l=e&ndb=1&id=37248.

Rex, Erica. (October 4, 2010). Scientist at work: Vance Vredenburg: Toiling to save a threatened frog. *The New York Times*. http://www.nytimes.com/2010/10/05/science/05frog.html?_r=3&scp=1&sq=Chytri%20diomycosis%20&st=cse.

Rifken, Jeremy. (1980). *Entropy: a new world view*. Viking, NY.

Rissler, Jane and Mellon, Margaret. (1996). *The ecological risks of engineered crops*. MIT Press, Cambridge, MA.

Rivlin, Michael A. (Fall 2001). Northern exposure. *Onearth*. 23(3). pg. 14-20.

Rockstrom, Johan et al. (September 24, 2009). A safe operating space for humanity. *Nature*. 461. pg. 472-5.

Rogers, Peter and Leal, Susan. (2010). *Running out of water: The looming crisis and solutions to conserve our most precious resource*. Palgrave Macmillan, NY.

Roseboro, Ken, Ed. (April 2008). *Oops, another unapproved GM corn gets into the food supply*. Organic Consumers Association, www.organicconsumers.org.

Rosenthal, Elisabeth. (August 23, 2010). In the fields of Italy, a conflict over corn. *The New York Times*.

http://www.nytimes.com/2010/08/24/world/europe/24modify.html?scp=1&sq=in%20the%20fields%20of%20italy&st=cse.

Rosenthal, Elisabeth. (January 21, 2011). For many species, no escape as temperature rises. *The New York Times*. http://www.nytimes.com/2011/01/22/science/earth/22kenya.html.

Rosenthal, Elisabeth. (November 11, 2012). U.S. to be world's top oil producer in five years. *The New York Times*. http://www.nytimes.com/2012/11/13/business/energy-environment/report-sees-us-as-top-oil-producer-in-5-years.html.

Royal Society & The Royal Academy of Engineering. (2004). *Nanoscience and nanotechnologies: Opportunities and uncertainties*. The Royal Society & The Royal Academy of Engineering, London. http://www.nanotec.org.uk/finalReport.htm.

Sabini, Meredith. (2002). *The Earth has a soul: The nature writings of C.G. Jung*.

Safina, Carl. (2012). *The view from Lazy Point: A natural year in an unnatural world*. Picadore, London.

Schellnhuber, Hans Joachim. (December 8, 2009). Tipping elements in the earth system. *Proceedings of the National Academy of Sciences*. 106(49).

Schlesinger, W. H. (1991). *Biogeochemistry: An analysis of global change*. Academic Press, Burlington, MA.

Schumacher, E. F. (1989). *Small is beautiful: Economics as if people mattered*. Harper Perennial, NY.

Sessions, George, Ed. (1995). *Deep ecology for the twenty-first century*. Shambhala, Boston, MA.

Sheikh, Pervaze A. (October 18, 2005). *The impact of hurricane Katrina on biological resources*. CRS Report for Congress. The Library of Congress, Washington, DC.

Shelford, Victor E. (1963). *The ecology of North America*. University of Illinois Press, Urbana, IL.

Small, Meredith and Broude, Sylvia. (2008). *Clear as a lake: A resource guide to invasive aquatic plants and non-toxic treatment alternatives*. Toxics Action Center, Boston, MA.

Solnit, Rebecca. (2009). *A paradise built in hell: The extraordinary communities that arise in disaster*. Viking Adult, NY.

Souder, William. (2000). *A plague of frogs*. Hyperion Press, NY.

Steele, J. H. and Schumacher, M. (2000). Ecosystem structure before fishing. *Fisheries Research*. 44. pg 201-5.

Steffen, Konrad. (2010). *Olaf Otto Becker: Above zero*. Hatje Cantz, Berlin, Germany.

Steingraber, Sandra. (2001). *Having faith: An ecologist's journey to motherhood*. Perseus Publishing, Cambridge, MA.

Stokstad, Erik. (November 3, 2006). Global loss of biodiversity harming ocean bounty. *Science*. 314(5800). pg. 745.

Teasdale, J. R., Coffman, C. B. and Mangum, R. W. (2007). Potential long-term benefits of no-tillage and organic cropping systems for grain production and soil improvement. *Agronomy Journal*. 99. pg. 1297-1305.

The Economist. (December 30, 2008). Troubled waters: A special report on the sea. *The Economist*. pg. 1-16.

Thurman, H. V. (1997). *Introductory oceanography*. Prentice Hall College, NJ.

Union of Concerned Scientists. (2005). *Environmental impacts of renewable energy technologies*. Cambridge, MA. http://www.ucsusa.org/clean_energy/technology_and_impacts/impacts/environmental-impacts-of.html.

Union of Concerned Scientists. (2008). *Re: USDA APHIS proposed rule regulating genetically engineered organisms*. APHIS-2008-0023. Regulatory Analysis and Development, USDA APHIS PPD, Riverdale, MD.

United Nations. (2006). *Water a shared responsibility: The United Nations world water development report 2*. World Water Assessment Program. http://www.unesco.org/water/wwap/wwdr/wwdr2/.

United Nations Environment Programme. (2007). *GEO4: Global environment outlook: Environment for development*. Progress Press Ltd., Valletta, Malta.

United Nations Environment Programme. (2008). *Organic agriculture and food security in Africa*. United Nations, NY.

Verity, Peter G., Smetacek, Victor and Smayda, Theodore J. (2002). Status, trends and the future of the marine pelagic ecosystem. *Environmental Conservation*. 29(2). pg. 207-37.

Wackernagle, Mathis and Rees, William E. (1996). *Our ecological footprint: Reducing human impact on the Earth*. New Society Publishers, Gabriola Island, BC, Canada.

Waring, R. H. and Running, S. W. (1998). *Forest ecosystems: An analysis at multiple scales*. Academic Press, Burlington, MA.

Washington, Richard, Bouet, Christel, Cautenet, Guy, Mackenzie, Elisabeth, Ashpole, Ian, Engelstaedter, Sebastian, Lizcano, Gil, Henderson, Gideon M., Schepanski, Kerstin and Tegen, Ina. (December 8, 2009). Dust as a tipping element: The Bodélé Depression, Chad. *Proceedings of the National Academy of Sciences*. 106(49).

Watson, J. D. (1965). *Molecular biology of the gene*. W. A. Benjamin, NY.

Watson, R. and Pauly, D. (2001). Systemic distortions in world fisheries catch trends. *Nature*. 414. pg. 534-6.

Weiner, Jonathan. (1990). *The next one hundred years: Shaping the fate of our living earth*. Bantam, NY.

White, Lynn Townsend, Jr. (1967). The historical roots of our ecological crisis. *Science*. 155(3767). Pg. 1203-1207. www.zbi.ee/~kalevi/lwhite.htm.

Wilson, E. O. (1984). *Biophilia*. Harvard University, Cambridge, MA.

Wilson, Edward O. (2002). *The future of life*. Vintage, NY.

Wilson, Harold F. (1935). The rise and decline of the sheep industry in northern New England. *Agricultural History*. 9(1). pg. 25.

Worldwatch Institute. (2006). *American energy*. Worldwatch Institute, Washington, DC.

Worm, Boris, Barbier, Edward B., Beaumont, Nicola, Duffy, J. Emmett, Folke, Carl, Halpern, Benjamin S., Jackson, Jeremy B. C., Lotze, Heike K., Micheli, Fiorenza, Palumbi, Stephen R., Sala, Enric, Selkoe, Kimberley A., Stachowicz, John J. and Watson, Reg. (November 3, 2006). Impacts of biodiversity loss on ocean ecosystem services. *Science*. 314(5800). pg. 787-90.

Worm, B., Lotze, H. K. Hillebrand, H. and Sommer, U. (2002). Consumer versus resource control of species diversity and ecosystem functioning. *Nature*. 417. pg. 848-51.

Worm, B. and Myers, R. A. (2003). Meta-analysis of cod-shrimp interactions reveals top-down control in oceanic food webs. *Ecology*. 84. pg. 162-73.

Worm, B. et al. (2009). Rebuilding global fisheries. *Science*. 325. pg. 578-85. http://www.sciencemag.org/content/325/5940/578.abstract?sid=f30dcece-e65b-4cf3-9492-3c8f41a920e3.

History, Economics, and Politics

Ahamed, Liaquat. (2009). *Lords of finance: The bankers who broke the world*. Penguin Books, NY.

Arvedlund, Erin. (2009). *Too good to be true: The rise and fall of Bernie Madoff*. Portfolio Hardcover, NY.

Baghai, Mehrdad and Quigley, James. (2011). *As one*. Portfolio/Penguin, NY.

Bajaj, Vikas. (December 30, 2009). Heart-stopping fall, breathtaking rally. *The New York Times*. http://www.nytimes.com/2009/12/31/business/31stox.html?_r=1&sq=vikas%20bajaj%20wall%20street&st=nyt&adxnnl=1&scp=7&adxnnlx=1269363936-M9iSw4fjtYxn5VoGwCzIqQ.

Baker-Said, Stephanie. (May 2008). Flight of the black swan. *Bloomberg Markets*.

Barlow, Maude. (2002). *Blue gold: The fight to stop the corporate theft of the world's water*. New Press, NY.

Barlow, Maude. (2008). *Blue covenant: The global water crisis and the coming battle for the right to water*. New Press, NY.

Bartlett, Bruce. (2006). *Imposter: How George W. Bush bankrupted America and betrayed the Reagan legacy*. Doubleday, NY.

Bartlett, Bruce. (2009). *The new American economy: The failure of Reaganomics and a new way forward*. Palgrave Macmillan, Hampshire, UK.

Bauman, Zygmunt. (2010). *Living on borrowed time: Conversations with Citlali Rovirosa-Madrazo*. Polity, UK.

Beck, Glenn. (2005). *The real America: Messages from the heart and heartland*. Pocket, NY.

Beck, Glenn. (2009). *Glenn Beck's common sense: The case against an out-of-control government, inspired by Thomas Paine*. Threshold Editions, NY.

Berman, Morris. (2001). *The twilight of American culture*. W. W. Norton & Company, NY.

Berman, Morris. (2006). *Dark ages America: The final phase of empire*. W. W. Norton & Company, NY.

Bernstein, Peter L. (2005). *Capital ideas: The improbable origins of Wall Street*. John Wiley and Sons, Hoboken, NJ.

Bishop, Matthew and Green, Michael. (2010). *The road from ruin: How to revive capitalism and put America back on top.* Crown Business, NY.

Bishop, Matthew and Green, Michael. (March 11, 2010). *The Road from Ruin*: Wake up, you can fix this financial mess. http://www.huffingtonpost.com/matthew-bishop/the-road-from-ruin-wake-u_b_494599.html.

Bloom, Howard. (2009). *The genius of the beast: A radical re-vision of capitalism.* Prometheus Books, Amherst NY.

Bogle, John C. (2005). *The battle for the soul of capitalism.* Yale University Press, New Haven, CT.

Bollier, David. (2002). *Silent theft the private plunder of our common wealth.* Routledge, NY.

Bonner, William, Wiggin, Addison and Agora. (2009). *The new empire of debt: The rise and fall of an epic financial bubble.* Wiley, NY.

Bookstaber, Richard. (2007). *Demon of our own design.* John Wiley and Sons, Hoboken, NJ.

Brack, H. G. (2008). *Handbook for ironmongers.* Pennywheel Press, Hulls Cove, ME.

Bradley, Bill, Ferguson, Niall, Krugman, Paul, Roubini, Nouriel, Soros, George, Wells, Robin et al. (June 11, 2009). The crisis and how to deal with it. *New York Review of Books.* 56(10). pg. 73-6. http://www.nybooks.com/articles/22756.

Briscoe, John. (2005). *India's water economy: Bracing for a turbulent future.* World Bank, New Delhi.

Brooks, Arthur C. (2010). *The battle: How the fight between free enterprise and big government will shape America's future.* Basic Books, NY.

Brown, Aaron. (2006). *The poker face of Wall Street.* John Wiley and Sons, Hoboken, NJ.

Brown, Lester R. (2003). *Plan B: Rescuing a planet under stress and a civilization in trouble.* Earth Policy Institute, Washington, DC.

Brown, Lester R. (2008). *Plan B 3.0: Mobilizing to save civilization.* W. W. Norton and Company, NY.

Brown, Lester R. (2011). *World on the edge: How to prevent environmental and economic collapse.* W. W. Norton & Company, NY.

Brown, Peter G. and Garver, Geoffrey. (2009). *Right relationship: Building a whole earth economy.* Berrett-Koehler Publishers, San Francisco, CA.

Brueggemann, John. (2010). *Rich, free, and miserable: The failure of success in America.* Rowman & Littlefield Publishers, Lanham, MD.

Brulle, Robert J. and Pellow, David N. (April 2006). Environmental justice: Human health and environmental inequalities. *Annual Review of Public Health.* 27. pg. 103-24.

Bryce, Robert. (2008). *Gusher of lies: The dangerous delusions of "energy independence".* Public Affairs Books, NY.

Bryce, Robert. (2010). *Power hungry: The myths of "green" energy and the real fuels of the future.* PublicAffairs, NY.

Bryson, Bill. (2003). *A short history of nearly everything.* Broadway Books, Random House, NY.

Burrough, Bryan and Helyar, John. (2008). *Barbarians at the gate: The fall of RJR Nabisco.* HarperBusiness, NY.

Campbell, Colin. (2009). *What is peak oil?* www.peakoil.net.

Campbell, Kurt M., Gulledge, Jay, McNeill, J.R., Podesta,John, Ogden, Peter, Fuerth, Leon, Woolsey, R. James, Lennon, Alexander T.J., Smith, Julianne, Weitz, Richard and Mix, Derek. (2007). *The age of consequences: The foreign policy and national security implications of global climate change.* Center for Strategic & International Studies and Center for a New American Security. Washington, DC. http://csis.org/files/media/csis/pubs/071105_ageofconsequences.pdf.

Capra, Fritjof. (1882). *The turning point: Science, society, and the rising culture.* Simon & Schuster, NY.

Capra, Fritjof. (2000). *The Tao of physics: An exploration of the parallels between modern physics and eastern mysticism.* 25th anniversary edition. Shambhala Publications, Inc., Boston, MA.

Carr, Nicholas. (2010). *The shallows: What the internet is doing to our brains.* W. W. Norton & Company, NY.

Cassidy, John. (2009). *How markets fail: The logic of economic calamities.* Farrar, Straus and Giroux, NY.

Charles, Daniel. (2002). *Lords of the harvest: Biotech, big money, and the future of food.* Basic Books, NY.

Chen, Shaohua and Ravallion, Martin. (Fall 2004). How have the world's poorest fared since the early 1980s? *The World Bank Research Observer.* 19(2). pg. 141-69.

Chomsky, Noam and Herman, Edward S. (2002). *Manufacturing consent: The political economy of the mass media.* Pantheon, NY.

Ciplet, David. (2009). *An industry blowing smoke: 10 reasons why gasification, pyrolysis & plasma incineration are not "green solutions".* Global Alliance for Incinerator Alternatives, Berkeley, Ca.

Coburn, Tom and Hart, John. (2012). *The debt bomb: A bold plan to stop Washington from bankrupting America.* Thomas Nelson, Nashville, TN.

Cohan, William. (2009). *House of cards: A tale of hubris and wretched excess on Wall Street.* Doubleday, NY.

Cohen, Stephen S. and Delong, J. Bradford. (2010).*The end of influence: What happens when other countries have the money.* Basic Books, NY.

Coll, Steve. (2012). *Private empire: ExxonMobil and American power.* Penguin Press HC, NY.

Collins, Daryl, Morduch, Jonathan, Rutherford, Stuart and Ruthven, Orlanda. (2009). *Portfolios of the poor: How the world's poor live on $2 a day.* Princeton University Press, Princeton, NJ.

Committee on Prospering in the Global Economy of the 21st Century: An Agenda for American Science and Technology, National Academy of Sciences, National Academy of Engineering, and Institute of Medicine. (2007). *Rising above the gathering storm: Energizing and employing America for a brighter economic future.* National Academies Press, Washington, DC.

Cooper, Michael and Walsh, Mary Williams. (December 4, 2010). Mounting debts by states stoke fears of crisis: Costs remain hidden: Analysts who predicted mortgage meltdown see a similarity. *The New York Times*. http://www.nytimes.com/2010/12/05/us/politics/05states.html.

Coyle, Diane. (2011). *The economics of enough: How to run the economy as if the future matters*. Princeton University Press, Princeton, NJ.

Cribb, Julian. (2010).*The coming famine: The global food crisis and what we can do to avoid it*. University of California Press, Berkeley, CA.

Daily Telegraph. (September 12, 2006). Thirst for cash that threatens a crash. *Daily Telegraph of London*. http://www.telegraph.co.uk/finance/2947119/Thirst-for-cash-that-threatens-a-crash.html.

Das, Satyajit. (2011). *Extreme money: Masters of the universe and the cult of risk*. FT Press, Upper Saddle River, NJ.

Davies, Howard. (2010). *The financial crisis: Who is to blame?* Polity, Cambridge, UK.

Davis, Mike. (1999). *Ecology of fear: Los Angeles and the imagination of disaster*. Vintage Books, London.

Davis, Mike. (2002). *Late Victorian holocausts: El Niňo famines and the making of the third world*. Verso, London.

Davis, Mike. (2007). *Planet of slums*. Verso, London.

Dawidoff, Nicholas. (March 29, 2009). The civil heretic. *The New York Times Magazine*. pg. 32-9, 54, 57-9.

Deffeyes, Kenneth S. (2001). *Hubbert's peak: The impending world oil shortage.* Princeton University Press, Princeton, NJ.

Dhillon, Navtej and Yousef, Tarik. (2009). *Generation in waiting: The unfulfilled promise of young people in the Middle East*. Brookings Institution Press, Washington, DC.

Diamond, Jared. (2005). *Collapse: How societies succeed or choose to fail*. Penguin Books, NY.

Drew, Christopher and Oppel, Richard A., Jr. (March 6, 2004). How industry won the battle of pollution control at E.P.A. *The New York Times*. http://www.nytimes.com/2004/03/06/politics/06LOBB.html?scp=7&sq=oppel%202004&st=cse.

Dunbar, Nicholas. (2011). *The devil's derivatives: The untold story of the slick traders and hapless regulators who almost blew up Wall Street... and are ready to do it again*. Harvard Business Review Press, Boston, MA.

Easterbrook, Greg. (2003). *The progress paradox: How life gets better while people feel worse*. Random House, NY.

Erlanger, Steven. (May 22, 2010). Europeans fear crisis threatens liberal benefits. *The New York Times*. http://www.nytimes.com/2010/05/23/world/europe/23europe.html?pagewanted=1&sq=aging population&st=nyt&scp=1.

Evans, Alex. (2009). *The feeding of the nine billion: Global food security for the 21st century*. Chatham House, London. http://www.humansecuritygateway.com/documents/CHATHAM_FeedingNineBillion_GlobalFood Security21stCentury.pdf.

Farrell, Diana, Key, Aneta Marcheva and Shavers, Tim. (March 2005). Mapping the global capital markets. *The McKinsey Quarterly*. http://www.mckinseyquarterly.com/Mapping_the_global_capital_markets_1579.

Ferguson , Niall. (2002). *The cash nexus: Money and power in the modern world, 1700-2000*. Basic Books, NY.

Ferguson, Niall. (2003). *Empire*. Basic Books, NY.

Ferguson, Niall. (2008). *The ascent of money: A financial history of the world*. The Penguin Press, NY.

Ferguson, Niall. (January 13, 2009a). MPBN-TV.

Ferguson, Niall. (November 3, 2009b). Charlie Rose Show. MPBN-TV.

Fox, Justin. (2009). *The myth of the rational market: A history of risk, reward, and delusion on Wall Street*. Harper Business/HarperCollins Publishers, NY.

Fraser, Steve. (2008). *Wall Street: America's dream palace*. Yale University Press, New Haven, CT.

Friedman, Thomas L. (2000). *The Lexus and the olive tree: Understanding globalization*. Anchor reprint edition, NY.

Friedman, Thomas L. (2006). *The world is flat 3.0: A brief history of the twenty-first century*. Farrar, Straus, and Giroux, NY.

Friedman, Thomas L. (2008). *Hot, flat, and crowded: Why we need a green revolution--and how it can renew America*. Farrar, Straus, and Giroux, NY.

Friedman, Thomas L. (March 7, 2009). The inflection is near? *The New York Times*. http://www.nytimes.com/2009/03/08/opinion/08friedman.html?_r=1.

Friedman, Thomas L. (April 4, 2010). Start-ups not bailouts. *The New York Times*.

Friedman, Thomas L. and Mandelbaum, Michael. (2011). *That used to be us: How America fell behind in the world it invented and how we can come back*. Farrar, Straus and Giroux, NY.

Fritze, Ronald H. (2009). *Invented knowledge: False history, fake science and pseudo-religions*. Reaktion Books, London.

Fukuyama, Francis. (1999). *The great disruption: Human nature and the reconstitution of social order*. The Free Press, NY.

Fukuyama, Francis. (2002). *Our posthuman future: Consequences of the biotechnology revolution*. Picador, NY.

Fukuyama, Francis. (January/February 2011). Left out. *The American Interest*. 6(3). pg. 22-8.

Fukuyama, Francis. (2008). *Blindside: How to anticipate forcing events and wild cards in global politics*. Brookings Institute Press, Washington, DC.

Fukuyama, Francis and Colby. (September/October 2009). What were they thinking? The role of economists in the financial debacle. *The American Interest*.

Gale Group, Inc. (2003). *Encyclopedia of espionage, intelligence, and security*. The Gale Group, Inc. www.fas.org/irp/cia/product/go_appendixa_032796.html.

Gardner, Gary. (2006). *Inspiring progress: Religions' contributions to sustainable development.* Worldwatch Institute, Washington, DC.

Garfinkle, Adam. (January/February 2011). Plutocracy & democracy: Terms of contention. *The American Interest*, 6(3). pg. 4-15.

Gasparino, Charles. (2009). *The sellout: How three decades of Wall Street greed and government mismanagement destroyed the global financial system.* Harper, NY.

Gelinas, Nicole. (2009). *After the fall: Saving capitalism from Wall Street and Washington.* Encounter Books, NY.

Giampetro, Mario and Pimentel, David. (1994). *The tightening conflict: Population, energy use, and the ecology of agriculture.* http://dieoff.org/page69.htm.

Gladwell, Malcolm. (2002). *The tipping point: How little things can make a big difference.* Back Bay Books, Boston, MA.

Gladwell, Malcolm. (April 22/29, 2002). Blowing up: How Nassim Taleb turned the inevitability of disaster into an investment strategy. *New Yorker*.

Glanz, James. (September 23, 2012). Power, pollution and the internet: Industry wastes vast amounts of electricity, belying image. *The New York Times*. pg. 1, 20-1.

Goodell, Jeff. (2006). *Big coal: The dirty secret behind America's energy future.* Houghton Mifflin, NY.

Gray, John. (April 9, 2009). The way of all debt. *The New York Review of Books.* 56(60). pg. 46-7.

Greenspan, Alan. (2007). *Adventures in a new world.* Penquin, NY.

Greenwald, Glenn. (2008). *Great American hypocrites: Toppling the big myths of Republican politics.* Three Rivers Press, NY.

Griffin, G. Edward. (2010). *The creature from Jekyll Island: A second look at the Federal Reserve.* American Media, Inc., NY.

Grimes, William. (August 7, 2010). Tony Judt, chronicler of history, is dead at 62. *The New York Times*. http://www.nytimes.com/2010/08/08/books/08judt.html?pagewanted=1&sq=tony judt dead&st=cse&scp=1.

Gross, Daniel. (September 4, 2004). The next shock: Not oil, but debt. *The New York Times.* http://www.nytimes.com/2004/09/05/business/yourmoney/05view.html?_r=1&scp=1&sq=The%20 next%20shock:%20Not%20oil,%20but%20debt&st=cse.

Halperin, Mark and Heilemann, John. (2010). *Game change: Obama and the Clintons, McCain and Palin, and the race of a lifetime.* Harper, NY.

Hartmann, Thom. (2009). *Threshold: The crisis of western culture.* Viking, NY.

Hartung, William D. (2010). *Profits of war: Lockheed Martin and the making of the military-industrial complex.* Nation Books, NY.

Hawken, Paul. (1993). *The ecology of commerce: A declaration of sustainability.* Harper Collins, NY.

Hayes, Brian. (2005). *Infrastructure: A field guide to the industrial landscape.* W. W. Norton, NY.

Hedges, Chris. (2008). *American fascists: The Christian right and the war on America*. Free Press, NY.

Hedges, Chris. (2009). *Empire of illusion: The end of literacy and the triumph of spectacle*. Nation Books, NY.

Hedges, Chris. (2010). *Death of the liberal class*. Nation Books, NY.

Heinberg, Richard. (2003). *The party's over*. New Society Publishers, Gabriola Island, BC, Canada.

Heinberg, Richard. (2007). *Peak everything: Waking up to the century of declines*. New Society Publishers, Gabriola Island, BC, Canada.

Heinberg, Richard. (December 3, 2007). What will we eat as the oil runs out? *Global Public Media*. http://www.energybulletin.net/node/38091.

Heinberg, Richard and Campbell, Colin. (2006). *The oil depletion protocol: A plan to avert oil wars, terrorism and economic collapse*. New Society Publishers, Gabriola Island, BC, Canada.

Heller, Anne C. (2009). *Ayn Rand and the world she made*. Nan A. Talese, NY.

Herbert, Bob. (February 6, 2010). Time is running out. *The New York Times*, Opinion.

Herrero, M. et al. (February 12, 2010). Smart investments in sustainable food production: Revisiting mixed crop-livestock systems. *Science*. 327(5967). pg. 822-5.

Hill, Steven. (2010). *Europe's promise: Why the European way is the best hope in an insecure age*. University of California Press, Berkeley, CA.

Huffington, Arianna. (2003). *Pigs at the trough: How corporate greed and political corruption are undermining America*. Crown Publishers, NY.

Huffington, Arianna. (2010). *Third World America: How our politicians are abandoning the middle class and betraying the American dream*. Crown Publishers, NY.

International Center for Technology Assessment. (2004/2005). *The real cost of gasoline: an analysis of the hidden external costs consumers pay to fuel their automobiles*. International Center for Technology Assessment, Washington, DC.

International Consortium of Investigative Journalists. (2010). *Looting the seas: How overfishing, fraud, and negligence plundered the majestic bluefin tuna*. ICIJ: A Project of the Center for Public Integrity. http://www.publicintegrity.org/treesaver/tuna/#-/treesaver/tuna/00-toc.html.

Jackson, Brooks and Jamieson, Kathleen Hall. (2007). *UnSpun: Finding facts in a world of disinformation*. Random House Trade Paperbacks, NY.

Jacobs, Jane. (1993). *Systems of survival: A dialog on the moral foundations of commerce and politics*. Random House, NY.

Jacoby, Susan. (2009). *The age of American unreason*. Vintage Books, NY.

Jacques, Martin. (2009). *When China rules the world: The end of the western world and the birth of a new global order*. The Penguin Press HC, NY.

James, Harold. (2009). *The creation and destruction of value: The globalization cycle*. Harvard University Press, Cambridge, MA.

Jan Honigsberg, Peter. (2009). *Our nation unhinged: The human consequences of the war on terror*. University of California Press, Berkeley, CA.

Jensen, Derrick. (2006). *Endgame, Vol. 1: The problem of civilization*. Seven Stories Press, NY.

Jensen, Derrick. (2006). *Endgame, Vol. 2: Resistance*. Seven Stories Press, NY.

Johnson, Chalmers A. (2004). *Blowback, second edition: The costs and consequences of American empire (American Empire Project)*. Holt Paperbacks, NY.

Johnston, David Cay. (2003). *Perfectly legal: The covert campaign to rig our tax system to benefit the super rich--and cheat everybody else*. Portfolio Hardcover, NY.

Johnston, David Cay. (2007). *Free lunch: How the wealthiest Americans enrich themselves at government expense (and stick you with the bill)*. Portfolio Hardcover, NY.

Johnson, Simon and Kwak, James. (2010). *13 bankers: The Wall Street takeover and the next financial meltdown*. Pantheon, NY.

Judt, Tony. (2010). *Ill fares the land*. Penguin Press HC, NY.

Kaiser, Robert G. (2010). *So damned much money: The triumph of lobbying and the corrosion of American government*. Vintage, London.

Karabell, Zachary. (2009). *Superfusion: How China and America became one economy and why the world's prosperity depends on it*. Simon & Schuster, NY.

Kanter, James. (January 26, 2011). China, once suspect on emissions, is rapidly becoming a clean-energy power. *The New York Times*. http://query.nytimes.com/gst/fullpage.html?res=9D05E0DF113EF935A15752C0A9679D8B63&ref=jameskanter.

Kelly, Kate. (2009). *Street fighters: The last 72 hours of Bear Stearns, the toughest firm on Wall Street*. Portfolio, NY.

Klare, Michael T. (2001). *Resource wars: The new landscape of global conflict*. Metropolitan Books, NY.

Klare, Michael. (2004). *Blood and oil*. Henry Holt, NY.

Klein, Naomi. (2008). *The shock doctrine: The rise of disaster capitalism*. Picador, NY.

Knoke, William. (1997). *Bold new world: The essential road map to the twenty-first century*. Kodansha America, NY.

Koppelaar, Rembrandt H.E.M. (September 2006). *World oil production and peaking outlook*. Peak Oil Netherlands Foundation, Amsterdam, Netherlands. http://peakoil.nl/wp-content/uploads/2006/09/asponl_2005_report.pdf.

Korin, Anne, and Luft, Gal. (2009). *Turning oil into salt: Energy independence through fuel choice*. Booksurge Publishing, Charleston, SC.

Kosman, Josh. (2009). *The buyout of America: How private equity will cause the next great credit crisis*. Portfolio, NY.

Kotke, William H. (2007). *The final empire: The collapse of civilization and the seed of the future*. Author House, Central Milton Keynes, UK.

Kotlikoff, Lawrence J. (July/August 2006). Is the United States bankrupt? *Federal Reserve Bank of St. Louis Review*. 88(4). pg. 235-49.

Krugman, Paul. (1994). *The age of diminished expectations: U. S. economic policy in the 1990s.* The MIT Press, Cambridge, MA.

Krugman, Paul. (2003). *The great unraveling: Losing our way in the new century.* W. W. Norton & Co., NY.

Krugman, Paul. (December 18, 2008). What to do? *New York Review of Books.* 55(20). http://www.nybooks.com/articles/22151.

Krugman, Paul. (June 1, 2009a). Reagan did it. *The New York Times.* http://www.nytimes.com/2009/06/01/opinion/01krugman.html?_r=1.

Krugman, Paul. (2009b). *The conscience of a liberal.* W. W. Norton & Co., NY.

Krugman, Paul. (2009c). *The return of depression economics and the crisis of 2008.* W. W. Norton & Co., NY.

Kugelman, Michael. (February 6, 2013). The global farmland rush. *The New York Times.* http://www.nytimes.com/2013/02/06/opinion/the-global-farmland-rush.html.

Kuhn, T. (1962). *The structure of scientific revolutions.* University of Chicago Press, Chicago, IL.

Kurlansky, Mark. (2003). *Salt: A world history.* Penguin Books, NY.

Kuttner, Robert. (February 6, 2011). Business doesn't need American workers. *Huffington Post.* http://www.huffingtonpost.com/robert-kuttner/business-doesnt-need-amer_b_819337.html.

Labaton, Stephen. (October 2, 2008). Up new debt. *The New York Times.*

Lancaster, John. (2010). *I.O.U.: Why everyone owes everyone and no one can pay.* Simon and Schuster, NY.

Lanton, Thomas Jr. (February, 7, 2010). Is debt trashing the Euro? Growing threats to dreams of greater unity. *The New York Times, Opinion.*

Lasch, Christopher. (1977). *Haven in a heartless world: The family besieged.* Basic Books, NY.

Lessig, Lawrence. (2001). *The future of ideas: The fate of the commons in a connected world.* Random House, NY.

Levitt, Steven D. and Dubner, Stephen J. (2005). *Freakonomics: A rogue economist explores the hidden side of everything.* Harper Collins, NY.

Levitt, Steven D. and Dubner, Stephen J. (2009). *SuperFreakonomics: Global cooling, patriotic prostitutes, and why suicide bombers should buy life insurance.* William Morrow, NY.

Lewis, Michael. (2007). *The blind side: Evolution of a game.* W. W. Norton & Co., NY.

Lewis, Michael. (2009). *Panic: The story of modern financial insanity.* W. W. Norton & Co., NY.

Lewis, Michael. (2010). *The big short: Inside the doomsday machine.* W. W. Norton & Co., NY.

Lewis, Michael. (2011). *Boomerang: Travels in the new third world.* W. W. Norton & Co., NY.

Linker, Damon. (2006). *The theocons.* Doubleday, NY.

Longman, Phillip. (2004). *The empty cradle: How falling birthrates threaten world prosperity and what to do about it.* Basic Books, NY.

Lowenstein, Roger. (2000). *When genius failed: The rise and fall of long-term capital management*. Random House, NY.

Lowenstein, Roger. (March 15, 2010). Who needs Wall Street? Society profits little from a dizzying casino. *The New York Times, Magazine*. pg. 15-6. http://www.nytimes.com/2010/03/21/magazine/21FOB-WWLN-t.html?ref=magazine.

Lowenstein, Roger. (2010). *The end of Wall Street*. Penguin Press HC, NY.

Luce, Edward. (2012). *Time to start thinking: America in the age of descent*. Atlantic Monthly Press, Boston, MA.

Lumb, Judy, Ed. (2009). *Fueling our future: A dialog about technology, ethics, public policy, and remedial action*. Quaker Institute for the Future Pamphlet 1. Producciones de la Hamaca, Caye Caulker, Belize.

Lynn, Barry. (2010). *Cornered: The new monopoly: Capitalism and the economics of destruction*. John Wiley and Sons, Hoboken, NJ.

MacDonald, James. (2003). *A free nation deep in debt: The financial roots of democracy*. Farrar, Straus and Giroux, NY.

MacDonald, Mia. (October 20, 2008). Opinion: Chinese farms a growing challenge. *Worldwatch Institute*. http://www.worldwatch.org/node/5916.

Maddock, Shane J. (2010). *Nuclear apartheid: The quest for American atomic supremacy from World War II to the present*. University of North Carolina Press, Chapel Hill, NC.

Madrick, Jeff. (2012). *Age of greed: The triumph of finance and the decline of America, 1970 to the present*. Vintage, NY.

Mallaby, Sebastian. (2010). *More money than God: Hedge funds and the making of a new elite*. Penguin Press, NY.

Mandelbrot, Benoit and Judson, Richard L. (2006). *The (mis)behavior of markets: A fractal view of financial turbulence*. Basic Books, NY.

Mann, Thomas E. and Ornstein, Norman J. (2012). *It's even worse than it looks: How the American constitutional system collided with the new politics of extremism*. Basic Books, NY.

Manning, Robert D. (2001). *Credit card nation: The consequences of America's addiction to credit*. Basic Books, NY.

Markoff, John. (February 20, 2013). Nanotubes seen as alternative when silicon chips hit their limits. *The New York Times*. http://www.nytimes.com/2013/02/20/technology/nanotubes-seen-as-an-alternative-to-silicon-circuits.html.

Martin, Douglas. (May 22, 2013). Heinrich Rohrer, physicist, dies at 79; helped open door to nanotechnology. *The New York Times*. http://www.nytimes.com/2013/05/22/science/heinrich-rohrer-physicist-who-won-nobel-dies-at-79.html?_r=0.

McDougall, Walter A. (1997). *Promised land, crusader state: The American encounter with the world since 1776*. Houghton Mifflin, NY.

McLuhan, Marshall and Fiore, Quentin. (1967). *The medium is the massage: An inventory of effects*. Bantam Books, NY.

Medoff, James and Harless, Andrew. (1996). *The indebted society: Anatomy of an ongoing disaster*. Little Brown & Co, NY.

Members of the 2005 "Rising Above the Gathering Storm" Committee, National Academy of Sciences, National Academy of Engineering, and Institute of Medicine. (2010). *Rising above the gathering storm, revisited: Rapidly approaching category 5*. National Academies Press, Washington, DC.

Miller, Judith, Engelberg, Stephen and Broad, William. (2001). *Biological weapons and America's secret war: Germs*. Simon & Schuster, NY.

Moore, Kathleen Dean and Nelson, Michael P., Eds. (2010). *Moral ground: Ethical action for a planet in peril*. Trinity University Press, San Antonio, TX.

Morgenson, Gretchen. (February 7, 2010). This crisis won't stop moving. *The New York Times, Opinion*.

Morgenson, Gretchen. (March 5, 2010). The swaps that swallowed your town. *The New York Times*. http://www.nytimes.com/2010/03/07/business/07gret.html?scp=2&sq=morgenson&st=nyt.

Morgenson, Gretchen and Rosner, Joshua. (2011). *Reckless endangerment: How outsized ambition, greed, and corruption led to economic Armageddon*. Times Books, NY.

Morris, Charles R. (2009). *The two trillion dollar meltdown: Easy money, high rollers, and the great credit crash*. Public Affairs Books, NY.

Moyers, Bill. (March 24, 2005). Welcome to Doomsday. *The New York Review of Books*. 52(5). http://www.nybooks.com/articles/article-preview?article_id=17852.

Moyo, Dambisa. (2011). *How the west was lost: Fifty years of economic folly—and the stark choices ahead*. Farrar, Straus & Giroux, NY.

Murray, Charles. (2013). *Coming apart: The state of white America, 1960-2010*. Crown Forum, NY.

National Energy Policy Development Group. (2001). *National energy policy: Reliable, affordable and environmentally sound energy for America's future*. U. S. Government Printing Office, Washington, DC. http://www.ne.doe.gov/pdfFiles/nationalEnergyPolicy.pdf.

National Resources Defense Council. (2012). *Going in reverse: The tar sands threat to central Canada and New England*. NRDC report R:12-06-A. National Resources Defense Council, NY. http://www.nrdc.org/energy/files/Going-in-Reverse-report.pdf.

Nef, John. (1932). *The rise of the British coal industry. Vol. 16*. Cambridge University Press, Cambridge, UK.

Neuwirth, Robert. (2004). *Shadow cities: A billion squatters, a new urban world*. Routledge, NY.

Nocera, Joe. (April 17, 2010). A Wall Street invention that let the crisis mutate. *The New York Times*. http://www.nytimes.com/2010/04/17/business/17nocera.html?dbk.

Noll, Mark A. and Harlow, Luke E. (2001). *The old religion in a new world: The history of North American Christianity*. Wm. B. Eerdmans Publishing Company, Grand Rapids, MI.

Noll, Mark A. and Harlow, Luke E. (2007). *Religion and American politics: From the colonial period to the present*. Oxford University Press, NY.

Ouroussoff, Alexandra. (2010). *Wall Street at war: The secret struggle for the global economy.* Polity, Cambridge, UK.

Paarlberg, Robert. (2008). *Starved for science: How biotechnology is being kept out of Africa.* Harvard University Press, Cambridge, MA.

Page, Benjamin I. and Jacobs, Lawrence R. (2009). *Class war? What Americans really think about economic inequality.* University of Chicago Press, Chicago, IL.

Panzer, Michael J. (2009). *When giants fall: An economic roadmap for the end of the American era.* John Wiley and Sons, Hoboken, NJ.

Participant Media. (2010). *Waiting for "Superman": How we can save America's failing schools.* PublicAffairs, NY.

Partnoy, Frank. (2009). *Infectious greed: How deceit and risk corrupted the financial markets.* John Wiley and Sons, Hoboken, NJ.

Patel, Raj. (2008). *Stuffed and starved: The hidden battle for the world food system.* Melville House, Brooklyn, NY.

Patterson, Scott. (2010). *The quants: How a new breed of math whizzes conquered Wall Street and nearly destroyed it.* Crown Business, NY.

Patterson, Scott and Lucchetti, Aaron. (May 8, 2008). Boom in "dark pool" trading networks is causing headaches on Wall Street. *Wall Street Journal.*

Patterson, Scott and Raghaven, Anita. (September 7, 2007). August ambush: How market turmoil waylaid the "Quants." *Wall Street Journal.*

Paulson, Henry. (2010). *On the brink: Inside the race to stop the collapse of the global financial system.* BusinessPlus, NY.

Paulson, Henry M. Jr. (December 5, 2012). How cities can save China. *The New York Times.* http://www.nytimes.com/2012/12/05/opinion/how-cities-can-save-china.html?_r=0.

Perkins, John. (2009). *Hoodwinked: An economic hit man reveals why the world financial markets imploded--and what we need to do to remake them.* Broadway Business, NY.

Pettifor, Ann. (2006). *The coming first world debt crisis.* Palgrave Macmillan, NY.

Phillips, Kevin P. (2004). *American dynasty: Aristocracy, fortune, and the politics of deceit in the house of Bush.* Viking, NY.

Phillips , Kevin P. (2006). *American theocracy: The peril and politics of radical religion, oil, and borrowed money in the 21st century.* Viking Adult, NY.

Phillips, Kevin. (2008a). *Bad money, Reckless finance, failed politics, and the global crisis of American capitalism.* Viking Adult, NY.

Phillips, Kevin. (April 30, 2008b). *Swimming with the sharks: The U.S. economy.* Alternative Radio. www.alternativeradio.org/programs/PHIK003.shtml.

Phillips, Melanie. (2006). *Londonistan.* Encounter Books, NY.

Pollack, Kenneth. (2008). *A path out of the desert: A grand strategy for America in the Middle East.* Random House, NY.

Popper, K. (1959). *The logic of scientific discovery.* Harper and Row, NY.

Porter, Eduardo. (2011). *The price of everything: Solving the mystery of why we pay what we do.* Portfolio Hardcover, NY.

Posner, Richard A. (2009). *A failure of capitalism: The crisis of '08 and the descent into depression.* Harvard University Press, Cambridge, MA.

Pozen, Robert. (2009). *Too big to save? How to fix the U.S. financial system.* Public Affairs Books, NY.

Prestowitz, Clyde V. (2004). *Rogue nation: American unilateralism and the failure of good intentions.* Basic Books, NY.

Prestowitz, Clyde V. (2006). *Three billion new capitalists: The great shift of wealth and power to the East.* Basic Books, NY.

Prestowitz, Clyde V. (2010). *The betrayal of American prosperity Free market delusions, America's decline, and how we must compete in the post-dollar era.* Free Press, NY.

Prins, Nomi. (2004). *Other people's money.* The New Press, NY.

Pryor, Frederic L. (2010). *Capitalism reassessed.* Cambridge University Press, Cambridge, UK.

Qiu, Jane. (May 2006). Unfinished symphony. *Nature.* 441. pg. 143-5.

Rachman, Gideon. (2011). *Zero-sum future: American power in an age of anxiety.* Simon & Schuster, NY.

Rajan, Raghuram. (2010). *Fault lines: How hidden fractures still threaten the world economy.* Princeton University Press, Princeton, NJ.

Ramo, Joshua Cooper. (2009). *The age of the unthinkable: Why the new world disorder constantly surprises us and what we can do about it.* Hachette Book Group, NY.

Ravitch, Diane. (November 11, 2010). The myth of charter schools. *The New York Review of Books.* 57(17). Pg. 22-4.

Reich, Robert B. (2008). *Supercapitalism: The transformation of business, democracy, and everyday life.* Vintage, London.

Reich, Robert B. (2010). *Aftershock: The next economy and America's future.* Knopf, NY.

Reinhart, Carmen M. and Rogoff, Kenneth (2009). *This time is different: Eight centuries of financial folly.* Princeton University Press, Princeton, NJ.

Rich, Frank. (January 3, 2010). The state of the union is comatose. *The New York Times, Opinion.*

Rigazio, John. (2010). *America is now a socialist country.* AuthorHouse, Bloomington, IN.

Ritholtz, Barry. (2009). *Bailout nation: How greed and easy money corrupted Wall Street and shook the world economy.* Wiley, Hoboken, NJ.

Roberts, Paul. (2005). *The end of oil: On the edge of a perilous new world.* Mariner Books, NY.

Robertson, Campbell. (July 29, 2010). Gulf of Mexico has long been dumping site. *The New York Times.* http://www.nytimes.com/2010/07/30/us/30gulf.html.

Rothkopf, David. (2009). *Superclass: The global power elite and the world they are making.* Farrar, Straus and Giroux, NY.

Roubini, Nouriel and Setser, Brad. (2004). *Bailouts or bailins: Responding to financial crisis in emerging markets.* Peterson Institute, Washington, DC.

Russell, Howard S. (1976). *A long, deep furrow: Three centuries of farming in New England*. University Press of New England, Hanover, NH.

Sandel, Michael J. (1998). *Democracy's discontent: America in search of a public philosophy*. Belknap Press of Harvard University Press, Cambridge, MA.

Saul, John Ralston. (2005). *The collapse of globalism: And the reinvention of the world*. Overlook Hardcover, NY.

Schama, Simon. (1988). *The embarrassment of riches*. University of California Press, Berkeley, CA.

Schindler, Jorg, Dipl.-Kfm. (February 2008). *Crude oil – the supply outlook*. Energy Watch Group/Ludwig-Boelkow-Foundation. Ottobrunn, Germany.

Schlosser, Eric. (2001). *Fast food nation: The dark side of the all-American meal*. Houghton Mifflin, NY.

Schultz, Ellen E. (2011). *Retirement heist: How companies plunder and profit from the nest eggs of American workers*. Portfolio/Penguin, NY.

Schultz, Stefan. (September 1, 2010). 'Peak oil' and the German government: Military study warns of a potentially drastic oil crisis. *Spiegel Online*. http://www.spiegel.de/international/germany/0,1518,715138,00.html.

Schwartz, Peter and Randall, Doug. (2003). *An abrupt climate change scenario*. US Department of Defense, Washington, DC.

Schwartz, Nelson D. and Dash, Eric. (February 25, 2010). Banks bet Greece defaults on debt they helped hide. *The New York Times*.

Shiller, Robert J. (2004). *The new financial order: Risk in the 21st century*. Princeton University Press, Princeton, NJ.

Shiller, Robert J. (2006). *Irrational exuberance*. Princeton University Press, Princeton, NJ.

Shiller, Robert J. (2008). *The subprime solution: How today's global financial crisis happened, and what to do about it*. Princeton University Press, Princeton, NJ.

Shiller, Robert J. (January 31, 2010). Stuck in neutral? Reset the mood. *The New York Times*. http://www.nytimes.com/2010/01/31/business/economy/31view.html?scp=2&sq=Robert+shiller&st=nyt.

Shulman, Seth. (January 8, 2007). *Undermining science: Suppression and distortion in the Bush administration*. University of California Press.

Shulman, Seth. (May 2002). *Trouble on the endless frontier: Science, invention and the erosion of the technological commons*. A report for the New America Foundation.

Shulman, Seth. (1999). *Owning the future*. Houghton Mifflin Company, Boston.

Simmons, Matthew R. (2006). *Twilight in the desert: The coming Saudi oil shock and the world economy*. John Wiley and Sons, Hoboken, NJ.

Simmons, Matthew R. (February 24, 2010). Twin threats to resource scarcity: Oil & water. *Marsh's National Oil Companies' Conference 2010*. Dubai, United Arab Emirates. http://www.simmonsco-intl.com/SearchResult.aspx?ID=1263.

Skidelsky, Robert. (April 17, 2008). Gloomy about globalization. *The New York Review of Books.* 55(6). pg. 60-4.

Smick, David M. (2008). *The world is curved: Hidden dangers to the global economy.* Portfolio/Penguin, NY.

Smil, V. (2000). *Feeding the world: A challenge for the twenty-first century.* MIT Press, Cambridge, MA.

Smil, Vaclav. (2003). *Energy at the crossroads.* The MIT Press, Cambridge, MA.

Smil, Vaclav. (September-October 2008). Water news: Bad, good, and virtual. *American Scientist.* 95(5). pg. 406-7.

Smith, Laurence C. (2010). *The world in 2050: Four forces shaping civilization's northern future.* Dutton Adult, NY.

Solomon, Steven. (2010). *Water: The epic struggle for wealth, power, and civilization.* Harper Collins Publishers, NY.

Sorkin, Andrew Ross. (2009). *Too big to fail: The inside story of how Wall Street and Washington fought to save the financial system---and themselves.* Viking, NY.

Soros, George. (1998). *The crisis of global capitalism.* Public Affairs, NY.

Soros, George. (2008a). *The new paradigm for financial markets: The credit crisis of 2008 and what it means.* Public Affairs, NY.

Soros, George. (December 4, 2008b). The crisis and what to do about it. *The New York Times Review of Books.* 55(19). http://www.nybooks.com/articles/22113.

Soros, George. (2009). *The crash of 2008 and what it means: The new paradigm for financial markets.* Public Affairs, NY.

Soros, George. (2010a). *The Soros lectures at the Central European University.* PublicAffairs, NY.

Soros, George. (August 19, 2010b). The euro & the crisis. *The New York Review of Books.* 57(13). pg. 28-9.

Soros, George. (November 11, 2010c). The real danger to the economy. *The New York Review of Books.* 57(17). pg. 16.

Starobin, Paul. (2009). *After America: Narratives for the next global age.* Viking Adult, NY.

Starr, Paul. (1984). *The social transformation of American medicine: The rise of a sovereign profession and the making of a vast industry.* Basic Books, NY.

Stauber, John and Rampton, Sheldon. (2002). *Toxic sludge is good for you!: Lies, damn lies and the public relations industry.* Common Courage Press, Monroe, ME.

Stearns, Peter. (2001). *Consumerism in world history.* Routledge, NY.

Stern, Nicholas. (2007). *The economics of climate change: The Stern review.* Cambridge University Press, Cambridge, UK.

Stern, Nicholas. (2009). *A blueprint for a safer planet: How to manage climate change and create a new era of progress and prosperity.* Bodley Head, NY.

Stewart, James B. (September 21, 2009). A reporter at large: Eight days: The battle to save the American financial system. *The New Yorker*. pg. 58-81.

Stiglitz, Joseph E. (2002). *Globalization and its discontents*. W. W. Norton & Co., NY.

Stiglitz, Joseph E. (2004). *The roaring nineties: A new history of the world's most prosperous decade*. W. W. Norton & Co., NY.

Stiglitz, Joseph E. (2007). *Making globalization work*. W. W. Norton & Co., NY.

Stiglitz, Joseph E. (2010). *Freefall: America, free markets, and the sinking of the world economy*. W. W. Norton & Co., NY.

Stiglitz, Joseph E. (2010). *The Stiglitz report: Reforming the international monetary and financial systems in the wake of the global crisis.* The New Press, NY.

Stokes, Doug and Raphael, Sam. (2010). *Global energy security and American hegemony (themes in global social change)*. Johns Hopkins University Press, Baltimore, MD.

Story, Louise. (February 9, 2011). New questions raised in mortgage financing. *The New York Times*. http://www.nytimes.com/2011/02/10/business/10mortgage.html.

Summers, Lawrence H. (October 3, 2004). The U.S. current account deficit and the global economy. *The Per Jacobsson Lecture*. www.perjocobsson.org/2004/100304.pdf.

Suskind, Ron. (2004). *The price of loyalty: George W. Bush, the White House, and the education of Paul O'Neill*. Simon and Schuster, NY.

Suskind, Ron. (2009). *The way of the world: A story of truth and hope in an age of extremism*. Harper Perennial, NY.

Taibbi, Matt. (2010). *Griftopia: Bubble machines, vampire squids, and the long con that is breaking America.* Spiegel & Grau, NY.

Tainter, Joseph A. (1990). *The collapse of complex societies*. Cambridge University Press, Cambridge, UK.

Taleb, Nassim Nicholas. (2010). *The black swan: The impact of the highly improbable*. Random House Trade Paperbacks, NY.

Taylor, John B. (2009). *Getting off track: How government actions and interventions caused, prolonged, and worsened the financial crisis*. Hoover Institution Press, Stanford, CA.

Terborgh, John. (Sept. 20, 2001). The age of giants. *The New York Review of Books*. 48(14). pg. 46. http://www.nybooks.com/articles/article-preview?article_id=14509.

Tester, Jefferson. (2005). *Sustainable energy: Choosing among options*. The MIT Press, Cambridge, MA.

The 9/11 Commission. (2004). *The 9/11 Commission report: Final report of the National Commission on terrorist attacks upon the United States*. W. W. Norton & Company, NY.

The New York Times. (2011). *Can Europe be saved? By Paul Krugman (Eurotrashed, how deep is their love?, the will to drill)*. The New York Times Magazine, NY.

Tibman, Joseph. (2009). *The murder of Lehman Brothers: An insider's look at the global meltdown*. Brick Tower Books, NY.

Tomasky, Michael. (April 9, 2009). Washington: Will the lobbyists win? *The New York Review of Books*. 56(6). pg. 18-22.

Trenor, Casson. (April 2010). *Carting away the oceans*. Greenpeace, Washington, DC. http://www.greenpeace.org/usa/Global/usa/report/2010/5/carting-away-the-oceans.pdf.

Triffin, Robert. (1961). *Gold and the dollar crisis*. Yale University Press, New Haven, CT.

Troy, Gil. (2007). *Morning in America: How Ronald Reagan invented the 1980's*. Princeton University Press, Princeton, NJ.

Tye, Larry. (2002). *The father of spin: Edward L. Bernays and the birth of public relations*. Picador, NY.

United States Government. (2005). *Energy policy act of 2005. Public law 109-58*. http://www.epa.gov/oust/fedlaws/publ_109-058.pdf.

Volcker, Paul. (April 10, 2005). An economy on thin ice. *The Washington Post*. http://pqasb.pqarchiver.com/washingtonpost/access/819763781.html?FMT=ABS&FMTS=ABS:FT&date=Apr+10%2C+2005&author=Paul+A.+Volcker&pub=The+Washington+Post&edition=&startpage=B.07&desc=An+Economy+On+Thin+Ice.

Volcker, Paul. (June 24, 2010). 'The time we have is growing short.' *The New York Review*. 57(11). pg. 12, 14.

Voltaire, Lex. (2010). *How Republicans can legally pay no taxes, change America, and save the world*. CreateSpace, Seattle, WA.

Walker, David M. (2010). *Comeback America: Turning the country around and restoring fiscal responsibility*. Random House, NY.

Wall Street Journal. (October 16-17, 2010). Throwing away our food. *The Wall Street Journal*. pg. C12.

Wallace, Henry A. (April 9, 1944). Wallace defines "American Fascism"; The Vice President says it pollutes public opinion, encourages intolerance and presents a challenge to our democratic way of life. *The New York Times*. pg. SM7.

Warshofsky, Fred. (1977). *Doomsday: The science of catastrophe*. Reader's Digest Press, NY.

Wilkinson, Richard and Pickett, Kate. (2009). *The spirit level: Why greater equality makes societies stronger*. Bloomsbury Press, NY.

Wolf, Martin. (2008). *Fixing global finance (forum on constructive capitalism)*. The Johns Hopkins University Press, Baltimore, MD.

Wolff, Richard. (2009). *Capitalism hits the fan*. Movie. Media Education Foundation, Northampton, MA. http://www.capitalismhitsthefan.com/.

Wolin, Sheldon S. (2008). *Democracy incorporated: Managed democracy and the specter of inverted totalitarianism*. Princeton University Press, Princeton, NJ.

Wolman, William and Colamosca, Anne. (1997). *The Judas economy: The triumph of capital and the betrayal of work*. Addison -Wesley, NY.

World Wildlife Federation. (June 1, 2005). *The history of whaling and the International Whaling Commission (IWC)*. http://www.panda.org/about_our_earth/all_publications/?13796/The-History-of-Whaling-and-the-International-Whaling-Commission-IWC.

Worldwatch Institute. (2006). *State of the world 2006: Special focus China and India.* Worldwatch Institute, Washington, DC.

Worldwatch Institute. (2006). *Vital signs 2006-2007: The trends that are shaping our future.* Worldwatch Institute, Washington, DC.

Worldwatch Institute. (2007). *Vital signs 2007 – 2008: The trends that are shaping our future.* Worldwatch Institute, Washington, DC.

Worldwatch Institute. (2009). *Vital signs 2009.* Worldwatch Institute, Washington, DC. http://www.worldwatch.org/vs2009.

Yergin, Daniel. (1994). *The prize.* Buccaneer Books, Cutchogue, NY.

Zakaria, Fareed. (2008). *The post-American world.* W. W. Norton & Co., NY.

Zuckerman, Gregory. (2009). *The greatest trade ever: The behind-the-scenes story of how John Paulson defied Wall Street and made financial history.* Broadway Business, NY.

Zuckerman, Gregory. (2013). *The frackers: The outrageous inside story of the new billionaire wildcatters.* Portfolio Hardcover, NY.

Cataclysmic Climate Change

Ackerman, Frank. (2009). *Can we afford the future? The economics of a warming world.* Zed Books, NY.

Ackerman, Frank and Stanton, Elizabeth A. (2008). *The cost of climate change: What we'll pay if global warming goes unchecked.* NRDC, Washington, DC. http://www.nrdc.org/globalwarming/cost/contents.asp.

Ackerman, Frank, Stanton, Elizabeth A., DeCanio, Stephen J., Goodstein, Eban, Howarth, Richard B., Norgaard, Richard B., Norman, Catherine S. and Sheeran, Kristen A. (2009). *The economics of 350: The benefits and costs of climate stabilization.* Economics for Equity and the Environment, Somerville, MA. http://www.sei-us.org/climate-and-energy/Economics-of-350-Final.pdf.

Aleklett, Kjell, Höök, Mikael, Jakobsson, Kristofer, Lardelli, Michael, Snowden, Simon and Söderbergh, Bengt. (March 2010). The peak of the oil age - analyzing the world oil production reference scenario in world energy outlook 2008. *Energy Policy.* 38(3). pg. 1398-414. http://www.tsl.uu.se/uhdsg/Publications/PeakOilAge.pdf.

Aldy, Joseph E. and Stavins, Robert N. (2009). *Post-Kyoto international climate policy: Summary for policymakers.* Cambridge University Press, Cambridge, UK.

Archer, David, Buffett, Bruce and Brovkin, Victor. (December 8, 2009). Ocean methane hydrates as a slow tipping point in the global carbon cycle. *Proceedings of the National Academy of Sciences.* 106(49).

Bamber, Jonathan L., Riva, Riccardo E. M., Vermeersen, Bert L. A. and LeBrocq, Anne M. (May 2009). Reassessment of the potential sea-level rise from a collapse of the West Antarctic ice sheet. *Science.* 324(5929), pg. 901-3. http://www.sciencemag.org/cgi/content/abstract/324/5929/901.

Barbier, Edward B. *A global green new deal: Rethinking the economic recovery.* Cambridge University Press, Cambridge, UK.

Bindoff, Nathaniel L. and Willibrand, Jurgen. (2007). Observations: Oceanic climate change and sea level. In: *Climate change 2007: The physical science basis. Contribution of Working Group I to the Fourth Assessment Report of the Intergovernmental Panel on Climate Change.* Solomon, S., et al. eds. Cambridge University Press, Cambridge, UK. http://ipcc-wg1.ucar.edu/wg1/Report/AR4WG1_Print_Ch05.pdf.

Bowermaster, Jon. (May 15, 2007). Global warming changing Inuit lands, lives, Arctic expedition shows. *National Geographic News.* http://news.nationalgeographic.com/news/2007/05/070515-inuit-arctic.html.

Braasch, Gary. (2009). *Earth under fire: How global warming is changing the world.* University of California Press, Berkeley, CA.

Brander, K. M. (December 11, 2007). Global fish production and climate change. *Proceedings of the National Academy of Sciences.* pg. 19709-14.

Broder, John M. (July 11, 2013). U.S. warns that climate change will cause more energy breakdowns. *The New York Times.* pg. A12. http://www.nytimes.com/2013/07/11/us/climate-change-will-cause-more-energy-breakdowns-us-warns.html?_r=0.

Calvin, William H. (2008). *Global fever: How to treat climate change.* University of Chicago Press, Chicago.

Christian-Smith, Juliet, Gleick, Peter H., Reilly, William K. and Cooley, Heather. (2012). *A twenty-first century U.S. water policy.* Oxford University Press, US.

Climate Change Research Center. *New England's changing climate, weather and air quality.* Institute for the Study of Earth, Oceans and Space, University of New Hampshire, Durham, NH.

Cohen, Dave. (2007). *The perfect storm.* Association for the Study of Peak Oil and Gas – USA. http://www.aspo-usa.com/archives/index.php?option=com_content&task=view&id=243&Itemid=91.

Cohen, Stewart J. and Waddell, Melissa W. (2009). *Climate change in the 21ˢᵗ century.* McGill-Queen's University Press, Montreal, Canada.

Conniff, Richard. (November 13, 2008). The greenhouse gas that nobody knew. *Yale Environment 360.* Yale School of Forestry and Environmental Studies, Yale University, New Haven, CT. http://www.e360.yale.edu/content/feature.msp?id=2085.

De Decker, Kris. (March 3, 2008). The ugly side of solar panels. *Low-tech Magazine.* http://www.lowtechmagazine.com/2008/03/the-ugly-side-o.html.

De Decker, Kris. (June 16, 2009). The monster footprint of digital technology. *Low-tech Magazine.* http://www.lowtechmagazine.com/2009/06/embodied-energy-of-digital-technology.html.

Donoghue, Andrew. (October 24, 2008). Solar panels linked to "powerful" greenhouse gas. *BusinessGreen.* http://www.businessgreen.com/business-green/news/2229052/solar-panels-linked-powerful.

Douglas, Bruce C. (1997). Global sea rise: A redetermination. *Surveys in Geophysics.* 18. pg. 279-292. http://www.springerlink.com/content/p364381652174757/.

Doyle, Alister. (March 15, 2010). CO2 at new highs despite economic slowdown. *Thomson Reuters.* http://www.reuters.com/article/idUSTRE62E2KJ20100315.

Dyer, Gwynne. (2008). *Climate wars.* Random House, NY.

Dyson, Freeman. (June 12, 2008a). The question of global warming. *The New York Review of Books*. 55(10). http://www.nybooks.com/articles/21494.

Dyson, Freeman. (September 25, 2008b). The question of global warming: An exchange. *The New York Review of Books*. 55(14). http://www.nybooks.com/articles/21811.

Earth Institute News. (March 23, 2006). *Glacial earthquakes point to rising temperatures in Greenland*. http://www.earthinstitute.columbia.edu/news/2006/story03-23-06.php.

Environmental Protection Agency. (2009). *Inventory of U. S. greenhouse gas emissions and sinks: 1990 – 2007*. EPA, Washington, DC. http://epa.gov/climatechange/emissions/downloads09/GHG2007entire_report-508.pdf.

Eriksson, Mats and Jianchu, Xu. (July 2008). Climate change impact on the Himalayan water tower. *Stockholm Water Front*. pg. 11. http://www.siwi.org/documents/Resources/Water_Front_Articles/2008/Two_Eyes_on_Asia_2-08.pdf.

Ewing B., Moore, D., Goldfinger, S., Oursler, A., Reed, A. and Wackernagel, M. (2010). *The ecological footprint atlas 2010*. Global Footprint Network, Oakland, CA. http://www.footprintnetwork.org/images/uploads/Ecological%20Footprint%20Atlas%202010.pdf.

Fairbridge, Rhodes W. (1960). Recent world-wide sea-level changes and their possible significance to New England archaeology. *Bulletin of the Massachusetts Archaeological Society*. 21(3-4). pg. 49-51.

Fischer, Douglas. (March 10, 2005). The great experiment. *Inside Bay Area*.

Flannery, Tim. (2006). *The weather makers: How man is changing the climate and what it means for life on earth*. Atlantic Monthly Press, Boston, MA.

Food and Agriculture Organization of the United Nations (FAO). (November 2007). *Climate change and food security*. United Nations joint press kit for Climate Change Conference, Bali, December 3-14, 2007. www.un.org/climatechange/pdfs/bali/fao-bali07-6.pdf.

French, Kenneth R., Baily, Martin N., Campbell, John Y., Cochrane, John H., Diamond, Douglas W., Duffie, Darrell, Kashyap, Anil K., Mishkin, Frederic S., Rajan, Raghuram G., Scharfstein, David S., Shiller, Robert J., Shin, Hyun Song, Slaughter, Matthew J., Stein, Jeremy C. and Stulz, Rene M. (2010). *The Squam Lake report: Fixing the financial system*. Princeton University Press, Princeton, NJ.

Friedlingstein, Pierre and Solomon, Susan. (2005). Contributions of past and present human generations to committed warming caused by carbon dioxide. *PNAS*. 102(31). pg. 10832-6.

Gardiner, Stephen, Caney, Simon, Jamieson, Dale and Shue, Henry, Eds. (2010). *Climate ethics: Essential readings*. Oxford University Press. Oxford, UK.

Giddens, Anthony. (2009). *Politics of climate change*. Polity, UK.

Gillis, Justin. (September 21, 2010). Extreme heat bleaches coral, and scientists see global threat. *The New York Times*. pg. A1, A9.

Gillis, Justin. (November 13, 2010). As glaciers melt, science seeks data on rising seas. *The New York Times*. http://www.nytimes.com/2010/11/14/science/earth/14ice.html?_r=1.

Gillis, Justin. (May 5, 2011a). Global warming reduces expected yields of harvests in some countries, study says. *The New York Times*. http://www.nytimes.com/2011/05/06/science/earth/06warming.html.

Gillis, Justin. (December 16, 2011b). As permafrost thaws, scientists study the risks. *The New York Times*. http://www.nytimes.com/2011/12/17/science/earth/warming-arctic-permafrost-fuels-climate-change-worries.html?scp=1&sq=as%20permafrost%20thaws,%20scientists%20study%20the%20risks&st=cse.

Gillis, Justin. (January 22, 2013). How high could the tide go? *The New York Times*. http://www.nytimes.com/2013/01/22/science/earth/seeking-clues-about-sea-level-from-fossil-beaches.html?pagewanted=all.

Gillis, Justin. (March 8, 2013b). Global temperatures highest in 4,000 years: Warming over longer period, study shows. *The New York Times*. http://www.nytimes.com/2013/03/08/science/earth/global-temperatures-highest-in-4000-years-study-says.html.

Gleick, Peter H. (1993). *Water in crisis: A guide to the world's fresh water resources*. Oxford University Press, US.

Gleick, Peter H. (2000). *The world's water 2000-2001: The biennial report on freshwater resources*. Island Press, Washington, DC.

Gleick, Peter H. (2004). *The world's water 2004-2005: The biennial report on freshwater*. Island Press, Washington, DC.

Gleick, Peter H. (2006). *The world's water 2006-2007: The biennial report on freshwater resources (world's water (quality))*. Island Press, Washington, DC.

Gleick, Peter H. (2008). *The world's water 2008-2009: The biennial report on freshwater*. Island Press, Washington, DC.

Gleick, Peter H. (2011). *Bottled and sold: The story behind our obsession with bottled water*. Island Press, Washington, DC.

Gleick, Peter H. (2011). *The world's water volume 7: The biennial report on freshwater resources (world's water (quality))*. Island Press, Washington, DC.

Gleick, Peter H., Garcetti, Gil, Carter, Jimmy and Annan, Kofi. (2007). *Water is key: A better future for Africa*. Balcony Press, Glendale, CA.

Gore, Al. (2006). *An inconvenient truth: The planetary emergency of global warming and what we can do about it*. Rodale Books, Emmaus, PA.

Gore, Al. (2006). *Earth in the balance: Ecology and the human spirit*. Rodale Books, Emmaus, PA.

Gore, Al. (2008). *Our purpose: the Nobel Peace Prize lecture 2007*. Rodale Books, Emmaus, PA.

Gore, Al. (2007). *The assault on reason*. Penguin, NY.

Gore, Al. (2009). *Our choice: A plan to solve the climate crisis*. Rodale Books, Emmaus, PA.

Grossman, Karl. (1997). *The wrong stuff: The space program's nuclear threat to our planet*. Common Courage Press, Monroe, ME.

Guinotte, John and Fabry, Victoria J. (2009). The threat of acidification on ocean ecosystems. *Current: The Journal of Marine Education*. pg. 2-7.

Halpern, Benjamin S. et al. (2008). A global map of human impact on marine ecosystems. *Science*. 319. pg. 948-52. http://www.sciencemag.org/content/319/5865/948.full.pdf.

Halpern, Benjamin S., Longo, Catherine, Hardy, Darren, McLeod, Karen L., Samhouri, Jameal F., Katona, Steven K., *et al.* (August 2012). An index to assess the health and benefits of the global ocean. *Nature*. 488. pg. 615-20.
http://www.nature.com/nature/journal/v488/n7413/full/nature11397.html.

Hansen, James. (2009). *Storms of my grandchildren: The truth about the coming climate catastrophe and our last chance to save humanity*. Bloomsbury USA, NY.

Hansen, James. (May 10, 2012). Game over for the climate. *The New York Times*. pg. A25.
http://www.nytimes.com/2012/05/10/opinion/game-over-for-the-climate.html.

Hansen, James, Sato, Makiko, Kharecha, Pushker, Beerling, David, Berner,Robert, Masson-Delmotte, Valerie, Pagani, Mark, Raymo, Maureen, Royer, Dana L. and Zachos, James C. (2008). Target atmospheric CO_2: Where should humanity aim? *The Open Atmospheric Science Journal*. 2. pg. 217-31.
http://www.bentham.org/open/toascj/articles/V002/217TOASCJ.pdf.

Hansen, James, Nazarenko, Larissa, Ruedy, Reto, Sato, Makiko, Willis, Johs, Del Genio, Anthony, Koch, Dorothy, Lacis, Andrew, Lo, Ken, Menon, Surabi, Novakov, Tica, Perlwitz, Judith, Russell, Gary, Schmidt, Gavin A. and Tausnev, Nicholas. (June 3, 2005). Earth's energy imbalance: Confirmation and implications. *Science*. 308(5727). pg. 1431-5.

Hassol, Susan Joy. (2004). *Impacts of a warming Arctic—Arctic climate impact assessment*. Cambridge University Press, Cambridge, UK.

Hill, Trevor and Symmonds, Graham. (2013). *The smart grid for water: How data will save our water and your utility.* Advantage Media Group, Charleston, SC.

Hoag, Hannah. (July 10, 2008). The missing greenhouse gas. *Nature Reports Climate Change*.
http://www.nature.com/climate/2008/0808/full/climate.2008.72.html.

Hofmann, Matthias and Rahmstorf, Stefan. (December 8, 2009). On the stability of the Atlantic meridional overturning circulation. *Proceedings of the National Academy of Sciences*. 106(49).

Hoggan, James. (2009). *Climate coverup: A crusade to deny global warming.* Vancouver Greystone Books, BC, Canada.

Holdren, John P. (Summer 2007). *Recent findings from climate science and their implications for policy*. Presentation. Energy Technology Innovation Policy research group, Belfer Center for Science and International Affairs, Harvard Kennedy School, Cambridge, MA.
http://belfercenter.ksg.harvard.edu/publication/17258/recent_findings_from_climate_science_and_their_implications_for_policy.html.

Holland, M. M., Bitz, C. M. and Tremblay, B. (2006). Future abrupt reductions in the summer Arctic sea ice. *Geophys. Res. Lett.* 33.
http://europa.agu.org/?view=article&uri=/journals/gl/gl0623/2006GL028024/2006GL028024.xml&t=gl,2006,holland.

Homer-Dixon, Thomas. (August 22, 2010). Disaster at the top of the world. *The New York Times.* http://www.nytimes.com/2010/08/23/opinion/23homer-dixon.html?_r=1&scp=1&sq=disaster%20at%20the%20top%20of%20the%20world&st=cse.

Intergovernmental Panel on Climate Change (IPCC). 2011. *IPCC special report on renewable energy sources and climate change mitigation.* Prepared by Working Group III of the Intergovernmental Panel on Climate Change [O. Edenhofer, R. Pichs-Madruga, Y. Sokona, K. Seyboth, P. Matschoss, S. Kadner, T. Zwickel, P. Eickemeier, G. Hansen, S. Schlömer, C. von Stechow (eds)]. Cambridge University Press, Cambridge, UK. http://srren.ipcc-wg3.de/report.

International Scientific Steering Committee. (2005). *International symposium on the stabilization of greenhouse gas concentrations.*

Jackson, J. B. C. et al. (2001). Historical overfishing and the recent collapse of coastal ecosystems. *Science.* 293. pg. 629-37.

Kerry, Emanuel. (August 4, 2005). Increasing destructiveness of tropical cyclones over the past 30 years. *Nature.* 436. pg. 686-8. http://www.nature.com/nature/journal/v436/n7051/full/nature03906.html.

Knickerbocker, Brad. (February 20, 2007). Human's beef with livestock: A warmer planet. *Christian Science Monitor.* http://www.csmonitor.com/2007/0220/p03s01-ussc.html.

Kolbert, Elizabeth. (2006). *Field notes from a catastrophe: Man, nature and climate change.* Bloomsbury USA, NY.

Krupa, S. V. (2003). Effects of atmospheric ammonia (NH_3) on terrestrial vegetation: A review. *Environmental Pollution.* 124. pg. 179-221.

Krupp, Fred and Horn, Miriam. (2008). *Earth: The sequel: The race to reinvent energy and stop global warming.* W. W. Norton, NY.

Kump, Lee R. (February 27, 2009). Perspectives: Atmospheric Science: Tipping pointedly colder. *Science.* 323(5918). pg. 1175-6.

Leggett, Jeremy. (2001). *The carbon war: Global warming and the end of the oil era.* Routledge, NY.

Levenson, Thomas. (1990). *Ice time: Climate, science, and life on Earth.* Perennial Library edition, Harper & Row Publishers, NY.

Levermann, Anders , Schewe, Jacob, Petoukhov, Vladimir and Held, Hermann. (December 8, 2009). Basic mechanism for abrupt monsoon transitions. *Proceedings of the National Academy of Sciences.* 106(49).

Linden, Eugene. (2006). *The winds of change: Climate, weather, and the destruction of civilizations.* Simon & Schuster Paperbacks, NY.

Lobell, David B. and Field, Christopher B. (March 16, 2007). Global scale climate-crop yield relationships and the impacts of recent warming. *Environmental Research Letters.* 2(1).

Lomborg, Bjorn. (2001). *The skeptical environmentalist: Measuring the real state of the world.* Cambridge University Press, Cambridge, UK.

Lomborg, Bjorn, Ed. (2010). *Smart solutions to climate change: Comparing costs and benefits.* Cambridge University Press, Cambridge, UK.

Lovell, Bryan. (2009). *Challenged by carbon: The oil industry and climate change*. Cambridge University Press, Cambridge, UK.

Lovelock, James. (2000). *Gaia: A new look at life on earth*. Oxford University Press, NY.

Lovelock, James. (2006). *The revenge of Gaia: Why the earth is fighting back – and how we can still save humanity*. Basic Books, NY.

Lovelock, James. (2009). *The vanishing face of Gaia: A final warning*. Basic Books, NY.

Lynas, Mark. (2008). *Six degrees: Our future on a hotter planet*. National Geographic Society, Washington, DC.

Malhi, Yadvinder, Aragão, Luiz E. O. C., Galbraith, David, Huntingford, Chris, Fisher, Rosie, Zelazowski, Przemyslaw, Sitch, Stephen, McSweeney, Carol and Meir Patrick. (December 8, 2009). Exploring the likelihood and mechanism of a climate-change-induced dieback of the Amazon rainforest. *Proceedings of the National Academy of Sciences*. 106(49).

McGowan, John A., Cayan, Daniel R. and Dorman, LeRoy M. (July 10, 1998). Climate-ocean variability and ecosystem response in the northeast Pacific. *Science*. 281(5374). pg. 210-7.

Molden, David. (2007). *Water for food, water for life: A comprehensive assessment of water management in agriculture*. Earthscan Publications Ltd., London.

Monbiot, George. (2009). *Heat: How to stop the planet from burning*. South End Press, Cambridge, MA.

Newell, Peter and Paterson, Matthew. (2010). *Climate capitalism: Global warming and the transformation of the global economy*. Cambridge University Press, Cambridge, UK.

Nordhaus, William. (2008). *A question of balance: Weighing the options on global warming policies*. Yale University Press, New Haven, CT.

Notz, Dirk. (December 8, 2009). The future of ice sheets and sea ice: Between reversible retreat and unstoppable loss. *Proceedings of the National Academy of Sciences*. 106(49).

Ogden, Joan M. and Williams, Robert H. (October 1989). *Solar hydrogen: Moving beyond fossils fuels*. World Resources Institute, Washington, DC.

Orr, David W. (2009). *Down to the wire: Confronting climate change*. Oxford University Press, Oxford, UK.

Owen, James. (February 23, 2007). Arctic expedition to spotlight warming impact on Inuit groups. *National Geographic News*. http://news.nationalgeographic.com/news/2007/02/070223-arctic-warming.html.

Pittock, A. Barrie. (2009). *Climate change: The science, impacts and solutions*. CSIRO Publishing, Victoria, Australia.

Pollack, Henry. (2009). *A world without ice*. Penguin, NY.

Portier, C. J., Thigpen, Tart K., Carter, S. R., Dilworth, C. H., Grambsch, A. E., Gohlke, J., Hess, J., Howard, S. N., Luber, G., Lutz, J. T., Maslak, T., Prudent, N., Radtke, M., Rosenthal, J. P., Rowles, T., Sandifer, P. A., Scheraga, J., Schramm, P. J., Strickman, D., Trtanj, J. M. and Whung, P-Y. (2010). *A human health perspective on climate change: A report outlining the research needs on the human health effects of climate change*. Environmental Health Perspectives/National

Institute of Environmental Health Sciences, Research Triangle Park, NC. http://www.niehs.nih.gov/health/docs/climatereport2010.pdf.

Prather, M. J. and Hsu, J. (2008). NF$_3$, the greenhouse gas missing from Kyoto. *Geophys. Res. Lett.* 35. http://www.agu.org/pubs/crossref/2008/2008GL034542.shtml.

Rogers, J., Averyt, K., Clemmer, S., Davis, M., Flores-Lopez, F., Frumhoff, P., Kenney, D., Macknick, J., Madden, N., Meldrum, J., Overpeck, J., Sattler, S. Spanger-Siegfried, E. and Yates, D. (2013). *Water-smart power: Strengthening the U.S. electricity system in a warming world.* Union of Concerned Scientists, Cambridge, MA. http://www.ucsusa.org/assets/documents/clean_energy/Water-Smart-Power-Full-Report.pdf.

Romm, Joseph R. (2007). *Hell and high water: Global warming--the solution and the politics--and what we should do.* William Morrow, NY.

Romm, Joseph R. (2010). *Straight up: America's fiercest climate blogger takes on the status quo media, politicians, and clean energy solutions.* Island Press, Washington DC.

Roosevelt, Margot. (July 8, 2008). A climate threat from flat TVs, chips. *Los Angeles Times.* http://articles.latimes.com/2008/jul/08/nation/na-climate8.

Rosenthal, Elisabeth. (April 12, 2010). Europe finds clean energy in trash, but U.S. lags. *The New York Times.*

Rosenthal, Elisabeth. (January 27, 2013). Your biggest carbon sin may be air travel. *The New York Times.* http://www.nytimes.com/2013/01/27/sunday-review/the-biggest-carbon-sin-air-travel.html.

Rosenthal, Elizabeth and Lehren, Andrew W. (June 21, 2012). In rising use of air-conditioning, hard choices. *The New York Times.* pg. A1, A9.

Rosenthal, Elizabeth and Revkin, Andrew C. (February 2, 2007). Panel issues bleak report on climate change. *The New York Times.* http://www.nytimes.com/2007/02/02/science/earth/02cnd-climate.html.

Ruddiman, William. (2005). *Plows, plagues, and petroleum: How humans took control of climate.* Princeton University Press, Princeton, NJ.

Santer, B. D. et al. (September 25, 2007). Identification of human-induced changes in atmospheric moisture content. *Proceedings of the National Academy of Sciences.* www.pnas.org/cgi/doi/10.1073/pnas.0702872104.

Schwartz, Peter and Randall, Doug. (October 2003). *Imagining the unthinkable: An abrupt climate change scenario and its implications for United States national security.* EMS, Washington, D.C.

Searchinger, Timothy, et al. (2002). Use of US croplands for biofuels increases greenhouse gases through emissions from land-use change. *Science.* 319. pg. 1238-40.

Seckler, D., Amarasinghe, U., Molden, D., de Silva, R. and Barker, R. (1998). *World water demand and supply, 1990 to 2025: Scenarios and issues.* International Water Management Institute, Colombo, Sri Lanka.

Shah, Sonia. (October 15, 2009). The spread of new diseases: The climate connection. *Yale e360.* http://www.e360.yale.edu/content/feature.msp?id=2199.

Shellenberger, H. J., et al. eds. (2006). *Avoiding dangerous climate change*. Cambridge University Press, Cambridge, UK. www.defra.gov.uk/environment/climatechange/research /dangerous-cc/pdf/avoid-dangercc.pdf.

Shellenberger, Michael and Nordhaus, Ted. (2004). The death of environmentalism: Global warming politics in a post-environmental world. *Proceedings of the Environmental Grantmakers Association.*

Siegel, Arnold. (October 1956). Automatic programming of numerically controlled machine tools. *Control Engineering*. 3(10). pg. 65-70.

Smid, Peter, (2008). *CNC Programming Handbook* (3rd ed.). Industrial Press, NY.

Smith, Kevin. (2007). *The carbon neutral myth: Offset indulgences for your climate sins*. Carbon Trade Watch, Transnational Institute, Amsterdam, The Netherlands. http://www.carbontradewatch.org/pubs/carbon_neutral_myth.pdf.

Smith, P., Martino, D., Cai, Z., Gwary, D., Janzen, H., Kumar, P., McCarl, B., Ogle, S., O'Mara, F., Rice, C., Scholes, B. and Sirotenko, O. (2007). Agriculture. In: *Climate change 2007: Mitigation*. Contribution of Working Group III to the *Fourth Assessment Report of the Intergovernmental Panel on Climate Change*. Cambridge University Press, Cambridge, UK.

Stix, Gary. (September 2006). A climate repair manual. *Scientific American*. http://www.scientificamerican.com/article.cfm?id=a-climate-repair-manual

Stone, R. (2007). A world without corals? *Science*. 316(5825). pg 678-681.

U.S. Global Change Research Program (USGCRP). (2009). *Global climate change impacts in the United States*. http://www.globalchange.gov/publications/reports/scientific-assessments/us-impacts.

Union of Concerned Scientists. (2007). *Smoke, mirrors & hot air: How Exxon Mobile uses big tobacco's tactics to manufacture uncertainty on climate science*. Union of Concerned Scientists, Cambridge, MA. http://www.ucsusa.org/assets/documents/global_warming/exxon_report.pdf.

Vanhoenacker, Mark. (January 27, 2013). Enjoying snow, while we still have it. *The New York Times*. http://www.nytimes.com/2013/01/27/opinion/sunday/enjoying-snow-while-we-still-have-it.html.

Vine, Edward, Crawley, Drury and Centolella, Paul, Eds. (1991). *Energy efficiency and the environment: Forging the link*. American Council for an Energy-Efficient Environment, Washington, DC.

Verhulst, S. L., Nelen, V., Hond, E. D., Koppen, G., Beunckens, C., Vael, C., Schoeters, G. and Desager, K. (January 2009). Intrauterine exposure to environmental pollutants and body mass index during the first 3 years of life. *Environmental Health Perspectives*. 117(1). pg. 122–6. http://www.ncbi.nlm.nih.gov/pmc/articles/PMC2627855/?tool=pmcentrez.

Volk, Tyler. (2008). *CO_2 rising: The world's greatest environmental challenge*. MIT Press, Cambridge, MA.

Walker, Gabrielle and Kline, David, Sir. (2008). *The hot topic: What we can do about global warming*. Harcourt Trade Publishers, NY.

Williams, Chris. (2010). *Ecology and socialism: Solutions to capitalist ecological crisis.* Haymarket books, Chicago.

Wines, Michael. (March 13, 2013). Monarch migration plunges to lowest level in decades. *The New York Times.* http://www.nytimes.com/2013/03/14/science/earth/monarch-migration-plunges-to-lowest-level-in-decades.html.

World Wildlife Federation. (2004). *Living planet report 2004.* World Wildlife Foundation.

Zedillo, Ernesto, Ed. (2008). *Global warming: Looking beyond Kyoto.* Yale Center for the Study of Globalization/Brookings Institution Press, Washington, DC.

Zittel, Werner and Schindler, Jörg. (2007). *Crude oil: The supply outlook.* Report to the Energy Watch Group. EWG-Series No 3/2007. Ludwig-Bölkow-Stiftung, Ottobrunn, Germany. http://www.energywatchgroup.org/fileadmin/global/pdf/EWG_Oilreport_10-2007.pdf.

Chemical Fallout and Biological Monitoring

Abad, E., et al. (1999). Dioxin like compounds from municipal waste incinerator emissions: Assessment of the presence of polychlorinated naphthalenes. *Chemosphere.* 38(1). pg. 109-20.

Adams, Mike. (January 26, 2009). India's waterways a toxic stew of pharmaceutical chemicals dumped from big pharma factories. *Natural News.* http://www.naturalnews.com/025415.html.

Adolfsson-Erici M., M. Peterson, J. Parkkonen, J. Sturve. (2002). Triclosan, a commonly used bactericide found in human milk and in the aquatic environment in Sweden. *Chemosphere.* Vol. 46(9-10). pg. 1485-1489.

Agency for Toxic Substances and Disease Registry. (2000). *Toxicological profile for polychlorinated biphenyls (PCBs).* US Department of Health and Human Services, Public Health Service, Atlanta, GA.

Agency for Toxic Substances and Disease Registry. (2008). *ATSDR Studies on chemical releases in the Great Lakes Region.* US Department of Health and Human Services, Atlanta, GA. http://www.atsdr.cdc.gov/grtlakes/aocreport/2008.html.

Agency for Toxic Substances and Disease Registry. (2007). *Toxicological profile for Ethylbenzene.* US Department of Health and Human Services, Atlanta, GA. http://pubs.acs.org/doi/abs/10.1021/ja9755065.

Agency for Toxic Substances and Disease Registry. (2007). *Toxicological profile for styrene.* US Department of Health and Human Services, Atlanta, GA. http://www.atsdr.cdc.gov/tfacts53.html.

Agency for Toxic Substances and Disease Registry (ATSDR). (2010). *ATSDR ToxProfiles CD-ROM 2010.* Division of Toxicology and Environmental Medicine, Agency for Toxic Substances and Disease Registry, Atlanta, GA.

Al-Yakoob, Sami, Saeed, Talat and Al-Hashash, Huda. (1993). Polycyclic aromatic hydrocarbons in edible tissue of fish from the Gulf after the 1991 oil spill. *Marine Pollution Bulletin.* 27. pg. 297-301.

Allsopp, Michelle, Walters, Adam, Santillo, David and Johnston, Paul. (2006). *Plastic debris in the world's oceans.* Greenpeace, Netherlands. http://www.unep.org/regionalseas/marinelitter/publications/docs/plastic_ocean_report.pdf.

Amlund et al. (2007). Accumulation and elimination of methylmercury in Atlantic cod (*Gadus morhua* L.) following dietary exposure. *Aquatic Technology*. 83. pg. 323-30.

Anderson H. A., Falk C., Hanrahan L., et al. (1998). Profiles of Great Lakes critical pollutants: A sentinel analysis of human blood and urine. *Environmental Health Perspectives*. 106(5). pg. 279-289.

Andrews, David and Wiles, Richard. (2009). *Off the books: Industry's secret chemicals*. Environmental Working Group, Washington, DC. http://www.ewg.org/files/secret-chemicals.pdf.

Arctic Monitoring and Assessment Programme (AMAP). (2002). *AMAP assessment 2002: Persistent organic pollutants in the Arctic*. Oslo, Norway.

Ashby, J., Lefevre, P. A., Odum, J. et al. (1997). Synergy between synthetic estrogens? *Nature*. 385. pg. 385-494.

Ashford, Nicholas and Miller, Claudia. (1998). *Chemical exposures: Low levels and high stakes*. Van Nostrand Reinhold, NY.

Austen, Ian. (October 14, 2010). Canada declares BPA, a chemical in plastics, to be toxic. *The New York Times*. pg. A10.

Balmer, M. E., Poiger, T., Droz, C., Romanin, K., Bergqvist, P. A., Muller, M. D., et al. (2004). Occurrence of methyl triclosan, a transformation product of the bactericide triclosan, in fish from various lakes in Switzerland. *Environmental Science & Technology*. 38(2). pg. 390-95.

Barnes, Kimberlee K., Kolpin, Dana W., Meyer, Michael T., Thurman, E. Michael, Furlong, Edward T., Zaugg, Steven D. and Barber, Larry B. (2002). *Water quality data for pharmaceuticals, hormones, and other organic wastewater contaminants in U.S. streams, 1999-2000*. United States Geological Survey, Iowa City, IA. http://toxics.usgs.gov/pubs/OFR-02-94/

Benbrook, Charles. (July 1999). *Evidence of the magnitude of the Roundup ready soybean yield drag from university- based varietal trials in 1998*. Ag BioTech InfoNet Technical Paper Number 1. http://www.biotech-info.net/herbicide-tolerance.html.

Bergmann, Melanie and Klages, Michael. (October 2012). Increase of litter at the Arctic deep-sea observatory Hausgarten. *Marine Pollution Bulletin*. http://dx.doi.org/10.1016/j.marpolbul.2012.09.018.

Berntssen, et al. (2004). Maximum limits of organic and inorganic mercury in fish feed. *Aquaculture Nutrition*. 10. pg. 83-97.

Bittman, Mark. (February 6, 2013). The cosmetics wars. *The New York Times*. http://opinionator.blogs.nytimes.com/2013/02/05/the-cosmetics-wars/.

Boutron, C., Candelone, J. and Hong, S. (1994). Past and recent changes in the large-scale tropospheric cycles of lead and other heavy metals as documented in Antarctic and Greenland snow and ice: A review. *Geochimica et Cosmochimica Acta*. 58(15).

Boutron, C. and Lorius C. (1979). Trace metals in Antarctic snows since 1914. *Nature*. 277(5697). pg. C 551-554.

Boutron, C. and Wolff, E.W. (1989). Heavy metal and sulphur emissions to the atmosphere from human activities in Antarctica. *Atmospheric Environment*. 23(8).

Braun, M. L. (2001). *DES stories: Faces and voices of people exposed to Diethylstilbestrol*. VSW Press, Rochester, NY.

Broad, William J. (August 21, 2011). Laser advances raise fears of terrorist nuclear ability. *The New York Times*. pg. A1, A12. http://www.nytimes.com/2011/08/21/science/earth/21laser.html?pagewanted=all.

Broder, John M. (February 7, 2011). E.P.A. standards for drinking water single out a new group of toxic chemicals. *The New York Times*. http://www.nytimes.com/2011/02/03/science/earth/03epa.html.

Buckley, Christine. (August 10, 2010). Lobster dieoffs linked to chemicals in plastics. *UConn Today*. http://today.uconn.edu/?p=18691.

Bush, B., Seegal, R. F. and Fitzgerald, E. (1990). Human monitoring of PCB urine analysis. In: Hutzinger, O. and Fiedler, H. Eds. *Organohalogen compounds Vol 1: Dioxin '90--EPRI Seminar, Toxicology, Environment, Food, Exposure-Risk*. Ecoinforma Press, Bayreuth, Germany.

Calafat, Antonia M., Ye, Xiaoyun, Wong, Lee-Yang, Reidy, John A. and Needham, Larry L. (2008). Exposure of the U.S. population to bisphenol A and 4-tertiary-Octylphenol: 2003-2004. *Environmental Health Perspectives*. 116(1). pg. 39-44.

Casco Bay Estuary Partnership. (2006). *Casco Bay plan: 2006 update*. Casco Bay Estuary Partnership, Portland, ME. http://www.cascobay.usm.maine.edu/pdfs/CB%20PlanUpdate06.pdf.

Casco Bay Estuary Partnership. (2008). *2007 toxic pollution in Casco Bay: Sources and impacts*. Casco Bay Estuary Partnership, Portland, ME. http://www.cascobay.usm.maine.edu/toxicsreport07.html.

Casco Bay Estuary Project. (1996). *Casco Bay plan*. Casco Bay Estuary Project, Portland, ME. http://www.cascobay.usm.maine.edu/pdfs/CB%20PlanUpdate06.pdf.

Casselman, Ben and Gold, Russell. (January 21, 2010). Drilling tactic unleashes a trove of natural gas and a backlash. *The Wall Street Journal*.

Castro, Kathy. (November 2010). Extensive research implicates bacteria in lobster shell disease. *Commercial Fisheries News*. 37(3). https://fish-news.com/cfn/editorial/editorial_11_10/Extensive_research_implicates_bacteria_in_lobster_shell_disease.html.

Centre for Environment, Fisheries & Aquaculture Science. (1997). *Monitoring and surveillance of non-radioactive contaminants in the aquatic environment and activities regulating the disposal of wastes at sea, 1994*. Science Series. Aquatic Environment Monitoring Report No. 47. CEFAS, Ministry of Agriculture, Fisheries and Food, Great Britain.

Christopher, S. J., Vander Pol, S. S., Pugh, R. S., Day, R. D. and Becker, P. R. (June 22, 2002). Determination of mercury in the eggs of common murres (*Uria aalge*) for the seabird tissue archival and monitoring project. *Journal of Analytical Atomic Spectrometry*. Royal Society of Chemistry. Cambridge, UK.

Chuanchuen, R., Beinlich, K., Hoang, T. T., Becher, A., Karkhoff-Schweizer, R. R. and Schweizer, H. P. (2001). Cross-resistance between triclosan and antibiotics in *Pseudomonas aeruginosa* is mediated by multidrug efflux pumps: Exposure of a susceptible mutant strain to triclosan selects nfxB mutants overexpressing MexCD-OprJ. *Antimicrobial Agents and Chemotherapy*. 45(2). pg. 428-432.

Colborn, T. and Carroll, L. E. (2007). Pesticides, sexual development, reproduction, and fertility: Current perspective and future. *Human and Ecological Risk Assessment*. 13(5). pg. 1078-110. http://www.informaworld.com/smpp/content~db=all?content=10.1080/10807030701506405.

Conway, Gordon. (1999). *The doubly green revolution: Food for all in the twenty-first century*. Cornell University Press, Ithaca, NY.

Cox-Foster, D. L., Conlan, S., Holmes, E. C., Palacios, G., Evans, J. D., et al. (2007). A metagenomic survey of microbes in honey bee colony collapse disorder. *Science*. 318. pg. 283-7.

Cranor, C. F. (1993). *Regulating toxic substances*. Oxford University Press, NY.

Dahl, R. (1997). Can You Keep a Secret? *Environmental Health Perspectives*. 103(10). pg. 914-5.

Darnerud, Per Ola, et al. (March 2001). Polybrominated diphenyl ethers: Occurrence, dietary exposure, and toxicology. *Environmental Health Perspectives*. 109. Supplement 1. pg. 49-68.

Danelski, D. and Beeman, D. (April 27, 2003). Special report: Growing concerns: While scientists debate the risks, a study finds the rocket-fuel chemical in inland lettuce. *The Press-Enterprise*.

de Wit, Cynthia A. (2002). An overview of brominated flame retardants in the environment. *Chemosphere*. Vol. 46(5). pg. 583-624.

Dean, Cornelia and Nuwer, Rachel. (October 18, 2011). Salmon-killing virus seen for first time in the wild on the Pacific Coast. *The New York Times*. pg. A10.

Deese, Heather and Schmitt, Catherine. (February/March 2011). Fathoming: MSX strikes Maine oysters. *Working Waterfront*. 24(1). pg. 9.

Denison, R. (2009). Hiding a toxic nanomaterial's identity: TSCA's disappearing act. http://blogs.edf.org/nanotechnology/2009/07/14/hiding-a-toxic-nanomaterialsidentity-tscas-disappearing-act/.

Dennis, I. F., Clair, T. A., Driscoll, C. T., Kamman, N., Chalmers, A., Shanley, J., Norton, S. A. and Kahl, S. (2005). Distribution patterns of mercury in lakes and rivers of northeastern North America. *Ecotoxicology*. 14. pg. 113-25.

Department of Ecology. (2006). *Washington State polybrominated diphenyl ether (PBDE) chemical action plan: Final plan*. Department of Ecology Publications Distributions Office, Olympia, WA. http://www.ecy.wa.gov/biblio/0507048.html.

Department of Environmental Protection. (2008). *Dioxin monitoring program: Including data on dioxin-like PCBs collected in the surface water ambient toxics monitoring program*. Department of Environmental Protection, Augusta, ME. http://www.maine.gov/dep/blwq/docmonitoring/dioxin/2008/2008%20Dioxin%20Monitoring%20Program%20Report.pdf.

DeSorbo, C. R. and Evers, D. C. (2005). *Evaluating exposure of Maine's bald eagle population to mercury: Assessing impacts on productivity and spatial exposure patterns*. Report to Maine Department of Environmental Protection, no. BRI 2005-08. BioDiversity Research Institute, Gorham, ME.

Dewan, Shaila. (January 7, 2009). Hundreds of coal ash dumps lack regulation. *The New York Times*. www.nytimes.com/2009/01/07/us/07sludge.html.

Diaz, R. J. and Rosenberg, R. (2008). Spreading dead zones and consequences for marine ecosystems. *Science*. 321. pg. 926-9.

DiFranco, J., Bacon, L., Mower, B. and Courtemanch, D. (1995). *Fish tissue contamination in Maine lakes: Data report*. Maine Department of Environmental Protection, Augusta, ME.

Driscoll, Charles T., et al. (2007). Mercury contamination in forest and freshwater ecosystems in the northeastern United States. *Bioscience*. 57(1). www.biosciencemag.org.

Dufault, Renee, LeBlanc, Blaise, Schnoll, Roseanne, Cornett, Charles, Schweitzer, Laura, Patrick, Lyn, Hightower, Jane, Wallinga, David and Lukiw, Walter. (January 26, 2009). Mercury from chlor-alkali plants: Measured concentrations in food product sugar. *Environmental Health*. 8(2). http://www.ehjournal.net/content/8/1/2.

Eilperin, Juliet. (March 20, 2009). Major decline found in some bird groups but conservation has helped others. *Washington Post*. pg. A02.

Eisenhardt, S., B. Runnebaum, et al. (2001). Nitromusk compounds in women with gynecological and endocrine dysfunction. *Environmental Research*. Elsevier Ltd., London, UK. Vol. 87(3). pg. 123-30.

Environmental Defence, Canada. (September 22, 2008). *Sudbury Human Health Assessment Briefing*. www.environmentaldefence.ca.

Environment News Service. (September 6, 2001). *EU lawmakers vote broad fire retardant ban*. www.ens-newswire.com/.

Environmental Health Strategy Center. (2010). *That's a killer look: A study of chemicals in personal care products*. Alliance for a Clean and Healthy Maine. http://www.cleanandhealthyme.org/LinkClick.aspx?fileticket=7uqB2t8lR7U%3d&tabid=36.

Environmental Protection Agency. (1994). *Nonpoint source pollution: The nation's largest water quality problem*. EPA-841-F-94-005. EPA, Washington, DC. http://www.epa.gov/owow/nps/facts/point1.htm.

Environmental Protection Agency. (2003). *Trichloroethylene health risk assessment: Synthesis and characterization: An EPA Science Advisory Board Report*. www.epa.gov/sab/pdf/ehc03002.pdf.

Environmental Protection Agency. (2003). *Toxics release inventory (TRI) explorer database*. www.epa.gov/triexplorer.

Environmental Protection Agency. (2006). *List of lists: Consolidated list of chemicals subject to the Emergency Planning and Community Right-To-Know Act (EPCRA) and Section 112(r) of the Clean Air Act*. EPA 550-B-01-003. Chemical Emergency Preparedness and Prevention Office, Office of Solid Waste and Emergency Response, EPA. http://www.epa.gov/emergencies/docs/chem/title3_Oct_2006.pdf.

Environmental Protection Agency. (2008). *Hypoxia in the northern Gulf of Mexico: An update by the EPA science advisory board*. EPA-SAB-09-003. Environmental Protection Agency, Washington, DC.

Environmental Protection Agency (USEPA). (2009a). *The national study of chemical residues in lake fish tissue.* EPA-823-R-09-006. U.S. Environmental Protection Agency, Office of Water, Washington, DC. http://www.epa.gov/waterscience/fish/study/data/finalreport.pdf.

Environmental Protection Agency (USEPA). (2009b). *EPA releases report on fish contamination in U.S. lakes and reservoirs.* EPA-823-F-09-008. U.S. Environmental Protection Agency, Office of Water, Washington, DC. http://www.epa.gov/fishadvisories/study/data/factsheet.pdf.

Environmental Working Group. (2004). *Mother's milk: Record levels of toxic fire retardants found in American mother's breast milk.* www.ewg.org/reports/mothersmilk/.
　　This online article contains 115 references on this subject in the bibliography.

Environmental Working Group. (2013). *The dirty dozen: 12 hormone altering chemicals and how to avoid them.* http://www.ewg.org/research/dirty-dozen-list-endocrine-disruptors?inlist=Y&utm_source=201208endocrinedd&utm_medium=email&utm_content=image&utm_campaign=toxics and http://www.keep-a-breast.org/.

Erdogrul, O, Covaci, A. et al. (July 2004). Levels of organohalogenated persistent pollutants in human milk from Kahramanmaras region, Turkey. *Environment International.* 30(5). pg. 659-66.

Eriksson, Johan, et al. (February 2003). Photochemical transformations of Tetrabromobisphenol A and related phenols in water. *Chemosphere.* 54(1). pg. 117-26.

Eriksson, Per Ola, et al. (September 2001). Brominated flame retardants: A novel class of developmental neurotoxicants in our environment? *Environmental Health Perspectives.* 109(9). pg. 903-908.

European Parliament and the Council of the European Union. (2003). Directive 2002/96/EC of the European Parliament and of the Council of 27 January 2003 on waste electrical and electronic equipment (WEEE). *Official Journal of the European Union.*

European Union. (2002). Directive 2002/95/EC of the European Parliament and of the Council of 27 January 2003 on the restriction of the use of certain hazardous substances in electrical and electronic equipment. *Official Journal of the European Union.*

Evans, Alex. (August 8, 2008). The global fertiliser crisis. *Global Dashboard.* http://www.globaldashboard.org/2008/08/08/the-global-fertiliser-crisis/.

Evers, D. C. (2005). *Mercury connections: The extent and effects of mercury pollution in northeastern North America.* Biodiversity Institute, Gorham, ME.

Evers, D. C. and Clair, T. A. (2005). Mercury in northeastern North America: A synthesis of existing databases. *Ecotoxicology.* 14(1,2). pg. 7-14.

Evers, D. C., Han, Y-J., Driscoll, C. T., Kamman, N. C., Goodale, M. W., Lambert, K. F., Holsen, T. M., Chen, C. Y., Clair, T. A. and Butler, T. (2007). Biological mercury hotspots in the northeastern U.S. and southeastern Canada. *BioScience.* 57(1). pg. 29-43.

Evers, D. C., Lane, O. P., Savoy, L. and Goodale, W. (2004). *Assessing the impacts of methylmercury on piscivorous wildlife using a wildlife criterion value based on the Common Loon, 1998-2003.* Report to Maine Department of Environmental Protection no. BRI 2004-05. BioDiversity Research Institute, Gorham, ME.

Field, J. A. and Sierra-Alvarez, R. (September 2008). Microbial transformation and degradation of polychlorinated biphenyls. *Environ. Pollut.* 155(1). pg. 1-12.

Food and Drug Administration (USFDA). (2004-2005). Exploratory survey data on perchlorate in food. http://www.fda.gov/Food/FoodSafety/FoodContaminantsAdulteration/ChemicalContaminants/Perchlorate/ucm077685.htm.

Food Standards Agency. (May 4, 2006). *Dioxins and dioxin-like PCBs in processed fish and fish products: Food survey information sheet 07/06.* Food Standards Agency, London, UK. http://www.food.gov.uk/multimedia/pdfs/fsis0706.pdf.

Food Standards Australia New Zealand (FSANZ). (November 2005). *Report on a survey of chemical residues in domestic and imported aquacultured fish.* Canberra, AU and Wellington, NZ. http://www.foodstandards.gov.au/_srcfiles/Chemical%20Residues%20in%20Fish%20Survey.pdf.

Fox River Watch. (2009). *PCBs, dioxins, furans, mercury – they travel together.* http://www.foxriverwatch.com/dioxins_pcb_pcbs_1.html.

Freese, B. (2002). Alternatives to open-air biopharming. In: *Manufacturing drugs and chemicals in crops.* Friends of the Earth and Genetically Engineered Food Alert.

Furl, C. and Meredith, C. (2008). *Quality assurance project plan: PBT monitoring: Measuring perfluorinated compounds in Washington rivers and lakes.* Washington State Department of Ecology, Olympia, WA. http://www.ecy.wa.gov/biblio/0803107.html.

Gabrielli, P., Barbante, C., Planchon, F., Ferrari, C., Delmonte, B. and Boutron, C. (2003). Changes in the occurrence of heavy metals in Antarctic ice during the last climatic cycles. *Geophysical Research Abstracts.* 5.

Gandhi, Randu and Snedeker, Suzanne. (2000). *Consumer concerns about hormones in food.* Fact Sheet #37. Cornell University Program on Breast Cancer and Environmental Risk Factors, Ithaca, NY. http://envirocancer.cornell.edu/factsheet/diet/fs37.hormones.pdf

George-Ares, Anita and Clark, James R. (April 2000). Aquatic toxicity of two Corexit dispersants. *Chemosphere.* 40(8). pg. 897-906. http://www.sciencedirect.com/science?_ob=ArticleURL&_udi=B6V74-3YGDD0T-28&_user=10&_coverDate=04%2F30%2F2000&_rdoc=1&_fmt=high&_orig=search&_sort=d&_docanchor=&view=c&_searchStrId=1358169642&_rerunOrigin=google&_acct=C000050221&_version=1&_urlVersion=0&_userid=10&md5=4d62a6c3405178c695393b267882c76d.

George-Ares, Anita and Clark, James R. (2010). Acute aquatic toxicity of three Corexit products: An overview. Abstract for the May 23, 2011 International Oil Spill Conference, Portland, Oregon. http://www.iosc.org/papers/00020.pdf.

Gilbert, Steven G. Ed. (2008). *Scientific consensus statement on environmental agents associated with neurodevelopmental disorders.* Collaborative on Health and the Environment's Learning and Developmental Disabilities Initiative. Institute for Children's Environmental Health. Freeland, WA. http://www.iceh.org/pdfs/LDDI/LDDIStatement.pdf.

Ginsberg, Gary and Rice, Deborah. (2005). The NAS perchlorate review: Questions remain about the perchlorate RfD. *Environmental Health Perspectives.* 113. pg. 1117-1119. http://ehp03.niehs.nih.gov/article/fetchArticle.action?articleURI=info:doi/10.1289/ehp.8254.

Goldberg, Edward D. et al. (1978). The mussel watch. *Environment Conservation.* 5(2). pg. 101-125.

Golomb, D., Barry, E., Fisher, G., Varanusupakul, P., Koleda, M. and Rooney, T. (2001). Atmospheric deposition of polycyclic aromatic hydrocarbons near New England coastal waters. *Atmospheric Environment.* 35. pg. 6245-58. http://www.cascobay.usm.maine.edu/pdfs/Golumb.pdf.

Goodale, W. (2008). *Preliminary findings of contaminant screening in Maine birds: 2007 field season.* BioDiversity Research Institute, Gorham, ME. http://www.briloon.org/pub/doc/2008Contaminant.pdf.

Goodman, Amy. (September 3, 2009). Fracking and the environment: Natural gas drilling, hydraulic fracturing and water contamination. *Democracy Now! The War and Peace Report.* http://www.democracynow.org/2009/9/3/fracking_and_the_environment_natural_gas.

Goodman, Amy. (February 23, 2010). Congress to investigate safety of natural gas drilling practice known as hydraulic fracturing. *Democracy Now! The War and Peace Report.* http://www.democracynow.org/2010/2/23/congress_to_investigate_safety_of_natural.

Gopal, Sriram and Deller, Nicole. (February 2003). Precision bombing, widespread harm: Environmental and legal concerns about "precision bombing". *Science for Democratic Action.* 11(2). pg. 1-3, 5, 8-10.

Gottholm, B. W. and Turgeon, D. D. (1992). *Toxic contaminants in the Gulf of Maine.* NOAA, National Ocean Service, Office of Ocean Resources of Maine Conservation and Assessment, Rockville, MD.

Govan, Emilia L. (January 1993). The threat at home: The toxic legacy of the U.S. military. *Bulletin of the Atomic Scientists.* 49(1). pg. 49.

Graham, F. (1970). *Since silent spring.* Houghton Mifflin, Boston, MA.

Grossman, Elizabeth. (September 15, 2007). *High tech trash: Digital devices, hidden toxics, and human health.* Shearwater Press, Douglas, Isle of Man.

Grossman, Elizabeth. (2009). *Chasing molecules: Poisonous products, human health, and the promise of green chemistry.* Shearwater, Washington, DC.

Gulf of Maine Council. (2004). *Gulfwatch contaminants monitoring program: Mussels as bioindicators.* http://www.gulfofmaine.org/gulfwatch/mussels.asp.

Gulf Watch. (2003). *Monitoring chemical contaminants in Gulf of Maine coastal waters.* www.gulfofmaine.org.

Gumy, C., Chandsawangbhuwana, C., Dzyakanchuk, A. A., Kratschmar, D. V., Baker, M. E. and Odermatt, A. (October 28, 2008). Dibutyltin disrupts glucocorticoid receptor function and impairs glucocorticoid-induced suppression of cytokine production. *PLoS ONE.* 3(10). http://www.environmentalhealthnews.org/ehs/newscience/a-widely-used-understudiedchemical-alters-inflammation/.

Gurian-Sherman, D. (2009a). *Failure to yield: Evaluating the performance of genetically engineered crops.* Union of Concerned Scientists, Cambridge, MA.

Gurian-Sherman, D. and Gurwick, Noel. (2009b). *No sure fix: Prospects for reducing nitrogen fertilizer pollution through genetic engineering*. Union of Concerned Scientists, Cambridge, MA.

Gustafson, Bob. (February/March 2011). Are salmon pen pesticides killing lobsters? *Working Waterfront*. 24(1). pg. 3.

Hale, Robert C., et al. (July 12, 2001). Persistent pollutants in land-applied sludges. *Nature*. 412. pg. 140-141.

Halldorsson, T. I., Thorsdottir, I., Meltzer, H. M., Nielsen, F. and Olsen, S. F. (October 2008). Linking exposure to polychlorinated biphenyls with fatty fish consumption and reduced fetal growth among Danish pregnant women: A cause for concern? *Am J Epidemiol*. 168(8). pg. 958-65.

Hansen, Michael, Halloran, Jean M., Groth, Edward and Lefferts, Lisa Y. (September 1997). Potential public health impacts of the use of recombinant bovine somatotropin in dairy production. *Consumers Union*. http://www.consumersunion.org/food/bgh-codex.htm.

Hanson, Paul W. (February 1987). *Acid rain and waterfowl: The case for concern in North America*. Izaak Walton League of America, Arlington, VA.

Harriss, R. C., Browell, E. V., Sebacher, D. I., Gregory, G. L., Hinton, R. R., Beck, S. M., McDougal D. S. and Shipley, S. T. (1984). Atmospheric transport of pollutants from North America to the North Atlantic. *Nature*. 308. pg. 722-724. http://www.nature.com/nature/journal/v308/n5961/abs/308722a0.html.

Hawley, Charles. (July 30, 2010). A quarter century after Chernobyl: Radioactive boar on the rise in Germany. *Spiegel Online*. http://www.spiegel.de/international/zeitgeist/0,1518,709345,00.html.

Hayashi, Takehiko, Kamo, Masashi and Tanaka, Yoshinari. (September 2009). Population-level ecological effect assessment: Estimating the effect of toxic chemicals on density-dependent populations. *Ecological Research*. 24(5). pg. 945-54. http://www.springerlink.com/content/a256365w6301614v/.

Hayes, T. B., Case, P., Chui, S., Chung, D., Haeffele, C., Haston, K., Lee, M., Mai, V. P., Marjuoa, Y., Parker, J. and Tsui. M. (April 2006). Pesticide mixtures, endocrine disruption, and amphibian declines: Are we underestimating the impact?. *Environmental Health Perspectives*. 114 Suppl 1. pg. 40–50. http://www.ncbi.nlm.nih.gov/pmc/articles/PMC1874187/?tool=pmcentrez.

Hayes, Tyrone, Haston, Kelly, Tsui, Mable, Hoang, Anhthu, Haeffele, Cathryn and Vonk, Aaron. (2003). Atrazine-induced hermaphroditism at 0.1 ppb in American Leopard Frogs. *Environmental Health Perspectives*. 111(4). http://www.ehponline.org/members/2003/5932/5932.html.

Hearne, Shelley A. (1984). *Harvest of unknowns: Pesticide contamination in imported foods*. Natural Resources Defense Council, Inc., NY.

Heck, W. W., et al. (1982). *Ozone impacts on the productivity of selected crops: Effects of air pollution on farm commodities*. Izaak Walton League of America, Washington, D.C.

Hemminger, Pat. (October 2005). Damming the flow of drugs into drinking water. *Environmental Health Perspectives*. 113(10). http://ehp.niehs.nih.gov/members/2005/113-10/spheres.html.

Hites, R. A., Foran, J. A., Carpenter, D. O, Hamilton, M. C., Knuth, B. A. and Schwager, S. J. (2004). Global assessment of organic contaminants in farmed salmon. *Science*. 303(5655). pg. 226-9.

Hopkin, Michael. (January 9, 2004). Farmed salmon harbour pollutants. Study may undermine salmon's status as a 'healthy' food. *Nature*.

Hooper, Kim and McDonald, Thomas A. (May 2000). The PBDEs: An emerging environmental challenge and another reason for breast-milk monitoring programs. *Environmental Health Perspectives*. 108(5). pg. 387-392.

Hynes, H. P. (1989). *The recurring silent spring*. Pergamon Press, Elmsford, NY.

Illinois Department of Public Health. (1999). *Fact sheet, nitrates in drinking water*. www.idph.state.il.us/envhealth/factsheets/NitrateFS.htm

Institute for Agriculture and Trade Policy (IATP). (January 26, 2009). *Brand-name food products also discovered to contain mercury*.

Institute of Medicine of the National Academies. (March 4, 2005). *Veterans and agent orange update 2004*. www.iom.edu/?id=28146.

Ivahnenko, Tamara and Zogorski, J. S. (2006). *Sources and occurrence of chloroform and other trihalomethanes in drinking-water supply wells in the United States, 1986-2001*. USGS Publications, Washington, DC.

Jackson, T. J., Wade, T. L., McDonald, T. J., Wilkinson, D. L. and Brooks, J. M. (1994). Polynuclear aromatic hydrocarbon contaminants in oysters from the Gulf of Mexico (1986-1990). *Environmental Pollution*. 83. pg. 291-8.

Jaraczewska, K, Lulek, J., Covaci, A., Voorspoels, S., Kaluba-Skotarczak, A., Drews, K. and Schepens, P. (December 15, 2006). Distribution of polychlorinated biphenyls, organochlorine pesticides and polybrominated diphenyl ethers in human umbilical cord serum, maternal serum and milk from Wielkopolska region, Poland. *Science of the Total Environment*. 372(1). pg. 20-31.

Jarnberg, U., Asplund, L., de Wit, C., Grafstrom, A., Jaglund, P., Jansson, B., Lexen, K., Strandell, M., Olsson, M. and Jonsson, B. (1993). Polychlorinated biphenyls and polychlorinated naphthalenes in Swedish sediment and biota: Levels, patterns, and time trends. *Environ. Sci. Technol*. 27. pg. 1364-74.

Jedrychowski, W., Jankowski, J., Flak, E., Skarupa, A., Mroz, E., Sochacka-Tatara, E., Lisowska-Miszczyk, I., Szpanowska-Wohn, A., Rauh, V., Skolicki, Z., Kaim, I. and Perera, F. (June 2006). Effects of prenatal exposure to mercury on cognitive and psychomotor function in one-year-old infants: Epidemiologic cohort study in Poland. *Ann Epidemiol*. 16(6). pg. 439-47.

Jedrychowski, W., Perera, F., Jankowski, J., Rauh, V., Flak, E., Caldwell, K. L., Jones, R. L., Pac, A. and Lisowska-Miszczyk, I. (November 2007). Fish consumption in pregnancy, cord blood mercury level and cognitive and psychomotor development of infants followed over the first three years of life: Krakow epidemiologic study. *Environ Int*. 33(8). pg. 1057-62.

Jedrychowski, W., Perera, F., Rauh, V., Flak, E., Mróz, E., Pac, A., Skolicki, Z. and Kaim, I. (2007). Fish intake during pregnancy and mercury level in cord and maternal blood at delivery: An environmental study in Poland. *Int J Occup Med Environ Health*. 20(1). pg. 31-7.

Jensen, A. A. and Jørgensen, K. F. (1983). Polychlorinated terphenyls (PCTs) uses, levels and biological effects. *Science of the Total Environ*ment 27(2-3). pg. 231-50.

Johns, Robert. (February 25, 2013). New study finds pesticides leading cause of grassland bird declines. *American Bird Conservancy.* http://www.abcbirds.org/newsandreports/releases/130225.html.

Johnson, Kirk. (December 9, 2011). E.P.A. links tainted water in Wyoming to hydraulic fracturing for natural gas. *The New York Times.* pg. A19.

Jones, R. C. (2000). Avian reovirus infections. *Rev Sci Tech.* 19. pg. 614-25.

Jones, S. H., Chase, M., Sowles, J., Hennigar, P., Landry, N., Wells, P. G., Harding, G. C. H., Krahforst, C. and Brun, G. L. (2001). Monitoring for toxic contaminants in *Mytilus edulis* from New Hampshire and the Gulf of Maine. *Journal of Shellfish Research.* 20. pg. 1203-14.

Kajiwara, J., Todaka, T., Hirakawa, H., Hori, T., Yasutake, D., Nakagawa, R., Iida, T., Nagayama, J., Yoshimura, T. and Furue, M. (May 2009). Dioxin concentration in the preserved umbilical cord from Yusho patients. *Fukuoka Igaku Zasshi.* 100(5). pg. 179-82.

Kamman, N. C., Burgess, N. M., Driscoll, C. T., Simonin, H. A., Goodale, W., Linehan, J., Estabrook, R., Hutcheson, M., Major, A. Scheuhammer, A. M. and Scruton, D.A. (2005). Mercury in freshwater fish of northeast North America—A geographic perspective based on fish tissue monitoring databases. *Ecotoxicology.* 14(1,2). pg. 163-180.

Kannan, K., Koistinen, J., Beckmen, K., Evans, T., Gorzelany, J. F., Hansen, K. J., Jones, P. D., Helle, E., Nyman, M. and Giesy, J. P. (2001). Accumulation of perfluorooctane sulfonate in marine mammals. *Environmental Science and Technology.* 35. pg. 1593-1598.

KemI. (March 15, 1999). *KemI proposes a prohibition of flame retardants.* www.kemi.se/default_eng.htm.

Kilman, Scott. (August 29, 2011). Monsanto corn plant losing bug resistance. *The Wall Street Journal.*

Kinney, Chad A., Furlong, Edward T., Kolpin, Dana W., Burkhardt, Mark R., Zaugg, Steven D., Werner, Stephen L., Bossio, Joseph P. and Benotti, Mark J. (February 20, 2008). Bioaccumulation of pharmaceuticals and other anthropogenic waste indicators in earthworms from agricultural soil amended with biosolid or swine manure. *Environ. Sci. Technol.* 42(6), pg. 1863-70. http://pubs.acs.org/doi/abs/10.1021/es702304c?prevSearch=%255Bauthor%253A%2Bbenotti%25 2C%2BM.%255D&searchHistoryKey=.

Kolpin, D. W., Furlong, E. T., Meyer, M.T., Thurman, E. M., Zaugg, S. D., Barber, L. B., et al. (March 14, 2002). Pharmaceuticals, hormones, and other organic wastewater contaminants in U.S. streams, 1999-2000: a national reconnaissance. *Environmental Science and Technology.* 36(6). pg. 1202-11.

Kristof, Nicholas D. (August 26, 2012). Big Chem, big harm? *The New York Times.* pg. 11.

Kucklick, John R, et al. (no date) *Persistent organic pollutants in murre eggs from the Bering Sea and Gulf of Alaska.* National Institute of Standards, USGS, Alaska Biological Science Center, US Fish and Wildlife Service Alaska Maritime Wildlife Refuge. www.absc.usgs.gov/research/ammtap/Persistent_organic_pollutants_in_Murre_eggs.pdf

LaDlamme, Denise, Stone, Alex and Kraege, Carol. (2008). *Alternatives to Deca-BDE in televisions and computers and residential upholstered furniture.* Department of Ecology State of

Washington and Washington State Department of Health, Olympia, WA. http://www.ecy.wa.gov/pubs/0907041.pdf.

Lane, O. and Evers, D. (2006). *Methylmercury availability in New England estuaries as indicated by Saltmarsh-Sharp-Tail Sparrow, 2004–2005*. Report BRI 2006-01. BioDiversity Research Institute, Gorham, ME.

Lavalli, Kari. (October 2003). Ask the lobster doc: Shell disease and lobsters. *Commercial Fisheries News*.

Law, Robin, et al. (September 2003). Levels and trends of polybrominated diphenylethers and other brominated flame retardants in wildlife. *Environmental International*. 29(6). pg. 757-70.

Lederman, S. A., Jones, R. L., Caldwell, K. L., Rauh, V., Sheets, S. E., Tang, D., Viswanathan, S., Becker, M., Stein, J. L., Wang, R. Y. and Perera, F. P. (August 2008). Relation between cord blood mercury levels and early child development in a World Trade Center cohort. *Environmental Health Perspectives*. 116(8). pg. 1085-91.

Lee, Robert G. M., et al. (2004). PBDEs in the atmosphere of three locations in Western Europe. *Environmental Science and Technology*. 38(3).

Lewis, Michael A., Scott, Geoff I., Bearden, Dan W., Quarles, Robert L., Moore, James, Strozier, Erich D., Sivertsen, Scott K., Dias, Aaron R. and Sanders, Marion. (February 2002). Fish tissue quality in near-coastal areas of the Gulf of Mexico receiving point source discharges. *The Science of the Total Environment*. 284(1-3). pg. 249-261.

Lindstrom, A., Buerge, I. J., Poiger, T., Bergqvist, P. A., Muller, M. D. and Buser H. R. (2002). Occurrence and environmental behavior of the bactericide triclosan and its methyl derivative in surface waters and in wastewater. *Environmental Science & Technology*. 36(11). pg. 2322-29.

Lipton, Eric and Revkin, Andrew C. (November 19, 2001). With water and sweat, fighting the most stubborn fire. *The New York Times*.

Lo, S. C., Pripuzova, N., Li, B., Komaroff, A. L., Hung, G. C., Wang, R. and Alter, H. J. (September 2010). Detection of MLV-related virus gene sequences in blood of patients with chronic fatigue syndrome and healthy blood donors. *Proc Natl Acad Sci*. 107(36). pg. 15874-9. http://www.pnas.org/content/107/36/15874.long.

Lubick, Naomi. (December 2008). Drinking water contamination mapped. *Nature News*. http://www.nature.com/news/2008/081217/full/news.2008.1310.html.

MacInnis, Laura and Sherman, Debra. (March 19, 2009). European lab accidents raise biosecurity concerns. *Reuters*. http://www.reuters.com/article/idUSLJ556939.

Maekawa, A., Matsushima, Y., et al. (1990). Long-term toxicity/carcinogenicity of musk xylol in B6C3F1 mice. *Food Chemical Toxicology*. 28(8). pg. 581-6.

Magos, L. (2001). Review on the toxicity of ethylmercury, including its presence as a preservative in biological and pharmaceutical products. *J Appl Toxicol*. 21(1). pg. 1–5.

Magos, L., Brown, A. W., Sparrow, S., Bailey, E., Snowden, R. T. and Skipp, W. R. (1985). The comparative toxicology of ethyl- and methylmercury. *Arch Toxicol*. 57(4). pg. 260–7.

Maine Bureau of Health. (August 29, 2000). *Warning about eating freshwater fish*. Environmental Toxicology Program, Maine Bureau of Health, Augusta, ME. http://www.maine.gov/dhhs/eohp/fish/documents/2KFCA.pdf

Maine Bureau of Health. (2001). *Bureau of Health fish tissue action levels*. Maine Bureau of Health, Augusta, ME. http://www.maine.gov/dhhs/eohp/fish/documents/Action%20Levels%20Writeup.pdf.

Maine Bureau of Health and Maine Department of Environmental Protection. (2005). *Bromated flame retardants*. Augusta, ME.

Maine Center for Disease Control and Prevention. (June 3, 2009). *Warning about eating saltwater fish and lobster tomalley*. Environmental and Occupational Health Programs, Division of Environmental Health, Augusta, ME. http://www.maine.gov/dhhs/eohp/fish/saltwater.htm.

Maine Department of Environmental Protection. (2005). *Fish tissue contamination in the State of Maine*. http://www.maine.gov/dep/blwq/hg_pres.htm.

Maine Department of Environmental Protection. (2007). *3rd annual report on brominated flame retardants*. Augusta, ME.

Maine Department of Environmental Protection. (2008). *Surface water ambient toxic monitoring program: 2007 report*. Maine Department of Environmental Protection, Augusta, ME.

Maine Department of Environmental Protection. (2009). *Chemicals of high concern list*. Posted by Maine DEP July 17, 2009. http://www.maine.gov/dep/oc/safechem/highconcern/DEP.CHC.web.short_list_7_16_09.pdf

Maine Department of Human Services. (1997). *Maine 1997 fish consumption advisories*. Handout. Bureau of Health, Maine Department of Human Services, Augusta, ME.

Maine State Legislature. (2004). *LD 1790: An act to reduce contamination of breast milk and the environment from the release of brominated chemicals in consumer products*. Paper HP1312. 121st Maine Legislature, second special session, Augusta, ME. http://www.mainelegislature.org/legis/bills/display_ps.asp?PID=1456&snum=121&paper=&ld=1790.

McAllister, Philip E. (1990). *Viral hemorrhagic septicemia of fishes*. Fish Disease Leaflet 83. (Revision of Fish Disease Leaflet 6, Ken Wolf, 1966). US Department of the Interior, Fish and Wildlife Service.

McAvoy, D. C., et al. (2002). Measurement of triclosan in wastewater treatment systems. *Environmental Toxicology and Chemistry 2002*. pg. 1323-29.

McMurray, L. M., Oethinger, M. and Levy, S. B. (1998). Triclosan targets lipid synthesis. *Nature*. Vol. 394. pg. 531-532.

Medical News Today. (January 27, 2009). Mercury found in high fructose corn syrup used as food sweetener. *Medical News Today*. www.medicalnewstoday.com/articles/136879.php.

Melcher, Joan. (March 15, 2010). Poor deer season spurs chemical concerns. *Miller-McCune Online Magazine*. http://www.miller-mccune.com/print/?open=10902.

Miller, Georgia and Senjen, Rye. (2008). *Out of the laboratory and onto our plates: Nanotechnology in food & agriculture*. Friends of the Earth, Australia, Europe, and USA, Washington, DC. http://www.foe.org/pdf/nano_food.pdf.

Miller, Morton W. and Berg, George G., Eds. (1970). *Chemical fallout: Current research on persistent pesticides*. Charles C. Thomas, Publisher, Springfield, IL.

Mineau, Pierre and Whiteside, Melanie. (2013). Pesticide acute toxicity is a better correlate of U.S. grassland bird declines than agricultural intensification. *PLoS ONE*. 8(2). http://www.abcbirds.org/abcprograms/policy/toxins/Grassland_birds_PLOS_One_Feb_2013.pdf.

Ministries of Agriculture, Fish and Food. (1982). *Survey of lead in food: Second supplementary report*. Her Majesty's London Stationary Office, London.

Mittelstaedt, Martin. (April 1, 2010). BPA widespread in ocean water and sand. *The Globe and Mail*. http://www.theglobeandmail.com/life/health/bpa-widespread-in-ocean-water-and-sand/article1520625/.

Mittelstaedt, Martin. (September 24, 2010). The disturbing truth about cellphones. *The Globe and Mail*. http://www.theglobeandmail.com/life/the-disturbing-truth-about-cellphones/article1724983/.

Miyata, H., Aozasa, O., Nakao, T. and Ohta, S. (May 2009). Investigation of PCB pollution evaluation for Yusho victims using their preserved umblical cord. *Fukuoka Igaku Zasshi*. 100(5). pg. 183-91.

Mizota, K. and Ueda, H. (2006). Endocrine disrupting chemical atrazine causes degranulation through $G_{q/11}$ protein-coupled neurosteroid receptor in mast cells. *Toxicological Sciences*. 90(2). pg. 362-8. http://toxsci.oxfordjournals.org/cgi/content/full/90/2/362#BDY.

Moore, C., Moore, S., Leecaster, M. and Weisberg S. (2001). A comparison of plastic and plankton in the North Pacific central gyre. *Marine Pollution Bulletin*. 42(12). pg. 1297 - 300.

Moore, Charles. (2002). A comparison of neustonic plastic and zooplankton abundance in southern California's coastal waters and elsewhere in the North Pacific. *Algalita Marine Research Foundation*. http://www.mindfully.org/Plastic/Ocean/Marine-Debris-Panel3Ooct02.htm.

Morris, Steven, et al. (2004). Distribution and fate of HBCD and TBBPA brominated flame retardants in North Sea estuaries and aquatic food webs. *Environmental Science and Technology*. 38(21).

Mott, Lawrie and Broad, Martha. (March 15, 1984). *Pesticides in food: What the public needs to know*. Natural Resources Defense Council, Inc., San Francisco, CA.

Motzer, William E. (2001). Perchlorate: Problems, detection, and solutions. *Environmental Forensics*. 2(4). pg. 301-11.

Mouawad, Jad and Krauss, Clifford. (December 8, 2009). Dark side of a natural gas boom. *The New York Times*. http://www.nytimes.com/2009/12/08/business/energy-environment/08fracking.html?_r=1&adxnnl=1&ref=todayspaper&adxnnlx=1268316739-j+9YFP+MPdx/4g+wo26Lqw.

Mueller, Sue. (8/15/07). *Flame retardants blamed for cat thyroid disease*. Environmental Working Group. www.ewg.org/node/22439.

Murata, K., Dakeishi, M., Shimada, M. and Satoh, H. (September 2007). Usefulness of umbilical cord mercury concentrations as biomarkers of fetal exposure to methylmercury. *Nippon Eiseigaku Zasshi*. 62(4). pg. 949-59.

Naidenko, Olga, Sutton, Rebecca and Houlihan, Jane. (2008). *High levels of industrial chemicals contaminate cats and dogs*. Environmental Working Group. http://www.ewg.org/reports/pets.

National Atmospheric Deposition Program/Mercury Deposition Network. (2006). *Monitoring mercury deposition: A key tool to understanding the link between emissions and effects*. Illinois State Water Survey, Champaign, IL. http://nadp.sws.uiuc.edu/lib/brochures/mdn.pdf.

National Acid Precipitation Assessment Program. (1987). *Interim assessment; The causes and effects of acid deposition, Vol. 2, Emissions and control*. Government Printing Office, Washington, D.C.

National Institute of Health Sciences. (June 2002). *Tetrabromobisphenol A: Review of toxicological literature*.

National Research Council. (2000). *Toxicological effects of methylmercury*. Committee on the Toxicological Effects of Methylmercury. National Academy Press, Washington, DC. http://books.nap.edu/books/0309071402/html/index.html

National Research Council. (2004). *Safety of genetically engineered foods: Approaches to assessing unintended health effects*. National Academy Press, Washington, DC.

New York Times, The. (October 14, 2007). Pollution from Chinese coal casts a global shadow. *The New York Times*.

Noren, Koidu. (1983). Some aspects of the determination of organochlorine contaminants in human milk. *Archives of Environmental Contamination and Toxicology*. 12. pg. 277-283.

Oaks, J. L., Gilbert, M., Virani, M. Z., Watson, R. T., et al. (2004). Doclofenac residues as the cause of vulture population decline in Pakistan. *Nature*. 427(6975). pg. 630-3.

O'Conner, Tom. (1998). *State of the coast: Chemical contaminants in oysters and mussels*. National Oceanic and Atmospheric Administration (NOAA), Silver Spring, MD. http://oceanservice.noaa.gov/websites/retiredsites/sotc_pdf/CCOM.PDF.

Office of Technology Assessment. (1984). *Protecting the Nation's groundwater from contamination—Volume I*. OTA–O–233. U.S. Congress, Office of Technology Assessment, Washington, DC.

Organization of Economic Co-operation and Development. (2004). *Perfluorooctane Sulfonate (PFOS) and related chemical products*. www.oecd.org/document/58/0,2340,fr_2649_34375_238437_1-1-1-1,00.html.

Owens, A. J., et al. (1982). The atmospheric lifetimes of CFC 11 and CFC 12. *Geophysical Research Letters*. 9(6). pg 700-03.

Ozone Hole, The. (2008). *2008 sees fifth largest ozone hole*. www.theozonehole.com/noaa2008.htm.

Ozone Hole, The. (2009). *The ozone hole 2009*. www.theozonehole.com/ozone2009.htm.

Paarlberg, Robert. (2008). *Starved for science: How biotechnology is being kept out of Africa*. Harvard University Press, Cambridge, MA.

Pankow, J. F. and Cherry, J. A. (1996). *Dense chlorinated solvents and other DNAPLs in ground water—History, behavior, and remediation*. Waterloo Press, Portland, OR.

Pearson, R. G. and Percy, K. E. (1990). *The 1990 Canadian Long-Range Transport of Air Pollution and Acid Deposition Report, Part V, Terrestrial Effects*. Federal/Provincial Research and Monitoring Coordinating Committee, Ottawa, Canada.

Pellmar, T. C., Hogan, J. B., Benson, K. A. and Landauer, M. R. (1998). *Toxicological evaluation of depleted uranium in rats: Six month evaluation point*. AFRRI Special Publication 98-1. Armed Forces Radiobiology Research Institute.

Peterle, T. J. (November 1969). DDT in Antarctic snow. *Nature*. 224(5219). pg. 620. http://www.nature.com/nature/journal/v224/n5219/pdf/224620a0.pdf.

Peters, R. (2005). *Man-made chemicals in maternal cord blood*. B&O-A R 2005/129. TNO Report for GeenPeace International and World Wildlife Fund-UK. Netherlands Organisation for Applied Scientific Research, Apeldoorn, The Netherlands.

Pinkus, A. G. (June 25, 1997). Synthetic and Natural Phenols. Studies in Organic Chemistry #52 By J. H. P. Tyman. *Environmental Science and Technology*.

Planchon, F., Van de Velde, K., Rosman, K. J. R., Wolff, E. W., Ferrari, C. and Boutron, C. F. (2003). One-hundred fifty-year record of lead isotopes in Antarctic snow from Coats Land. *Geochim. Cosmochim. Acta*. 67(4). pg. 693-708.

Polishuk, Z. W., Ron, M., Wassermann, M., Cucos, S., Wassermann, D. and Lemesch, C. (March 1977). Organochlorine compounds in human blood plasma and milk. *Pesticides Monitoring Journal*. 10(4). pg. 121-129.

Polishuk, Z. W., Wassermann, D., Wassermann, M., Cucos, S. and Ron, M. (1977). Organochlorine compounds in mother and fetus during labor. *Environmental Research*. 13. pg. 278-284.

Pollack, Andrew. (September 20, 2012). Foes of modified corn find support in a study. *The New York Times*. pg. B5.

Pope, Carl. (February 23, 2009). Who is getting it done? *Sierra Club Insider*. Sierra Club USA.

Psomopoulos, C. S., Bourke, A. and Themelis, N. J. (2009). Waste-to-energy: A review of the status and benefits in USA. *Waste Management*. 29. pg. 1718-24. http://www.nmwda.org/news/documents/Tab3-Psomopoulosetal2009WTEstatusandbenefits2.pdf.

Public Employees for Environmental Responsibility. (September 1995). *Genetic genie: The premature commercial release of genetically engineered bacteria*. PEER, Washington, DC.

Puckett, Jim and Smith, Ted, Eds. (2002). *Exporting harm: The high-tech trashing of Asia*. Basel Action Network, Seattle, WA and Silicon Valley Toxics Coalition, San Jose, CA.

Quammen, David. (2012). *Spillover: Animal infections and the next human pandemic*. W. W. Norton & Co., NY.

Quraishi, Ash-har. (November 23, 2011). Tiny mussels invade Great Lakes, threaten fishing industry. *PBS NewsHour*. http://www.pbs.org/newshour/bb/environment/july-dec11/mussels_11-23.html.

Rachowicz, L. J., Knapp, R. A., Morgan, J. A. T., Stice, M. J., Vredenburg, V. T., Parker, J. M. and Briggs, C. J. (2006). Emerging infectious disease as a proximate cause of amphibian mass mortality. *Ecology.* 87(7). pg. 1671-83. http://web.me.com/vancevredenburg/Vances_site/Publications_files/Rachowicz%20et%20al%202006.pdf.

Ravishankara, A. R., Daniel, J. S. and Portmann, R. W. (2009). Nitrous oxide (N_2O): The dominant ozone-depleting substance emitted in the 21st century. *Science.* 326(5949). pg. 123-5.

Ritter, L., Solomon, K. R., Forget, J., Stemeroff, M. and O'Leary, C. (1995). *Persistent organic pollutants: An assessment report on: DDT-aldrin-dieldrin-endrin-chlordane heptachlor-hexachlorobenzene mirex-toxaphene polychlorinated biphenyls dioxins and furans.* The International Programme on Chemical Safety (IPCS).

Roan, S. (1989). *Ozone crisis.* John Wiley and Sons, NY.

Rogan, W. J. and Ragan, N. B. (July 2003). Evidence of effects of environmental chemicals on the endocrine system in children. *Pediatrics.* 112(1 Pt 2). pg. 247-52. http://pediatrics.aappublications.org/cgi/reprint/112/1/S1/247.

Rollin, H. B., Rudge, C. V., Thomassen, Y., Mathee, A. and Odland, J. Ø. (March 2009). Levels of toxic and essential metals in maternal and umbilical cord blood from selected areas of South Africa – results of a pilot study. *Journal of Environmental Monitoring.* 11(3). pg. 618-27.

Rollin, H. B., Sandanger, T. M., Hansen, L., Channa, K. and Odland, J. Ø. (December 2009). Concentration of selected persistent organic pollutants in blood from delivering women in South Africa. *Science of the Total Environment.* 408(1). pg. 146-52.

Ronald, Pamela and Adamchak, Raoul. (2008). *Tomorrow's table: Organic farming, genetics, and the future of food.* Oxford University Press, NY.

Rosenthal, Elisabeth. (August 24, 2010). In the fields of Italy a conflict over corn. *The New York Times.* pg. A4.

Rosman, K. J. R., Chisholm, W., Boutron, C. F., Hong, S., Edwards, P. R., Morgan, V. I. and Sedwick, P. N. (1998). Lead isotopes and selected metals in ice from Law Dome, Antarctica. *Annals of Glaciology.* 27(1). pg. 349-354.

Ross, Will. (August 5, 2008). Computers pile up in Ghana dump. *BBC News.* http://news.bbc.co.uk/2/hi/africa/7543489.stm.

Rossol, Monona. (2011). *Pick your poison: How our mad dash to chemical utopia is making lab rats of us all.* Wiley, Hoboken, NJ.

Ryan, P. A., Hafner, H. R. and Brown, S. G. (2003). *Deposition of air pollutants to Casco Bay.* Casco Bay Estuary Partnership Final Report STI-902150-2209-FR2. University of Southern Maine, Portland, ME. http://www.cascobay.usm.maine.edu/pdfs/SONOMA.pdf.

Saey, Tina Hesman. (May 24, 2008). Epic genetics: Genes' chemical clothes may underlie the biology behind mental illness. *Science News.* http://www.thefreelibrary.com/Epic+genetics%3a+genes'+chemical+clothes+may+underlie+the+biology...-a0179535436.

Sakamoto, M., Murata, K., Kubota, M., Nakai, K. and Satoh, H. (January 2010). Mercury and heavy metal profiles of maternal and umbilical cord RBCs in Japanese population. *Ecotoxicol Environ Saf.* 73(1). pg.1-6.

Schade, Michael. (2010). *A call for congressional action: Toxic Toys "R" Us: PVC toxic chemicals in toys and packaging: A report to the National Commission of Inquiry into Toxic Toys.* Center for Health, Environment & Justice (CHEJ) and Teamsters Office of Consumer Affairs, Washington, DC. http://www.toxictoysrus.com/documents/Toxic-Toys-Report_11.17.10.pdf.

Schapiro, Mark. (2007). *Exposed: The toxic chemistry of everyday products and what's at stake for American power.* Chelsea Green Publishing, White River Jct., VT.

Schoch, Nina and Jackson, Allyson. (2012). *Adirondack loons – sentinels of mercury pollution in New York's aquatic ecosystems.* BRI Report 2011-29. Biodiversity Research Institute, Gorham, ME. https://docs.google.com/a/briloon.org/viewer?url=http://www.briloon.org/uploads/centers/looncenter/AdkLoonsSofMP-lr.pdf.

Schubert, Charlotte. (October 13, 2001). Burned by flame retardants? *Science News.* 160. pg. 238-239.

Schultz, Terrie. (March 26, 2010). Toxic chemicals found in cord blood of newborns: Infants are exposed to harmful substances before birth. *suite101.com.* http://prenatal-health.suite101.com/article.cfm/toxic-chemicals-found-in-cord-blood-of-newborns.

Science Daily. (April 18, 2008). Early exposure to common weed killer impairs amphibian development. *ScienceDaily.* http://www.sciencedaily.com/releases/2008/04/080416091015.htm.

Scott, C. (2009). 30 Years of US-Approved Recombinant Cell Culture Products. *BioProcess Int. supplement 7.*

Shah, Sonia. (April 15, 2010). As pharmaceutical use soars, drugs taint water and wildlife. *Yale e360.* http://www.e360.yale.edu/content/print.msp?id=2263.

Sharp, Kathleen. (October 12, 2012). A drug to quicken the blood. *The New York Times.* pg. A23.

Shulman, Seth. (1992). *The threat at home: Confronting the toxic legacy of the U.S. military.* Beacon Press, Boston.

Shulman, Seth. (1993) *Biohazard: How the Pentagon's biological warfare research program defeats its own goals.* A report for the Center for Public Integrity. http://www.publicintegrity.org/assets/pdf/BIOHAZARD.pdf.

Silicon Valley Toxics Coalition. (2002). *Poison PCs and toxic TVs.* San Jose, CA. www.svtc.org.

Silicon Valley Toxics Coalition. (2009). *Toward a just and sustainable solar energy industry.* A Silicon Valley Toxics Coalition White Paper. http://www.svtc.org/site/DocServer/Silicon_Valley_Toxics_Coalition_-_Toward_a_Just_and_Sust.pdf?docID=821.

Skakkebaek, N. E. (2001). Hormone and endocrine disrupters in food and water: Possible impact on human health. *APMIS.* Supplement No. 13. (109).

Smith, Andrew E. and Frohmberg, Eric. (2008). *Evaluation of the health implications of levels of polychlorinated dibenzo-p-dioxins (dioxins) and polychlorinated dibenzofurans (furans) in fish from Maine rivers: 2008 update*. Maine Department of Health and Human Services, Augusta, ME. http://www.maine.gov/dhhs/eohp/fish/documents/FinalDraft_Eval_of_PCDD.pdf.

Smith, Jeffrey M. (2003). *Seeds of deception: Exposing industry and government lies about the safety of the genetically engineered foods you're eating*. Yes! Books, Fairfield, IA.

Smith, Jeffrey M. (2007). *Genetic Roulette: The documented health risks of genetically engineered foods*. Chelsea Green, White River Junction, VT.

Solomon, Gina, Weiss, Pilar, Owen, Beth and Citron, Annicke. (March 25, 2005). *Healthy milk, healthy baby: Chemical pollution and mother's milk*. Natural Resources Defense Council. www.nrdc.org/breastmilk/.

Solomon, Susan, et al. (2006). *Contrasts between Antarctic and Arctic ozone depletion*. Earth System Research Laboratory, NOAA. Fort Collins, CO.

Solomon, Susan, et al. (January 3, 2007). Contrasts between Antarctic and Arctic ozone depletion. *Proceedings of the National Academies of Science*. www.pnas.org/cgi/doi/10.1073/pnas.0604895104.

Speight, J. G., Ed. (1999). *The chemistry and technology of petroleum*. Marcel Dekker, NY, NY.

Spiro, T. G. and Thomas, V. M. (October 1994). Sources of dioxin. *Science*. 266(5184). pg. 349.

Stapleton, Heather M., et al. (2004). Debromination of polybrominated diphenol ether congeners BDE 99 and BDE 183 in the intestinal tract of the common carp (Cyprinus carpio). *Environmental Science and Technology*. 38(4).

Stapleton, H. M., Dodder, N. G., Offenberg, J. H., Schantz, M. and Wise, S. A. (2005). Polybrominated diphenyl ethers in house dust and clothes dryer lint. *Environ.Sci.Technol.* 39. pg. 925-31. http://pubs.acs.org/doi/abs/10.1021/es0486824.

Strum, Carol Van. (1983). *A bitter fog: Herbicides and human rights*, Sierra Club Books, San Francisco, CA.

Sturchio, Neil C., Caffee, Marc, Beloso, Abelardo D., Jr., Heraty, Linnea J., Böhlke, John Karl, Hatzinger, Paul B., Jackson, W. Andrew, Gu, Baohua, Heikoop, Jeffrey M. and Dale, Michael. (2009). Chlorine-36 as a tracer of perchlorate origin. *Environmental Science & Technology*, 43(18). pg. 6934-8. http://pubs.acs.org/doi/abs/10.1021/es9012195?prevSearch=sturchio&searchHistoryKey=.

Szlinder-Richert, J., Barska, I., Mazerski, J. and Usydus, Z. (May 2008). Organochlorine pesticides in fish from the southern Baltic Sea: Levels, bioaccumulation features and temporal trends during the 1995-2006 period. *Marine Pollution Bulletin*. 56(5). pg. 927–40.

Tan L, Nielsen, N. H., Young, D. C. and Trizna, Z. (2002). Use of antimicrobial agents in consumer products. *Archives of Dermatology*. 138(8). pg. 1082-6.

Tennant, P. A., Norman, C. G. and Vicory, A. H., Jr. (1992). The Ohio River Valley Water Sanitation Commission's Toxic Substances Control Program for the Ohio River. *Water Science and Technology*. 26(7–8). pg. 1779-88.

Thomas, Jim. (February 2, 2011). The sins of syn bio: How synthetic biology will bring us cheaper plastics by ruining the poorest nations on earth. *Slate Magazine*. http://www.slate.com/articles/technology/future_tense/2011/02/the_sins_of_syn_bio.html.

Thomas, V. M., Bedford, J. A. and Cicerone, R. J. (June 1, 1997). Bromine emissions from lead gasoline. *Geophysical Research Letters*. 24(11). pg. 1371-74.

Thornton, Joe. (2000). *Pandora's poison: Chlorine, health and a new environmental strategy*. MIT Press, Cambridge, MA.

Thornton, Joe. (2002). *Environmental impacts of polyvinyl chloride building materials*. Healthly Building Network, Washington, DC.
http://www.healthybuilding.net/pvc/Thornton_Enviro_Impacts_of_PVC.pdf.

Thurlow, William. (1979). *Matacil spray report*. Gander Environmental Group, Gander, Newfoundland.

Tomlinson, G. H. (date unknown). Dieback of red spruce; acid deposition, and changes in soil nutrient status - a review. In: Ulrich, B. and Pankrath, J., Eds. *Effects of accumulation of air pollutants in forest ecology*.

Tomy, Gregg T., et al. (2004). Bioaccumulation, biotransformation, and biochemical effects of brominated diphenyl ethers in juvenile lake trout (Savelinus manaycush). *Environmental Science and Technology*. 38(5).

Trasande, L., Attina, T. M. and Blustein, J. (September 2012). Association between urinary bisphenol A concentration and obesity prevalence in children and adolescents. *Journal of the American Medical Association*. 308(11). pg.1113-21.

Troesken, Werner. (2008). *The great lead water pipe disaster*. The MIT Press, Cambridge, MA.

Ueno, Daisuke, et al. (2004). Global pollution monitoring of polybrominated dipheyl ethers using skipjack tuna as a bioindicator. *Environmental Science and Technology*. 38(5).

Union of Concerned Scientists. (March 2009). Do bioplastics deserve a seat at your table? *Greentips*. http://www.ucsusa.org/publications/greentips/.

United Nations Environment Programme Ozone Secretariat. (1999). *The Montreal protocol on substances that deplete the ozone layer*. UNEP Ozone Secretariat, United Nations Environment Programme, Beijing.

United States Environmental Protection Agency. (1997). *Mercury study report to Congress. Volume VI: An ecological assessment of anthropogenic mercury emissions in the United States*. 452/R-97-008.

United States Environmental Protection Agency. (1997). *The incidence and severity of sediment contamination in surface waters of the United States: Volume 1: National sediment quality survey*. EPA 823-R-97-006. United States Environmental Protection Agency, Office of Science and Technology, Washington, DC.

United States Environmental Protection Agency. (2001). *Frequently asked questions about atmospheric deposition: A handbook for watershed managers*. EPA-453/R-01-009. http://www.epa.gov/oar/oaqps/gr8water/handbook/airdep_sept_1.pdf.

United States Environmental Protection Agency. (2000). *Guidance for assessing chemical contaminant data for use in fish advisories. Volume 2. Risk assessment and fish consumption limits*. EPA 823-B-00-007. USEPA, Washington, DC.
http://www.epa.gov/waterscience/fish/advice/volume2/v2cover.pdf.

United States Food and Drug Administration. (November 2010). *Protocol for interpretation and use of sensory testing and analytical chemistry results for re-opening oil-impacted areas closed to seafood harvesting due to the Deepwater Horizon oil spill.* http://www.fda.gov/Food/ucm217601.htm.

University of London School of Pharmacy. (2009). *State of the art report on mixture toxicity.* European Commission. http://ec.europa.eu/environment/chemicals/pdf/report_Mixture%20toxicity.pdf.

Urbina, Ian. (February 27, 2011). Regulation lax as gas wells' tainted water hits rivers. *The New York Times.* pg. 1, 22-3.

Urbina, Ian. (June 27, 2011). Behind veneer, doubt on future of natural gas. *The New York Times.* pg. 1, A12.

Urbina, Ian. (April 14, 2013). Think those chemicals have been tested? *The New York Times.* http://www.nytimes.com/2013/04/14/sunday-review/think-those-chemicals-have-been-tested.html?_r=0.

Vallelonga, P., Van de Velde, K., Candelone, J.-P., Morgan, V. I., Boutron, C. F. and Rosman, K. J. R. (2002). The lead pollution history of Law Dome, Antarctica, from isotopic measurements on ice cores: 1500 AD to 1989 AD. *Earth and Planetary Science Letters.* 204(1-2). pg. 291-306.

Vandal, G., Fitzgerald, W., Boutron, C. and Candelone, J. (April 1993). Variations in mercury deposition to Antarctica over the past 34,000 years. *Nature.* 362.

Vander Pol, et al. (2003). *Seabird tissue archival and monitoring project: egg collections and analytical results for 1999-2002.* National Institute of Standards and Technology; US Dept. of the Interior, US Fish and Wildlife Service, Alaska Maritime National Wildlife Refuge; US Dept. of the Interior, US Geological Society, Alaska Science Center.

Vives, Ingrid, et al. (2004). Polybromodiphenyl ether flame retardants in fish from lakes in European high mountains and Greenland. *Environmental Science and Technology.* 38(8).

Vogelmann, H. W., Badger, G. J., Bliss, M. and Klein, R. M. (August 1985). Forest decline on Camel's Hump Vermont. *Bulletin of the Torrey Botanical Club.* 112(3). pg. 274-87. http://www.jstor.org/pss/2996543.

Vokoun, J. C. and Perkins, C. R. (2008). *Second statewide assessment of mercury contamination in fish tissue from Connecticut lakes (2005-2006).* Completion Report submitted to Connecticut Department of Environmental Protection. University of Connecticut, Storrs, CT. http://www.nre.uconn.edu/pages/people/bios/VokounDocs/MercuryBass2005_06_Vokoun_Perkins2008.pdf.

Wade, Terry L. and Sweet, Stephen T. (May 2005). *Assessment of sediment contamination in Casco Bay.* Casco Bay Estuary Project, Portland, ME. http://www.cascobay.usm.maine.edu/pdfs/CCBay%20Report%20May2005.pdf.

Wargo, John. (1996). *Our children's toxic legacy: How science and law fail to protect us from pesticides.* Yale University Press, New Haven, CT.

Weschler. C. J. (2009). Changes in indoor pollutants since the 1950s. *Atmospheric Environment.* 43(1). pg. 153-69.

Westrick, J. J. (1990). National surveys of volatile organic compounds in ground and surface waters. In: Ram, N. M., Christman, R. F. and Cantor, K. P., Eds. *Significance and treatment of volatile organic compounds in water supplies*. Lewis Publishers, Chelsea, MI.

Westrick, J. J., Mello, J. W. and Thomas, R. F. (1984). The groundwater supply survey. *Journal of the American Water Works Association*. 76(5). pg. 52-9.

White, Louise, Wells, Peter, Jones, Steve, Krahforst, Christian, Harding, Gareth, Hennigar, Peter, Grun, Guy and Landry, Natalie. (2001). *Nine-year review of Gulfwatch in the Gulf of Maine: Trends in tissue contaminant levels in the blue mussel, Mytilus edulis L., 1993-2001*. Gulfwatch, Boscawan, NH.

Wilkening, K. E., Barrie, L. A. and Engle, M. (2000). Trans-Pacific air pollution. *Science*. 290. pg. 65-7.

Wilson, Duff. (2001). *Fateful harvest: The true story of a small town, a global industry, and a toxic secret*. Harper Collins, NY.

Wines, Michael. (September 17, 2013). Gas leaks in fracking disputed in study: Less methane found than U.S. believed. *The New York Times*. http://www.nytimes.com/2013/09/17/us/gas-leaks-in-fracking-less-than-estimated.html?_r=0.

Wolff, E. W. and Suttie, E. D. (1994). Antarctic snow record of southern hemisphere lead pollution. *Geophysical Research Letters*. 21. pg. 781-784.

Wolff, M. S., Teitelbaum, S. L., Windham, G., Pinney, S. M., Britton, J. A. and Chelimo, C. (2007). Pilot study of urinary biomarkers of phytoestrogens, phthalates, and phenols in girls. *Environmental Health Perspectives*. 115. pg. 116-21.

Wolkers, Hans, et al. (2004). Congener-specific accumulation and food chain transfer of polybrominated diphenyl ethers in two Arctic food chains. *Environmental Science & Technology*. 38(6). pg. 1667-74.

World Bank. (July 9, 2007). Bangladesh: *overusing pesticides in farming*. http://web.worldbank.org/WBSITE/EXTERNAL/COUNTRIES/SOUTHASIAEXT/0,,contentMDK:21204863~pagePK:146736~piPK:146830~theSitePK:223547,00.html.

Yates, D. and Evers, D. C. (2006). *Assessment of bats for mercury contamination on the North Fork of the Holston River, VA*. Report BRI 2006-9. BioDiversity Research Institute, Gorham, ME.

Yorifuji, T., Kashima, S., Tsuda, T. and Harada, M. (December 2009). What has methylmercury in umbilical cords told us? - Minamata disease. *Sci Total Environ*. 408(2). pg. 272-6.

Zeller, Tom, Jr. (April 14, 2010). A program to certify electronic waste recycling rivals an industry-U.S. plan. *The New York Times*. http://www.nytimes.com/2010/04/15/business/energy-environment/15ewaste.html.

Zogorski, John S., Carter, Janet M., Ivahnenko, Tamara, Lapham, Wayne W., Moran, Michael J., Rowe, Barbara L., Squillace, Paul J. and Toccalino, Patricia L. (2006). *Volatile organic compounds in the nation's ground water and drinking-water supply wells*. U.S. Department of the Interior, U.S. Geological Survey. Circular 1292. Reston, VA. http://pubs.usgs.gov/circ/circ1292/pdf/circular1292.pdf.

Health Effects and Human Body Burdens

Adams, J. B. (July 2004). *A review of the autism-mercury connection.* Conference Proceedings of the Annual Meeting of the Autism Society of America.

Agency for Toxic Substances and Disease Registry (ATSDR). (2004). *Interaction profile for: Persistent chemicals found in breast milk (chlorinated dibenzo-p-dioxins, hexachlorobenzene, p,p'-DDE, methylmercury, and polychlorinated biphenyls.* US Department of Health and Human Services, Public Health Service, Atlanta, GA. http://www.atsdr.cdc.gov/interactionprofiles/IP-breastmilk/ip03.pdf.

Allen, R. H., Gottlieb, M., Clute, E. et al. (1997). Breast cancer and pesticides in Hawaii: The need for further study. *Environmental Health Perspectives.* 105(suppl. 3). pg. 679-83.

American Academy of Pediatrics. (September 1999). Thimerosal in vaccines - An interim report to clinicians (RE9935). *American Academy of Pediatrics Policy Statement.* 104 (3). pg. 570-574.

American Chemical Society. (February 24, 2005). Perchlorate found in dairy and breast milk samples from across the country. *ScienceDaily.* http://www.sciencedaily.com/releases/2005/02/050222110959.htm.

Ames, B. N., Magaw, R. and Gold, L. S. (1987). Ranking possible carcinogenic hazards. *Science.* 236. pg. 271-80.

Amin-Zaki, Laman, Majeed, M. A., Greenwood, Michael R., Elhassani, Sami B., Clarkson, Thomas W. and Doherty, Richard A. (1981). Methylmercury poisoning in the Iraqi suckling infant: A longitudinal study over five years. *Journal of Applied Toxicology.* 1(4). pg. 210-14. http://www3.interscience.wiley.com/journal/112227996/abstract.

Anderson, H. A., Imm, P., Knobeloch, L., Turyk, M., Mathew, J., Buelow, C. and Persky, V. (September 2008). Polybrominated diphenyl ethers (PBDE) in serum: Findings from a US cohort of consumers of sport-caught fish. *Chemosphere.* 73(2). pg. 187-94.

Angell, Marcia. (June 23, 2011). The epidemic of mental illness: Why? *The New York Review.* http://www.nybooks.com/articles/archives/2011/jun/23/epidemic-mental-illness-why/?pagination=false.

Apostolidis S, Chandra, T., et al. (2002). Evaluation of carcinogenic potential of two nitro-musk derivatives, musk xylene and musk tibetene in a host-mediated in vivo/in vitro assay system. *Anticancer Research.* 22(5). pg/ 2657–62. http://www.geocities.com/Athens/Rhodes/8729/iiar/present.htm.

Arnold, Ryan S., Thom, Kerri A., Sharma, Saarika, Phillips, Michael, Johnson, J. Kristie, and Morgan, Daniel J. (January 2011). Emergence of *Klebsiella pneumoniae* Carbapenemase (KPC)-Producing Bacteria. *South Med J.* 104(1). pg. 40-5. http://www.ncbi.nlm.nih.gov/pmc/articles/PMC3075864/pdf/nihms-253051.pdf.

Association of Infection Control Professionals and Epidemiologists. (November 8, 2008). *National prevalence study of C. difficile in the US healthcare facilities.* www.apic.org.

Bakalar, Nicholas. (February 5, 2013). Most food illness from greens. *The New York Times.* http://well.blogs.nytimes.com/2013/02/04/most-food-illnesses-come-from-greens/.

Bakalar, Nicholas. (February 5, 2013). Pneumonia strain has spread. *The New York Times.* http://well.blogs.nytimes.com/2013/02/04/pneumonia-strain-has-spread/.

Barclay, Eliza. (March 10, 2010). What's best for kids: Bottled water or fountains? *National Geographic News*. http://news.nationalgeographic.com/news/2010/02/100303-bottled-water-tap-schools/.

Barr, Dana B., Bishop, Amanda and Needham, Larry L. (2007). Concentrations of xenobiotic chemicals in the maternal-fetal unit. *Reproductive Toxicology*. 23. pg. 260-6.

Barrett, J. R. (2005). Phthalates and baby boys: Potential disruption of human genital development. *Environmental Health Perspective*. 113(8). pg. 542–542. http://www.jstor.org/stable/3436340.

Barringer, Felicity. (January 24, 2009). Exposed to solvent, worker faces hurdles. *The New York Times*. http://www.nytimes.com/2009/01/25/us/25toxic.html?_r=1&scp=10&sq=tce&st=cse.

Bauman, M. L. and Kemper, T. L. (1994). *The neurobiology of autism*. Johns Hopkins University Press, Baltimore, MD.

Behar, Andrew, Fugere, Danielle and Passoff, Michael. (2013). Slipping through the cracks: An issue brief on nanomaterials in foods. *AsYouSow.org*. http://www.asyousow.org/publications/2013/SlippingThroughTheCracks-20130208.pdf.

Bell, Michelle L., McDermott, Aidan, Zeger, Scott L., Samet, Jonathan M. and Dominici, Francesca. (2004). Ozone and short-term mortality in 95 US urban communities, 1987-2000. *Journal of the American Medical Association*. 292. pg. 2372-8.

Belluck, Pam. (January 13, 2010). Obesity rates hit plateau in U.S., data suggest. *The New York Times*. http://www.nytimes.com/2010/01/14/health/14obese.html.

Benachour, Nora and Aris, Aziz. (December 2009). Toxic effects of low doses of bisphenol-A on human placental cells. *Toxicology and Applied Pharmacology*. 241(3). pg. 322-8.

Berglund, A. (1990). Estimation by a 24-hours study of the daily dose of intra-oral mercury vapor inhaled after release from dental amalgam. *Journal of Dental Research*. 69. pg. 1646-1651.

Berkson, Lindsey. (2001). *Hormone deception*. Contemporary/McGraw-Hill, NY.

Bernard, S., Enayati, A., Redwood, L., Roger, H., and Binstock, T. (2001). Autism: a novel form of mercury poisoning. *Medical Hypotheses*. 56(4). pg. 462-471.

Bernard, S., Enayati, A., Roger, H. Binstock, T. and Redwood, L. (2002). The role of mercury in the pathogenesis of autism. *Mol. Psychiatr*. 7. pg. S42-3.

Bertrand, J., Mars, A., Boyle, C., Bove, F., Yeargin-Allsopp, M. and Decoufle, P. (2001). Prevalence of autism in a United States population: The Brick Township, New Jersey, Investigation. *Pediatrics*. 108(5). pg. 1155-61.

Betts, K. S. (February 2002). Rapidly rising PBDE levels in North America. *Environmental Science &. Technology*. 36(3). g. 50A-2A. http://pubs.acs.org/doi/abs/10.1021/es022197w.

Betts, Kellyn. (July 15, 2006). Bacteria may break down popular flame retardant to produce toxics. *Environmental Science & Technology*. pg. 4329-30.

Bilefsky, Dan. (October 15, 2010). Hungary sludge a warning for other sites in Europe. *The New York Times*. http://query.nytimes.com/gst/fullpage.html?res=9904E3D9133EF936A25753C1A9669D8B63&scp=1&sq=hungary%20sludge%20a%20warning%20for%20other%20sites%20in%20Europe&st=Search.

Birnbaum, L. S. (1994a). Endocrine effects of prenatal exposures to PCBs, dioxins, and other xenobiotics: Implications for policy and future research. *Environmental Health Perspectives.* 102. pg.676-9.

Birnbaum, L. S. (1994b). The mechanism of dioxin toxicity: Relationship to risk assessment. *Environmental Health Perspectives.* 102(suppl 9). pg. 157-67.

Birnbaum, L. S. (2000). Health effects of dioxins: People are animals, and vice-versa! *Organohalogen Compounds.* 49. pg. 101-3.

Birnbaum, L. S. and Fenton, S. E. (2003). Cancer and developmental exposure to endocrine disruptors. *Environmental Health Perspectives* 111. pg. 389-94.

Blaxill M. F. (2002). Any changes in prevalence of autism must be determined. *British Medical Journal.* 324. pg. 296.

Blaxill, Mark F., Redwood, Lyn and Bernard, Sallie. (2004). Thimerosal and autism? A plausible hypothesis that should not be dismissed. *Medical Hypotheses.* 62. pg. 788-94. http://www.vce.org/mercury/thim/Blaxill_Redwood_Bernard.pdf.

Blount, B. C., Pirkle, J. L., Oserloh, J. D., Valentin-Blasini, L. and Caldwell, K. L. (2006a). Urinary perchlorate and thyroid hormone levels in adolescent and adult men and women living in the Unites States. *Environmental Health Perspectives.* 114(12). pg. 1865-71.

Blount, B. C., Valentin-Blasini, L., Osterloh, J. D., Mauldin, J. P. and Pirkle, J. L. (2006b). Perchlorate exposure of the US population, 2001-2002. *Journal of Exposure Science and Environmental Epidemiology.* 17(4). pg. 400-7.

Blount, Benjamin C., Rich, David Q., Valentin-Blasini, Liza, Lashley, Susan, Ananth, Cande V., Murphy, Eileen, Smulian, John C., Spain, Betty J., Barr, Dana B., Ledoux, Thomas, Hore, Paromita and Robson, Mark. (September 2009). Perinatal exposure to perchlorate, thiocyanate, and nitrate in New Jersey mothers and newborns. *Environmental Science & Technology.* 43(19). pg. 7543-9.

Bolger, M., Egan, S., South, P., Murray, C., Robin, L., Wood, G., Kim, H. and Beru, N. (2009). U.S. Food and Drug Administration's Program for chemical contaminants in food. In: *Intentional and unintentional contaminants in food and feed.* Al-Taher, Fadwa, Jackson, Lauren and DeVries, Jonathan W. eds. ACS Symposium Series, Vol. 1020. American Chemical Society, Washington, DC. http://pubs.acs.org/isbn/9780841269798.

Bor, Jonathan. (March 2010). The science of childhood obesity. *Health Affairs.* 29(3). pg. 393-7. http://content.healthaffairs.org/cgi/content/abstract/29/3/393.

Bornehag, C. G., Sundrell, J., Weschler, C. J., Sigsgaard, T., Lundgren, Björn, Hasselgren, Mikael and Hägerhed-Engman, Linda. (2004). The association between asthma and allergic symptoms in children and phthalates in house dust: A nested case-control study. *Environmental Health Perspectives.* 112. pg.1393-7.

Bradstreet, J., Geier, D. A., Kartzinel, J. J, Adams, J. B. and Gier, M. R. (2003). A case-control study of mercury burden in children with autistic spectrum disorders. *Journal of American Physicians and Surgeons.* 8(3). pg. 76-79.

Branswell, Helen. (March 19, 2008). Superbug found in Canadian pork products. *The Canadian Press.*

Braydich-Stolle, L., Hussain, S., Schlager, J. J. and Hofmann, M. (2005). *In vitro* cytotoxicity of nanoparticles in mammalian germline stem cells. *Toxicological Sciences*. 88(2). pg. 412-9.

Buttar, R. A. (May 6, 2004). *Autism: The misdiagnosis of our future generations, US Congressional Sub-committee on Wellness and Human Rights.* Washington, D.C.

CBS News. (March 10, 2008). *Meds found in water may be affecting humans.* CBS3.com, Philadelphia, PA. http://cbs3.com/national/medicine.found.water.2.673206.html.

California Department of Food and Agriculture. (2004). Data cited in: Sharp, R. *Rocket fuel contamination in California milk.* Environmental Working Group. http://www.ewg.org/reports/rocketmilk/.

Carlsen, E., Giwervman, A., Keiding, N. and Skakkeback, N. E. (1992). Evidence for decreasing quality of semen during the past 50 years. *British Medical Journal*. 305. pg. 609-13.

Centers for Disease Control and Prevention. (1999). Thimerosal in vaccines: A joint statement of the American Academy of Pediatrics and the Public Health Service. *Morbidity and Mortality Weekly Report*. 49(26). pg. 563–65. http://www.cdc.gov/mmwr/preview/mmwrhtml/mm4826a3.htm.

Centers for Disease Control and Prevention. (December 22, 2000). Blood lead levels in young children -- United States and selected states, 1996 - 1999. *Morbidity and Mortality Weekly Report*. 49(50). pg. 1133-7. http://www.cdc.gov/mmwr/preview/mmwrhtml/mm4950a3.htm.

Centers for Disease Control and Prevention. (June 1, 2001). Public health consequences and outbreak management. *Morbidity and Mortality Weekly Report*. www.cdc.gov/mmwr/preview/mmwrhtml/rr5009a1.htm.

Centers for Disease Control and Prevention. (2001). *National report on human exposure to environmental chemicals*. National Center for Environmental Health, Atlanta, GA. http://www.cdc.gov/exposurereport/.

Centers for Disease Control and Prevention. (2005). *Third national report on human exposure to environmental chemicals*. CDC, Atlanta, GA. http://www.cdc.gov/exposurereport/.

Centers for Disease Control and Prevention. (August 10, 2007). Dengue Hemorrhagic Fever – U.S.-Mexico border, 2005. *Morbidity and Mortality Weekly Report*. 56(31). pg. 785-89.

Centers for Disease Control and Prevention. (2009). *Fourth national report on human exposure to environmental chemicals*. Department of Health and Human Services, CDC, Atlanta, GA. http://www.cdc.gov/exposurereport/pdf/FourthReport.pdf.

Centers for Disease Control and Prevention. (2012). *Community Report from the Autism and Developmental Disabilities Monitoring (ADDM) Network: Prevalence of autism spectrum disorders (ASDs) among multiple areas of the United States in 2008*. U. S. Department of Health and Human Services, Washington, DC. http://www.cdc.gov/ncbddd/autism/documents/ADDM-2012-Community-Report.pdf.

Centers for Disease Control and Prevention. (2013). *Antibiotic resistance threats in the United States, 2013*. U. S. Department of Health and Human Services, Centers for Disease Control and Prevention, Washington, DC. http://www.cdc.gov/drugresistance/threat-report-2013/.

Chang, Kenneth. (February 5, 2013). Gluten-free, whether you need it or not. *The New York Times*. http://well.blogs.nytimes.com/2013/02/04/gluten-free-whether-you-need-it-or-not/.

Channel Islands Marine and Wildlife Institute. (2009). *Domoic acid information and history*. http://www.cimwi.org/stranded_domoic.html.

Charles, Dan. (May 5, 2008). *New form of fungus threatens wheat supply*. National Public Radio, All Things Considered. www.npr.org/templates/story/story.php?storyId=90201538.

Charney, E., Sayre, J. and Coulter, M. (February 1980). Increased lead absorption in inner city children: Where does the lead come from?. *Pediatrics*. 65(2). pg. 226-31.

Chauhan, A, Chauhan V., Brown, W. T., and Cohen, I. (Oct 8, 2004). Oxidative stress in autism: Increased lipid peroxidation and reduced serum levels of ceruloplasmin and transferrin - the antioxidant proteins. *Life Science*. 75(21). pg. 2539-49.

Chen, Min and von Mikecz, Anna. (2009). Nanoparticle-induced cell culture models for degenerative protein aggregation diseases. *Inhalation Toxicology*. 21 Suppl 1. pg. 110-4.

Clark, Rachel Ann, Snedeker, Suzanne and Devine, Carol. (1998). *Estrogen and breast cancer risk: What is the relationship?* Cornell University Fact Sheet #9. Cornell University Program on Breast Cancer and Environmental Risk Factors, Ithaca, NY. http://envirocancer.cornell.edu/factsheet/General/fs9.estrogen.pdf

Colborn, T. (June 2004). Neurodevelopment and endocrine disruption. *Environ. Health Perspect*. 112(9). pg. 944-9. http://ehp.niehs.nih.gov/members/2003/6601/6601.html.

Colborn, T. and Clement, C. (1992). Chemically induced alterations in sexual and functional development: The wildlife/human connection. *Advances in Modern Environmental Toxicology*. 21.

Colborn, Theo, Dumanoski, Dianne and Meyers, John Peter. (1996). *Our stolen future: Are we threatening our fertility, intelligence and survival? A scientific detective story*. Dutton, NY.

Colborn, T., vom Saal, F. S. and Soto A. M. (1993). Developmental effects of endocrine disrupting chemicals in wildlife and humans. *Environmental Health Perspectives*. 101. pg. 378-83.

Commission on Life Sciences. (1999). *Hormonally active agents in the environment*. National Academy Press, Washington, DC.

Cooper, R. L., Stoker, T. E., Tyrey, L., Goldman, J. M. and McElroy, W. K. (2000). Atrazine disrupts the hypothalamic control of pituitaryovarian function. *Toxicol. Sci*. 53. pg. 297-307.

Costa, L. G. and Giordano, G. (November 2007). Developmental neurotoxicity of polybrominated diphenyl ether (PBDE) flame retardants. *Neurotoxicology*. 28(6). pg. 1047–67. http://www.ncbi.nlm.nih.gov/pmc/articles/PMC2118052/?tool=pmcentrez.

Counter, S. A. (2003). Neurophysiological anomalies in brainstem responses of mercury-exposed children of Andean gold miners. *Journal of Occupational and Environmental Medicine*. 45(1). pg. 87–9.

Croen, L. A. and Grether, J. K. (2003). A response to Blaxill, Baskin and Spitzer on Croen et al., The changing prevalence of autism in California. *Journal of Autism and Developmental Disorders*. 33(2). pg. 227-9.

Croen, L. A., Grether, J. K., Hoogstrate, J. and Selvin, S. (2002). The changing prevalence of autism in California. *Journal of Autism and Developmental Disorders*. 32. pg. 207-15.

Dally, A. (August 1997). The rise and fall of pink disease. *Society of Historical Medicine*. 10(2). pg. 291-304.

Damstra, Terri, Barlow, Sue, Bergman, Aake, Kavlock, Robert and van der Kraak, Glen, Eds. (2002). *Global assessment of the state-of-the-science of endocrine disruptors.* WHO/PCS/EDC/02.2. International Programme on Chemical Safety, World Health Organization, the International Labour Organisation, and the United Nations Environment Programme. http://www.who.int/ipcs/publications/new_issues/endocrine_disruptors/en/index.html.

Danish Environmental Protection Agency. (1995). *Male reproductive chemicals with estrogenic effects.* Miljø project (290). Danish Ministry of Environment and Energy. Denmark.

Danzo, B. J. (1997). Environmental xenobiotics may disrupt normal endocrine function by interfering with physiological ligands to steroid receptors and binding proteins. *Environmental Health Perspectives.* 105. pg. 294-301.

DeBess, E., Cieslak, P. R., Marsden-Haug, N., Goldoft, M., Wohrle, R., Free, C., Dykstra, E., Nett, R. J., Chiller, T., Lockhart, S. R. and Harris, J. (July 23, 2010). Emergence of *Cryptococcus gattii* --- Pacific Northwest, 2004--2010. *Morbidity and Mortality Weekly.* 59(28). pg. 865-8.

Dentzer, Susan, Ed. (March 2010). Child obesity: The way forward. *Health Affairs.* 29(3). http://content.healthaffairs.org/content/vol29/issue3/index.dtl.

Deykin, E. Y. and MacMahon, B. (1979). Viral exposure and autism. *Am. J. Epidemiol.* 109. pg. 628-38.

Diamanti-Kandarakis, E. et al. (2009). Endocrine-disrupting chemicals: An Endocrine Society scientific statement. *Endocrine Reviews.* 30(4). pg. 293-342.

Donn, Jeff, Mendoza, Martha and Pritchard, Jason. (March 16, 2008). AP: Pharmaceuticals found in drinking water. *Valley Morning Star.* http://www.valleymorningstar.com/news/water_21605___article.html/drinking_pharmaceuticals.html.

Drexler, H. and Schaller, K. H. (1998). The mercury concentration in breast milk resulting from amalgam fillings and dietary habits. *Environ Res.* 77(2). pg. 124-9.

Dugas, J., Nieuwenhuijsen, M. J., Martinez, D., Iszatt, N., Nelson, P. and Elliott, P. (2010). Use of biocides and insect repellents and risk of hypospadias. *Occupational and Environmental Medicine.* 67. Pg. 196-200. http://oem.bmj.com/content/67/3/196.abstract?sid=65055a2b-010a-4be4-adc7-4bf978844ec6.

EDSTAC (Endocrine Disruptor Screening and Advisory Committee). (1998). *Final report.* U.S. Environmental Protection Agency, Washington, DC. http://epa.gov/endo/pubs/edspoverview/finalrpt.htm.

Efron, Edith. (1984). *The apocalyptics: Cancer and the big lie.* Simon and Schuster, NY.

Elixhauser, PhD., A and M. Jung. (April 2008). Clostridium difficile-associated disease in U.S. hospitals 1993-2005. *HCUP* (Healthcare Cost and Utilization Project) *Statistical Brief 50.* Agency for Healthcare Research and Quality, Washington, DC.

Endocrine Society, The. (June 2009). *Position statement: Endocrine-disrupting chemicals.* Chevy Chase, MD. http://www.endo-society.org/advocacy/policy/upload/Endocrine-Disrupting-Chemicals-Position-Statement.pdf.

Engel, S. M., Miodovnik, A., Canfield, R. L., Zhu, C., Silva, M. J., Calafat, A. M. and Wolff, M. S. (January 8, 2010). Prenatal phthalate exposure is associated with childhood behavior and executive functioning. *Environmental Health Perspectives.*

Environment News Service. (September 6, 2001). EU lawmakers vote broad fire retardant ban. *Environment News Service.* http://www.ens-newswire.com/.

Environment News Service. (September 1, 2009). Mercury found in blood of one-third of American women. *Environment News Service.* http://www.ens-newswire.com/ens/sep2009/2009-09-01-092.asp.

Environmental Health Tracking Project Team. (2000). *America's environmental health gap: Why the country needs a nationwide health tracking network.* The Pew Environmental Health Commission. http://healthyamericans.org/reports/files/healthgap.pdf.

Environmetal Protection Agency (USEPA). (1997). Endocrine disruptors glossary: Definition of endocrine disruptor, as defined by EDSTAC. U.S. Environmental Protection Agency, Office of Prevention, Pesticides, and Toxic Substances, Washington, DC. http://www.epa.gov/scipoly/oscpendo/history/glossary.htm.

Environmental Protection Agency (USEPA). (2005). *Guidelines for carcinogen risk assessment.* EPA/630/P–03/001F. United States Environmental Protection Agency, Washington DC. http://cfpub.epa.gov/ncea/cfm/recordisplay.cfm?deid=116283.

Environmental Protection Agency (USEPA). (2007). *Inorganic mercury: TEACH chemical summary: U.S. EPA, toxicity and exposure assessment for children's health.* United States Environmental Protection Agency, Washington DC. http://www.epa.gov/teach/chem_summ/mercury_inorg_summary.pdf.

Environmental Working Group. (2009). *Pollution in people: Cord blood contaminants in minority newborns.* Environmental Working Group. http://www.ewg.org/files/2009-Minority-Cord-Blood-Report.pdf.

Falk, F. Y., Ricci, A., Wolff, M. et al. (1992). Pesticides and polychlorinated biphenyl residues in human breast lipids and their relation to breast cancer. *Archives of Environmental Health.* 47. pg. 143-6.

Finkelstein, M. M. (2008). Diesel particulate exposure and diabetes mortality among workers in the Ontario construction trades. *Occupational and Environmental Medicine.* 65. pg. 215.

Fisher, J. S. (March 2004). Environmental anti-androgens and male reproductive health: Focus on phthalates and testicular dysgenesis syndrome. *Reproduction.* 127(3). pg. 305–15. http://www.reproduction-online.org/cgi/content/abstract/127/3/305.

Fombonne, E. (1999). The epidemiology of autism: A review. *Psychological Medicine.* 29(4). pg. 769–86.

Free Market News Network. (October 18, 2005). *Silver kills viruses, study finds.* www.freemarketnews.com/WorldNews.asp?nid=1401

Freedman, D. M., Stewart, P., Kleinerman, R. A., Wacholder, S., Hatch, E. E., Tarone, R. E., et al. (2001). Household solvent exposures and childhood acute lymphoblastic leukemia. *Am J Public Health.* 91. pg. 564-7.

Fuller, Thomas. (January 26, 2009). Spread of malaria feared as drug loses potency. *The New York Times*. http://www.nytimes.com/2009/01/27/health/27malaria.html?scp=7&sq=malaria%20africa&st=cse.

Fürst, P. (October 2006). Dioxins, polychlorinated biphenyls and other organohalogen compounds in human milk. Levels, correlations, trends and exposure through breastfeeding. *Molecular Nutrition & Food Research* 50(10). pg. 922–33. http://www3.interscience.wiley.com/journal/113376467/abstract?CRETRY=1&SRETRY=0.

Furst, Peter, Furst, Christiane and Wilmers, Klaus. (1994). Human milk as a bioindicator for body burden of PCDDs, PCDFs, organo-chlorine pesticides, and PCBs. *Environmental Health Perspectives Supplements*. 102(1). pg. 187-193.

Garland, Cedric F., Garland, F. C. and Gorham, E. D. (April 1992). Could sunscreens increase melanoma risk? *American Journal of Public Health*. 82. pg. 614-5. http://ajph.aphapublications.org/cgi/reprint/82/4/614?maxtoshow=&hits=10&RESULTFORMAT= 1&author1=garland%2C+c&andorexacttitle=and&andorexacttitleabs=and&andorexactfulltext=an d&searchid=1&FIRSTINDEX=0&sortspec=relevance&resourcetype=HWCIT.

Garrett, Laurie. (1995). *The coming plague*. The Penguin Group. Penguin Books USA, Inc. NY.

Geier, D. A., and Geier, M. R. (2003). Thimerosal in childhood vaccines, neurodevelopment disorders, and heart disease in the United States. *Journal of American Physicians and Surgeons*. 8(1). pg. 6-11.

Geier, D. A., and Geier, M. R. (Apr-Jun 2003). An assessment of the impact of thimerosal on childhood neurodevelopmental disorders. *Pediatric Rehabilitation*. 6(2). pg. 97-102.

Geier, D. A., King, P. G., Sykes, L. K. and Geier, M. R. (October 2008). A comprehensive review of mercury provoked autism. *Indian J Med Res*. 128(4). pg. 383-411. http://www.ncbi.nlm.nih.gov/pubmed/19106436.

Geiser, M., Rothen-Rutishauser, B., Kapp, N., Schurch, S., Kreyling, W., Schulz, H., et al. (2005). Ultrafine particles cross cellular membranes by nonphagocytic mechanisms in lungs and in cultured cells. *Environmental Health Perspectives*. 113. pg. 1555-60.

Ginsberg, G. L., Hattis, D. B., Zoeller, R. T. and Rice, D. C. (2007). Evaluation of the U.S. EPA/OSWER preliminary remediation goal for perchlorate in groundwater: Focus on exposure to nursing infants. *Environmental Health Perspectives*. 115. pg. 361-9.

Gofman, J. W. (1981). *Radiation and human health; a comprehensive investigation of the evidence relating low-level radiation to cancer and other diseases*. Sierra Club Books, San Francisco, CA.

Goodman, W. (September 2, 1994). Something is attacking male fetus sex organs. *The New York Times*. Sec. D. pg. 17.

Gordon, Amanda. (March 21, 2001). *First-ever government report a "wake-up call" on toxic exposure of average Americans*. http://www.pirg.org/envirohealth/media/toxicexposure.html.

Grady, Denise. (September 4, 2012). New virus tied to ticks poses puzzle for doctors. *The New York Times*. pg. D5.

Grady, Denise. (October 22, 2012). 'Worried sick': Meningitis risk haunts 14,000. *The New York Times*. pg. A1, A3. http://www.nytimes.com/2012/10/22/health/meningitis-risk-haunts-14000-people.html?_r=0.

Graef, J. W. (1994). Heavy metal poisoning. Chapter 396. In: Isselbacher, et al. Eds. *Harrison's Principles of Internal Medicine*. 13th Ed. McGraw-Hill. pg. 2465.

Greenwood, Michael R. (1985). Methylmercury poisoning in Iraq: An epidemiological study of the 1971 - 1972 outbreak. *Journal of Applied Toxicology*. 5(3). pg. 148-59. http://www3.interscience.wiley.com/journal/112394403/abstract.

Greer, M. A., Goodman, G., Pleuss, R. C. and Greer, S.E. (2002). Health effect assessment for environmental perchlorate contamination: The dose response for inhibition of thyroidal radioiodide uptake in humans. *Environmental Health Perspectives*. 110(9). pg. 927-37. http://www.ehponline.org/docs/2002/110p927-937greer/abstract.html.

Haslam, D. W. and James, W. P. (2005). Obesity. *Lancet*. 366. pg. 1197-209.

Hayakawa. R., Matsunaga, K. and Arima, Y. (1987). Depigmented contact dermatitis due to incense. *Contact Dermatitis Journal*. 16(5). pg. 272-4.

Hauser, P., Zamertkin, A. J., Martinez, P. et al. (1993). Attention deficit-hyperactivity disorder in people with generalized resistance to thyroid hormone. *New England Journal of Medicine*. 328. pg. 997-1000.

Hertz-Picciotto. I., Green, P. G., Delwiche, L., Hansen, R., Walker, C. and Pessah, I. N. (January 2010). Blood mercury concentrations in CHARGE Study children with and without autism. Environmental Health Perspectives. 118(1). pg. 161-6. http://www.ncbi.nlm.nih.gov/pmc/articles/PMC2831962/.

Hessler, Wendy and Patisaul, Heather. (February 2, 2010). Human placenta cells die after BPA exposure. *Environmental Health News*. http://www.environmentalhealthnews.org/ehs/newscience/human-placental-cells-die-after-bpa-exposure/.

Hites, R. A. (February 2004). Polybrominated diphenyl ethers in the environment and in people: A meta-analysis of concentrations. *Environmental Science & Technology*. 38(4). pg. 945–56. http://pubs.acs.org/doi/abs/10.1021/es035082g.

Hooper, Kim and McDonald, Thomas A. (May 2000). The PBDEs: An emerging environmental challenge and another reason for breast-milk monitoring programs. *Environmental Health Perspectives*. 108(5). pg. 387-392.

Hossain, Parvez, Kawar, Bisher and El Nahas, Meguid. (January 18, 2007). Obesity and diabetes in the developing world – a growing challenge. *The New England Journal of Medicine*. 356(3). pg. 213-5. http://content.nejm.org/cgi/reprint/356/3/213.pdf.

Hotez, Peter J. (2007). Neglected diseases and poverty in "the other America": The greatest health disparity in the United States? *PLoS Negl Trop Dis*. 1(3). pg. 149. http://gnntdc.sabin.org/files/press_releases/Neglected_Diseases_and_Poverty_in_The_Other_America.pdf.

Houlihan, Jane, Brody, Charlotte and Schwan, Bryony. (2002). *Not too pretty: Phthalates, beauty products & the FDA*. Environmental Working Group, Washington, DC. http://www.ewg.org/files/nottoopretty_final.pdf.

Howard, V. (1997). Synergistic effects of chemical mixtures—can we rely on traditional toxicology? *Ecologist*. 27. pg. 192-4.

Hoyer, A., Grandjean, P., Jorgensen, T. et al. (1998). Organochlorine exposure and risk of breast cancer. *Lancet*. 352. pg. 1816-20.

Iguchi, Taisen and Katsu, Yoshinao. (December 2008). Commonality in signaling of endocrine disruption from snail to human. *BioScience*. 58(11). pg. 1061-7. http://caliber.ucpress.net/doi/abs/10.1641/B581109.

Illinois Department of Public Health. (1999). *Nitrates in drinking water*. Environmental Health Fact Sheet, Division of Environmental Health, Springfield, IL. http://www.idph.state.il.us/envhealth/factsheets/NitrateFS.htm.

Ilonka, A., Meerts, T. M., van Zanden, J. J., Luijks, E. A. C., van Leeuwen-Bol, I., Marsh, G., Jakobsson, E., Bergman, Å. and Brouwer, A. (2000). Potent competitive interactions of some brominated flame retardants and related compounds with human transthyretin in vitro. *Toxicological Sciences*. 56. pg. 95-104.

Institute of Medicine. (2003). *Dioxins and dioxin-like compounds in the food supply – strategies to decrease exposure*. National Academies Press, Washington, DC.

International Organization for Economic Cooperation and Development. (April 16, 1997). *European Workshop on the Impact of Endocrine Disruptors on Human Health and Wildlife. Report on proceedings*. DGX11. EUR 17459.

Jaakkola, Jouni J. K., Ieromnimon, Antonia and Jaakkola, Maritta S. (October 2006). Interior surface materials and asthma in adults: A population-based incident case-control study. *American Journal of Epidemiology*. 164(8). pg. 742-49.

Jacob, Anila. (April 2009). *CDC scientists find rocket fuel chemical in infant formula: Powdered cow's milk formula contains thyroid toxin*. Environmental Working Group. http://www.ewg.org/node/27772/.

Jacobson, J. L. and Jacobson, S.W. (1996). Intellectual impairment in children exposed to polychlorinated biphenyls in utero. *New England Journal of Medicine*. 335. pg. 783-9.

Johnson, R. J., Perez-Pozo, S. E., Sautin, Y. Y., Manitius, J., Sanchez-Lozada, L. G., Feig, D. I., Shafiu, M., Segal, M., Glassock, R. J., Shimada, M., Roncal, C. and Nakagawa, T. (2009). Hypothesis: Could excessive fructose intake and uric acid cause type 2 diabetes? *Endocr. Rev*. 30. pg. 96-116.

Johnson, Steven. (2006). *The ghost map: The story of London's most terrifying epidemic—and how it changed science, cities, and the modern world*. Riverhead, NY.

Joshi, Mohit. (December 1, 2008). Chemical in paint may play role in promoting obesity. *OneIndia*. http://www.topnews.in/healthcare/diseases/obesity.

Kaiser Daily HIV/AIDS Report. (February 23, 2009). *About 42% of pregnant women in Swaziland are living with HIV*. www.kaisernetwork.org/daily_reports/rep_index.efm?hint=1&DR_ID=57103.

Kay, Jane. (July 10, 2003). Widely used flame retardant feared to be a health hazard found in women's breast tissue, fish – bill would ban it. *San Francisco Chronicle*.

Kearney, P. M., Whelton, M., Reynolds, K., Muntner, P., Whelton, P. K. and He, J. (2005). Global burden of hypertension: Analysis of worldwide data. *Lancet*. 365. pg. 217-23.

Kelland, Kate. (November 17, 2011). Europe in the grip of drug resistant superbugs. *ibtimes.com*.

302

Kerper, L. E., Ballatori, N. and Clarkson, T. W. (1992). Methylmercury transport across the blood–brain barrier by an amino acid carrier. *American Journal of Physiology.* 262(5 Pt 2). pg. R761–5.

Kirk, A. B. (2006). Environmental perchlorate: Why it matters. *Analytical Chimica Acta.* 567(1). pg. 4-12.

Kirk, A. B., Martinelango, P. K., Tian, K., Dutta, A., Smith, E. E. and Dasgupta, P. K. (April 2005). Perchlorate and iodide in dairy and breast milk. *Environmental Science & Technology.* 39(7). pg. 2011-7.

Kirk, A. B., Smith, E. E, Tian, K., Anderson, T. A. and Dasgupta, P K. (2003). Perchlorate in milk. *Environ Sci Technol.* 37(21). pg. 4979-81.

Kodama, Hirokadzu and Ota, Hideo. (March/April 1980). Transfer of polychlorinated biphenyls to infants from their mothers. *Archives of Environmental Health.* 35(2). pg. 95-100.

Kodavanti, P. R. (2006). Neurotoxicity of persistent organic pollutants: Possible mode(s) of action and further considerations. *Dose Response.* 3(3). pg. 273–305. http://www.ncbi.nlm.nih.gov/pmc/articles/PMC2475949/?tool=pmcentrez.

Kolata, Gina. (June 23, 2011). Scientists find 2 unusual traits blended in German E. coli strain. *The New York Times.* pg. A5.

Krimsky, Sheldon. (1991). *Biotechnics and society: The rise of industrial genetics.* Praeger Publishers, Westport, CT.

Krimsky, Sheldon. (2000). *Hormonal chaos: The scientific and social origins of the environmental endocrine hypothesis.* The Johns Hopkins University Press, Baltimore, MD.

Kristof, Nicholas D. (December 6, 2009). Cancer from the kitchen? *The New York Times.* http://www.nytimes.com/2009/12/06/opinion/06kristof.html?_r=1.

Kristof, Nicholas D. (October 17, 2013). This is your brain on toxins. *The New York Times.* pg. A29. http://www.nytimes.com/2013/10/17/opinion/kristof-this-is-your-brain-on-toxins.html.

Kumarasamy, K. K., Toleman, M. A., Walsh, T. R., et al. (August 2010). Emergence of a new antibiotic resistance mechanism in India, Pakistan, and the UK: A molecular, biological, and epidemiological study. *Lancet Infect Dis.* 10(9). pg. 597-602. http://www.thelancet.com/journals/laninf/article/PIIS1473-3099(10)70143-2/fulltext.

Lal, Rattan, Hansen, David O. and Uphoff, Norman, Eds. (2002). *Food security and environmental quality in the developing world.* CRC Press, Boca Raton, FL.

Landrigan, Philip J. (January 16, 2010a). What causes autism? Exploring the environmental contribution. *Current Opinion in Pediatrics.* http://journals.lww.com/co-pediatrics/Abstract/publishahead/What_causes_autism__Exploring_the_environmental.99878.aspx

Landrigan, P. L., Needleman, H. L. and Landrigan, M. (2002). *Raising healthy children in a toxic world: 101 smart solutions for every family.* Rodale Press, Kutztown, PA.

Landrigan, Philip J., Satlin, Lisa M. and Boffetta, Paolo. (2010b). *Why are we subsidizing childhood obesity?* Mount Sinai School of Medicine, NY. http://www.mountsinai.org/static_files/MSMC/Files/NYT%20OpEds/MS_OpEd%20Ad_Obesity_final.pdf.

Li, N., Sioutas, C., Cho, A., Schmitz, D., Misra, C., Sempf, J., Wang, M., Oberley, T., Froines, J. and Nel, A. (2003). Ultrafine particulate pollutants induce oxidative stress and mitochondrial damage. *Environmental Health Perspectives*. 111(4). pg. 455-60.

Lorber, M. (January 2008). Exposure of Americans to polybrominated diphenyl ethers. *J Expo Sci Environ Epidemiol*. 18(1). pg. 2–19.
http://www.nature.com/jes/journal/v18/n1/abs/7500572a.html.

Lunder, S. and Jacob, A. (September 1, 2008). *Fire retardants in toddlers and their mothers*. Environmental Working Group. http://www.ewg.org/reports/pbdesintoddlers.

Mahaffey, K. R., Clickner, R. P. and Bocurow, C. C. (2004). Blood organic mercury and dietary mercury intake: National health and nutrition examination survey, 1999 and 2000. *Environmental Health Perspectives*. 112(5). pg. 562-570.

Mahaffey, K. R. (2005). Mercury exposure: Medical and public health issues. *Transactions of the American Clinical and Climatological Association*. 116. pg. 127-154.
http://www.pubmedcentral.nih.gov/articlerender.fcgi?artid=1473138.

Main, Katharina M., Mortensen, Gerda K., Kaleva,Marko M., Boisen, Kirsten A., Damgaard, Ida N., Chellakooty, Marla, Schmidt, Ida M., Suomi, Anne-Maarit, Virtanen, Helena E., Petersen, Jørgen H., Andersson, Anna-Maria, Toppari, Jorma and Skakkebæk, Niels E. (September 2005). Human breast milk contamination with phthalates and alterations of endogenous reproductive hormones in infants three months of age. *Environmental Health Perspectives*. 114(2). pg. 270-6.

Marsee, Kevin, Woodruff, Tracey J., Axelrad, Daniel A., Calafat, Antonia M. and Swan, Shanna H. (June 2006). Estimated daily phthalate exposures in a population of mothers of male infants exhibiting reduced anogenital distance. *Environmental Health Perspectives*. 114(6). pg. 805-9.

McGinnis, W. R. (2004). Oxidative stress in autism. *Alternative Therapies in Health Medicine*. 10(6). pg. 22-36.

McGrath, Susan. (March-April 2010). Pandora's water bottle. *Audubon*. pg. 33-9.
http://www.audubonmagazine.org/currents/currents1003.html.

McKee, Maggie. (February 23, 2005). Perchlorate found in breast milk across US. *New Scientist*.
http://www.newscientist.com/article/dn7057.

McKenna, Maryn. (2010). *Superbug: The fatal menace of MRSA*. Free Press, NY.

McLachlan, J. A., Ed. (1980). *Estrogens in the environment*. Elsevier North-Holland, NY.

McLachlan, J. A., Ed. (1985). *Estrogens in the environment II: Influences on development*. Elsevier North-Holland, NY.

McLachlan, J. A. and Korach, K. S., Eds. (1995). *Estrogens in the environment III: Global health implications*. National Institute of Environmental Health Services, US National Institutes of Health, Washington, DC.

McNeil, Donald G., Jr. (June 24, 2008). Tropical diseases add to burden among the poor in the U.S. *The New York Times*.
http://www.nytimes.com/2008/06/24/health/24glob.html?_r=3&scp=1&sq=Tropical+diseases&st=nyt&oref=slogin.

McNeil, Donald G., Jr. (August 11, 2010). Antibiotic-resistant bacteria moving from South Asia to U. S. *The New York Times*.

McNeil, Donald G., Jr. (October 15, 2010). Virus deadly in livestock is no more, U.N. declares. *The New York Times*. http://www.nytimes.com/2010/10/16/science/16pest.html.

MedIndia.com. (December 21, 2008). *Common infant virus may trigger Type 1 Diabetes.*; www.medindia.net/news/Common-Infant-Virus-may-Trigger-Type-1-Diabetes-45425-2.htm

Melzer, David, Rice, Neil, Depledge, Michael H., Henley, William E. and Galloway, Tamara S. (May 2010). Association between serum perfluorooctanoic acid (PFOA) and thyroid disease in the U.S. National Health and Nutrition Examination Survey. *Environmental Health Perspectives*. 18(5). Pg. 686-92. http://ehsehplp03.niehs.nih.gov/article/info:doi%2F10.1289%2Fehp.0901584.

Minogue, Kristen. (January 23, 2009). *Drugs from genetically engineered animals poised to debut in US*. Medill School, Northwestern University, Chicago. http://news.medill.northwestern.edu/chicago/news.aspx?id=112645.

Morland, K. B., Landrigan, P. J., Sjödin, A., Gobeille, A. K., Jones, R. S., McGahee, E. E., Needham, L. L. and Patterson, D. G. (December 2005). Body burdens of polybrominated diphenyl ethers among urban anglers. *Environmental Health Perspectives*. 113(12). pg. 1689-92. http://www.ncbi.nlm.nih.gov/pmc/articles/PMC1314906/?tool=pmcentrez.

Mundy, Alicia and Tomson, Bill. (October 1, 2010). Industry fights altered salmon. *The Wall Street Journal*. pg. A7.

Naseri, Iman, Jerris, Robert C. and Steven E. Sobol. (January 2009). Nationwide trends in pediatric *staphylococcus aureus* head and neck infections. *Archives of Otolaryngology-Head & Neck Surgery*. 135(1). pg. 14-16. http://archotol.ama-assn.org/cgi/content/abstract/135/1/14.

National Environmental Trust. (2000). *Polluting our future: Chemical pollution in the U. S. that effects child development and learning*. The National Environmental Trust, Physicians for Social Responsibility, and the Learning Disabilities Association of America, Washington, DC. http://www.ehinitiative.org/pdf/polluting_report.pdf.

National Wildlife Federation. (1994). *Fertility on the brink: The legacy of the chemical age*. National Wildlife Federation, Washington, DC.

Needleman, H. L. and Gatsonis, C.A. (1990). Low-level lead exposure and the IQ of children: A meta-analysis of modern studies. *Journal of the American Medical Association*. 263. pg. 73-8.

Nelson, K. B. and Bauman, M. L. (2003). Thimerosal and autism? *Pediatrics*. 111(3). pg. 674–9.

Noedl, Harald, Se, Youry, Docheat, Duong and Fukuda, Mark. (2008). Evidence of Artemisinin-resistant malaria in Western Cambodia. *New England Journal of Medicine*. 359(24). pg. 2619-2620. http://content.nejm.org/cgi/content/full/359/24/2619.

Noren, Koidu. (1983). Some aspects of the determination of organochlorine contaminants in human milk. *Archives of Environmental Contamination and Toxicology*. 12. pg. 277-283.

Norén, K. and Meironyté, D. (2000). Certain organochlorine and organobromine contaminants in Swedish human milk in perspective of past 20-30 years. *Chemosphere*. 40(9-11). pg. 1111–23.

Onstot J, Ayling R, Stanley J. (1987). *Characterization of HRGC/MS Unidentified Peaks from the Analysis of Human Adipose Tissue. Volume 1: Technical Approach.* Washington, DC: U.S. Environmental Protection Agency Office of Toxic Substances (560/6-87-002a).

Palmer, R. F., Blanchard, S., Stein, Z., Mandell, D. and Miller, C. (2006). Environmental mercury release, special education rates, and autism disorder: An ecological study of Texas. *Health Place.* 12(2). pg. 203-9.

Parker, R. D., E. V. Buehler, et al. (1986). Phototoxicity, photoallergy, and contact sensitization of nitro musk perfume raw materials. *Contact Dermatitis.* 14(2). pg. 103-9.

Parker-Pope, Tara. (April 14, 2009). Stomach bug crystallizes an antibiotic threat. *The New York Times.* http://www.nytimes.com/2009/04/14/health/14well.html?emc=eta1.

Patterson, J. E., Weissberg, B. G., and Dennison, P. J. (1981). Mercury in human breath from dental amalgams. *Journal of Dental Research.* 60. pg. 1668-71.

Pearce, Elizabeth N., Leung, Angela M., Blount, Benjamin C., Bazrafshan, Hamid R., He, Xuemei, Pino, Sam, Valentin-Blasini, Liza and Braverman, Lewis E. Breast milk iodine and perchlorate concentrations in lactating Boston-area women. *Journal of Clinical Endocrinology & Metabolism.* 92(5). pg. 1673-7. http://jcem.endojournals.org/cgi/content/abstract/92/5/1673.

Perencevich, E. N., Wong, M. T. and Harris, A. D. (2001). National and regional assessment of the antibacterial soap market: A step toward determining the impact of prevalent antibacterial soaps. *American Journal of Infection Control.* 29(5). pg. 281-283.

Pichichero, M. E., Cernichiari, E., Lopreiato, J. and Treanor, J. (2002). Mercury concentrations and metabolism in infants receiving vaccines containing thimerosal: A descriptive study. *Lancet.* 360. pg. 1737-41.

Polishuk, Z. W., Ron, M., Wassermann, M., Cucos, S., Wassermann, D. and Lemesch, C. (March 1977). Organochlorine compounds in human blood plasma and milk. *Pesticides Monitoring Journal.* 10(4). pg. 121-129.

Polishuk, Z. W., Wassermann, D., Wassermann, M., Cucos, S. and Ron, M. (1977). Organochlorine compounds in mother and fetus during labor. *Environmental Research.* 13. pg. 278-84.

Pollack, Andrew. (February 26, 2010a). Rising threat of infections unfazed by antibiotics. *The New York Times Editorials.* www.nytimes.com/2010/02/27/business/27germ.html.

Pollack, Andrew. (September 20, 2010b). Panel leans in favor of engineered salmon. *The New York Times.* http://www.nytimes.com/2010/09/21/business/energy-environment/21salmon.html?_r=1&scp=1&sq=panel%20leans%20in%20favor%20of%20engineered%20salmon&st=cse.

Rabin, Roni Caryn. (January 20, 2009). Children's staph infections increasingly resistant to drugs. *The New York Times.* http://www.nytimes.com/2009/01/21/health/research/21staph.html?_r=1&scp=4&sq=rabin%20antibiotic&st=cse.

Rabin, Roni Caryn. (February 23, 2010). Rate of chronic health problems rises. *The New York Times.*

Randolph, Theron G. (1962). *Human ecology and susceptibility to the chemical environment.* Charles C. Thomas, Publisher, Springfield, IL.

Reuben, Suzanne H. (2010). *Reducing environmental cancer risk: What we can do now.* 2008-2009 Annual Report. President's Cancer Panel. National Cancer Institute, Washington, DC. http://deainfo.nci.nih.gov/advisory/pcp/pcp08-09rpt/PCP_Report_08-09_508.pdf.

Ritvo, E. R., Freeman, B. J., Pingree, C., Mason-Brothers, A., Jorde, L., Jenson, W. R., et al. (1989). The UCLA-University of Utah epidemiologic survey of autism: Prevalence. *American Journal of Psychiatry.* 146(2). pg. 194-9.

Robbins, Jim. (May 29, 2012). In wild animals, charting the pathways of disease. *The New York Times.* pg. D2.

Robbins, Jim. (July 15, 2012). Man-made epidemics. *The New York Times.* pg. 1, 6-7.

Robbins, Jim. (October 15, 2013). Moose die-off alarms scientists. *The New York Times.* pg. D3. http://www.nytimes.com/2013/10/15/science/earth/something-is-killing-off-the-moose.html.

Rodier, P. M. (2000). The early origins of autism. *Scientific American.* pg. 56-63.

Ronald, Pamela C. and McWilliams, James E. (May 14, 2010). Genetically engineered distortions. *The New York Times.* http://www.nytimes.com/2010/05/15/opinion/15ronald.html.

Rossignol, Daniel A. and Bradstreet, J. Jeffrey. (2008). Evidence of mitochondrial dysfunction in autism and implications for treatment. *American Journal of Biochemistry and Biotechnology.* 4(2). pg. 208-17. http://www.scipub.org/fulltext/ajbb/ajbb42208-217.pdf.

Rudel, R. A., Seryak, L. M. and Brody, J. G. (2008). PCB-containing wood floor finish is a likely source of elevated PCBs in resident's blood, household air and dust: A case study of exposure. *Environmental Health.* 7. pg. 2. http://www.ncbi.nlm.nih.gov/pmc/articles/PMC2267460/?tool=pmcentrez.

Saint Louis, Catherine. (February 5, 2013). Warning too late for some babies. *The New York Times.* http://well.blogs.nytimes.com/2013/02/04/warning-too-late-for-some-babies/.

Sanchez, C. A., Crump, K. S., Krieger, R. I., Khandaker, N. R. and Gibbs, J. P. (2005). Perchlorate and nitrate in leafy vegetables of North America. *Environmental Science & Technology.* 39(24). pg. 9391-7.

Sarantis, Heather, Naidenko, Olga V., Gray, Sean, Houlihan, Jan and Malkan, Stacy. (2010). *Not so sexy: The health risks of secret chemicals.* Campaign for Safe Cosmetics and Environmental Defence Canada, Toronto, Canada. http://www.environmentaldefence.ca/reports/pdf/FragranceReport.pdf.

Sasco, A. J. (2001). Epidemiology of breast cancer: An environmental disease? *APMIS.* 109. pg. 321-32.

Sayeed, Syed, McGiffert, Lisa and McCauley, Michael. (November 2008). *Hospital-acquired C. difficile infections (CDI).* Consumers Union Policy Brief. Consumers Union, NY. www.consumersunion.org/pub/Final%20Cdiff%20Policy%20Brief%2011-11-08.pdf

Schecter, Arnold, Cramer, P., Boggess, K., Stanley, J., Papke, O., Olson, J., Silver, A. and Schmitz, M. (2001). Intake of dioxins and related compounds from food in the U.S. population. *Journal of Toxicology and Environmental Health.* Part A. 63. pg. 101-118.

Schecter, A., Päpke, O., Tung, K.C., Staskal, D. and Birnbaum, L. (October 2004). Polybrominated diphenyl ethers contamination of United States food. *Environmental Science &. Technology.* 38(20). pg. 5306-11. http://pubs.acs.org/doi/abs/10.1021/es0490830.

Schettler, T., Solomon, G., Valenti, M. and Huddle, A. (1999). *Generations at risk.* MIT Press, Cambridge, MA.

Schober, S. E., Sinks, T. H., Jones, R. L., Bolger, P. M., McDowell, M., Osterloh, J., et al. (2003). Blood mercury levels in US children and women of childbearing age, 1999-2000. *Journal of the American Medical Association.* 289. pg. 1667-74.

Schreder, Erika. (2006). *Pollution in people: A study of toxic chemicals in Washingtonians.* Toxic-Free Legacy Coalition, Seattle, WA.

Schuiling, Jacqueline and van der Naald, Wytze. (2005). *A present for life: Hazardous chemicals in umbilical cord blood.* Greenpeace, Amsterdam, NL. http://www.greenpeace.org/international/en/news/features/poisoning-the-unborn111/.

Schwarz, Alan. (February 3, 2013). Drowned in a stream of prescriptions: Addict's parents couldn't halt flow of attention deficit drug. *The New York Times.* http://www.nytimes.com/2013/02/03/us/concerns-about-adhd-practices-and-amphetamine-addiction.html?pagewanted=all&_r=0.

Sears, Stephen D., Smith, Peter, Mills, Dora Anne, Robbins, Amy and Robinson, Sara. (2010). *Report to Maine Legislature: Lyme disease.* Maine Center for Disease Control and Prevention, Augusta, ME.

Senjen, Rye and Illuminato, Ian. (2009). *Nano and biocidal silver: Extreme germ killers present a growing threat to public health.* Friends of the Earth Australia and Friends of the Earth United States, Washington, DC. http://www.foe.org/sites/default/files/Nano-silverReport_US.pdf.

Silverlieb, Alan. (12/9/08). *Gulf War illness is real, new federal report says.* CNN.com; www.cnn.com/2008/HEALTH/11/17/gulf.war.illness.study/.

Skinner, Linda, Ed. (2003). *Seafood handbook: The comprehensive guide to sourcing, buying, and preparation.* Diversified Publications, Eastover, SC.

Slorach, Stuart A. and Vaz, Reggie. (1983). *Assessment of human exposure to selected organochlorine compounds through biological monitoring.* Swedish National Food Administration, Uppsala, Sweden.

Smith, E. P., Boyd, J., Frank, G. R., Takahashi, H., Cohen, R. M., Specker, B., Williams, T. C., Lubahn, D. B. and Korach, K. S. 1994. Estrogen resistance caused by a mutation in the estrogen-receptor gene in a man. *New England Journal of Medicine.* 331(16). pg. 1088-9.

Smith, Jeffrey. (April 26, 2010). Genetically modified soy linked to sterility, infant mortality. *NewsWithViews.com.* http://newswithviews.com/Smith/jeffrey130.htm.

Smith, Rick and Lourie, Bruce. (2010). *Slow death by rubber duck: The secret danger of everyday things.* Counterpoint, Berkeley, CA.

Spellberg, Brad. (2009). *Rising plague: The global threat from deadly bacteria and our dwindling arsenal to fight them.* Prometheus Books, Amherst, NY.

Stark, Jill. (September 8, 2008). Drug-resistant 'superbug' in Australia fueling fear of epidemic. *Melbourne Age*. www.organicconsumers.org/articles/article_14536.cfm.

Stein, Rob. (March 5, 2013). Infections with 'nightmare bacteria' are on the rise in U.S. hospitals. *National Public Radio*. http://www.npr.org/blogs/health/2013/03/05/173526084/infections-with-nightmare-bacteria-are-on-the-rise-in-u-s-hospitals.

Stern, A. H. and Smith, A. E. (2003). An assessment of the cord blood: Maternal blood methylmercury ratio: Implications for risk assessment. *Environmental Health Perspectives*. 111(12). pg. 1465-70.

Sternberg, Steve. (June 24, 2008). Analysis: U.S. poor are vulnerable to 'neglected' diseases. *USA Today*. http://www.usatoday.com/news/health/2008-06-23-neglected-diseases_N.htm

Stratton, K., Gable, A. and McCormick, M., Eds. (2001). *Immunization safety review: Thimerosal containing vaccines and neurodevelopmental disorders*. National Academy Press, Washington, DC.

Strom, Stephanie. (February 6, 2013). Study looks at particles used in food. *The New York Times*. http://www.nytimes.com/2013/02/06/business/nanoparticles-in-food-raise-concern-by-advocacy-group.html?_r=0.

Stromland, K., Nordin, V. and Miller, M. (1994). Autism in thalidomide embryopathy: A population study. *Developmental Medicine and Child Neurology*. 36. pg. 351-6.

Swan, S. H., Waller, K., Hopkins, B. and DeLorenzo, G. (1998). Trihalomethanes in drinking water and spontaneous abortion. *Journal of Epidemiology*. 9. pg. 134-40.

Swaminathan, Nikhil. (September 2008). Why is melamine in baby formula, your food -- and your pet's meals? *Scientific American*.

Swan, Shanna H., Main, Katharina M., Liu, Fan, Stewart, Sara L., Kruse, Robin L., Calafat, Antonia M., Mao, Catherine S., Redmon, J. Bruce, Ternand, Christine L., Sullivan, Shannon and Teague, J. Lynn. (August 2005). Decrease in anogenital distance among male infants with prenatal phthalate exposure. *Environmental Health Perspectives*. 113(8). pg. 1056-61.

Tan, Z. X., Lal, R. and Weibe, R. D. (2005). Global soil nutrient depletion and yield reduction. *Journal of Sustainable Agriculture*. 26(1). pg. 133, 36, 41.

Tavernise, Sabrina. (September 4, 2012). Farm use of antibiotics defies scrutiny. *The New York Times*. pg. D1.

Tavernise, Sabrina. (September 17, 2013). Antibiotic-resistant infections lead to 23,000 deaths a year, C.D.C. finds. *The New York Times*. pg. A13. http://www.nytimes.com/2013/09/17/health/cdc-report-finds-23000-deaths-a-year-from-antibiotic-resistant-infections.html.

The Fishery and Aquaculture Industry Research Fund. (2009). *Decontamination of persistent organic pollutants in fishmeal and fish oil*. Oslo, Norway. http://www.fiskerifond.no/index.php?current_page=prosjekter&subpage=archive&detail=1&id=611&gid=3

ToxicExposure.org coalition. (March 22, 2001). *First-ever government report: A "wakeup call" on toxic exposure of average Americans*. www.igc.org/igc/gateway/en/archive/arch032601.html.

Treffert D. A. Epidemiology of infantile autism. (1970). *Archives of General Psychiatry*. 22(5). pg. 431-8.

Tuller, David. (2010). Study links chronic fatigue to virus class. *The New York Times*. pg. D6. http://query.nytimes.com/gst/fullpage.html?res=9F0DE7D81039F937A1575BC0A9669D8B63&scp=2&sq=Study%20links%20chronic%20fatigue%20to%20virus%20class&st=cse.

UNAIDS. (2008). *2008 report on the global AIDS epidemic*. Joint United Nations Programme on HIV/AIDS, Geneva, Switzerland. www.unaids.org/en/KnowledgeCentre/HIVData/GlobalReport/2008/2008_Global_report.asp.

United Nations. (1996). *Sources and effects of ionizing radiation: United Nations Scientific Committee on the Effects of Atomic Radiation UNSCEAR 1996 report to the General Assembly, with scientific annex.* UN, NY.

US Agency for Toxic Substances and Disease Registry (1998). *Public health implications of persistent toxic substances in the Great Lakes and St. Lawrence River Basins*. Public Health Service, U.S. Department of Health and Human Services, Atlanta, GA.

US Environmental Protection Agency. (1997). *Special report on environmental endocrine disrupters: An effects assessment and analysis*. Technical Panel, Office of Research and Development, Office of Prevention, Pesticides and Toxic Substances, EPA, Washington, DC.

USA Today. (December 24, 2009). Two of three California homes had excessive formaldehyde levels. *USA Today*. http://content.usatoday.com/communities/greenhouse/post/2009/12/two-of-three-california-homes-had-excessive-formaldehyde-levels/1.

USDA. (2006). *Emerging Disease Notice. Viral hemorrhagic septicemia in the Great Lakes.* Centers for Epidemiology and Animal Health, USDA.

University of Maine. (2001). *The U. S. health care system: Best in the world, or just the most expensive?* Bureau of Labor Education, University of Maine, Orono, ME.

Valera, Beatriz, Dewailly, Éric and Poirier, Paul. (October 2009). Environmental mercury exposure and blood pressure among Nunavik Inuit adults. *Hypertension*. 54. pg. 981-986. http://hyper.ahajournals.org/cgi/content/abstract/HYPERTENSIONAHA.109.135046v1.

Van den Berg, M., Birnbaum, L., Bosveld, A. T., Brunstrom, B., Cook, P., Feeley, M., et al. (1998). Toxic equivalency factors (TEFs) for PCBs, PCDDs, PCDFs for humans and wildlife. *Environmental Health Perspectives*. 103. pg.775-92.

Varney, Sarah. (February 12, 2009). New safety law doesn't mean all is well in toyland. *National Public Radio Morning Edition*. http://www.npr.org/templates/story/story.php?storyId=100038395.

Velasquez-Manoff, Moises. (August 26, 2012). An immune disorder at the root of autism. *The New York Times*. pg. 1, 12.

Vinay, S. P., Raghn, K. G., and Sood, P. P. (1990). Dose and duration related methylmercury deposition, glycosidase inhibition, myelin degeneration and chelation therapy. *Cellular and Molecular Biology*. 36(5). pg. 609-623.

Vinter, Levi J., St. Laurent, R. and Segal, L. M. (December 9, 2008). *Ready or not? Protecting the public's health from diseases, disasters, and bioterrorism*. Trust for America's Health and the Robert Wood Johnson Foundation.

Vos, Joseph G., et al. (2003). Brominated flame retardants and endocrine disruption. *Pure Applied Chemistry*. 75(11-12). pg. 2039-46.

Wade, Nicholas. (June 12, 2010). A decade later, gene map yields few new cures. *The New York Times*. http://www.nytimes.com/2010/06/13/health/research/13genome.html.

Wallinga, David, Sorensen, Janelle, Mottl, Pooja and Yablon, Brian. (January 26, 2009). *Not so sweet: Missing mercury and high fructose corn syrup*. Institute for Agriculture and Trade Policy (IATP), Minneapolis, MN. http://www.healthobservatory.org/library.cfm?refid=105026.

Weinhold, B. (2006). Epigenetics: The science of change. *Environmental Health Perspectives*. 114. pg. A160-7. http://ehp03.niehs.nih.gov/article/fetchArticle.action?articleURI=info:doi/10.1289/ehp.114-a160.

Weise, Elizabeth. (September 23, 2003). Flame retardant found in breast milk. *USA Today*. pg. 1.

Weiss, Bernard, Clarkson, Thomas W. and Simon, William. (2002). Silent latency periods in methylmercury poisoning and in neurodegenerative disease. *Environmental Health Perspectives*. 110(Suppl 5). pg. 851-4.

Wenner, Melinda. (Nov. 30, 2007). Humans carry more bacterial cells than human ones. *Scientific American*; http://www.sciam.com/article.cfm?id=strange-but-true-humans-carry-more-bacterial-cells-than-human-ones

Westerdahl, Johan, Ingvar, Christian, Måsbäck, Anna and Olsson, Håkan. (June 2000). Sunscreen use and malignant melanoma. *International Journal of Cancer*. 87(1). Pg. 145-50. http://www3.interscience.wiley.com/journal/72504428/abstract?CRETRY=1&SRETRY=0.

Wild, S., Roglic, G., Green, A., Sicree, R. and King, H. (2004). Global prevalence of diabetes: Estimates for the year 2000 and projections for 2030. *Diabetes Care*. 27. pg. 1047-53.

Williams, Florence. (January 9, 2005). Toxic breast milk? *The New York Times*.

Wilson, Matthew and Rasku, Jay. (2005). *Refuse to use ChemLawn: Be truly green*. Toxica Action Center, Boston, MA.

Windham, G., Zhang, L., Gunier, R., Croen, L. and Grether, J. (2006). Autism spectrum disorders in relation to distribution of hazardous air pollutants in the San Francisco Bay area. *Environmental Health Perspectives*. 114. pg. 1438-44.

Wing, L. and Potter, D. (2002). The epidemiology of autistic spectrum disorders: Is the prevalence rising? *Mental Retardation and Developmental Disabilities Research Review*. 8. pg. 151-61.

Wolfe, Nathan. 2011. *The viral storm: The dawn of a new pandemic age*. Times Books, Henry Holt and Company, NY.

World Health Organization. (1982). *Nuclear power: Health implications of transuranium elements*. European Series No. 11, WHO Regional Publications, Copenhagen.

World Health Organization. (1985). *The quantity and quality of breast milk: Report on the WHO collaborative study on breast feeding*. World Health Organization, Switzerland.

World Health Organization. (1990). *Environmental health criteria 101: Methylmercury*. World Health Organization, Geneva, Switzerland.

Worth, Robert F. (October 14, 2010). Parched earth where Syrian farms thrived. *The New York Times*.

http://www.nytimes.com/2010/10/14/world/middleeast/14syria.html?scp=1&sq=Parched+earth+w
here+Syrian+farms+thrived&st=nyt.

Xu, Qi. (January 29, 2007). Facing up to "invisible pollution". *ChinaDialogue*.
http://www.chinadialogue.net/article/show/single/en/724-Facing-up-to-invisible-pollution-.

Yeargin-Allsopp, M., Rice, C., Karapurkan, T., Doernberg, N., Boyle, C. and Murphy, C. (2003).
Prevalence of autism in a US metropolitan area. *Journal of the American Medical Association*.
289. pg. 49-55.

Young, Saundra. (August 31, 2012). New virus found in Missouri; ticks suspected: Doctors don't
yet know whether virus exists anywhere else. *Scripps Media Inc*.
http://www.abc2news.com/dpp/news/national/new-virus-found-in-missouri-ticks-suspected.

Zilberberg, MD, et al. (2008). Increase in adult Clostridium difficile-related hospitalizations and
case-fatality rate, United States, 2000-2005. *Emerging Infectious Diseases*. 14.6. pg. 929-31.

Zinczenko, David and Goulding, Matt. (2010). *Drink this not that!: The no-diet weight loss
solution*. Rodale Books, Emmaus, PA.

Anthropogenic Radioactivity

This abbreviated bibliography on anthropogenic radioactivity includes a representative selection of
some of the most important information sources on the subject. For a more detailed summary of
the literature on anthropogenic radioactivity, see the Davistown Museum's Center for Biological
Monitoring archives, RADNET Nuclear Information on the Internet.

Aarkrog, A., Buch, E., Chen, Q. J., Christensen, G. C., Dahlgaard, H., Hansen, H., Holm, E. and
Nielsen, S. P. (July 1989). *Environmental radioactivity in the North Atlantic region. The Faroe
Islands and Greenland included. 1987.* Riso-R-564. Riso National Laboratory, Roskilde,
Denmark.

Aarkrog, A., Botter-Jensen, L., Chen, Q. J., Clausen, J., Dahlgaard, H., Hansen, H., Holm, E.,
Lauridsen, B., Nielsen, S. P., Strandberg, M. and Sogaard-Hansen, J. (February 1995).
Environmental radioactivity in Denmark in 1992 and 1993. Riso-R-756(EN). Riso National
Laboratory, Roskilde, Denmark.

Bertell, Rosalie. (2000). *Host response to depleted uranium.* International Institute of Concern for
Public Health, Toronto, Canada. http://iicph.org/host_response_to_du.

Baldash, Lawrence. (1979). *Radioactivity in America: Growth and decay of a science.* Johns
Hopkins University Press, Baltimore, MD.

Brack, H. G. (1984). *RADSCAN: Information sampler on long-lived radionuclides.* Pennywheel
Press, Hulls Cove, ME.

Brack, H. G. (1986). *A review of radiological surveillance reports of waste effluents in marine
pathways at the Maine Yankee Atomic Power Company at Wiscasset, Maine--- 1970-1984: An
annotated bibliography.* Pennywheel Press, Hulls Cove, ME.

Brack, H. G. (1993). *Legacy for our children: The unfunded costs of decommissioning the Maine
Yankee Atomic Power Station.* Pennywheel Press, Hulls Cove, ME.

Brack, H. G. (1998). *Patterns of noncompliance: The Nuclear Regulatory Commission and the Maine Yankee Atomic Power Company: Generic and site-specific deficiencies in radiological surveillance programs.* Center for Biological Monitoring, Hulls Cove, ME.

Brack, H. G., Ed. (2009). *Chernobyl fallout data: Annotated bibliography.* Extracted from Section 10 of RADNET. www.davistownmuseum.org/cbm/Rad7.html.

Bradley, David. (1983). *No place to hide: 1946/1984.* University Press of New England, Hanover, NH.

Bryant, P. M. and Jones, J. A. (December 1972). *The future implications of some long-lived fission product nuclides discharged to the environment in fuel reprocessing wastes.* NRPB-R8. National Radiological Protection Board, Harwell, Didcot, Berkshire, England.

Caldicott, Helen. (1978). *Nuclear madness: What you can do!* Bantam Books, Inc., NY. Reprinted in 1980.

Camplin, W. C. and Aarkrog, A. (1989). *Radioactivity in north European waters: Report of Working Group II of CEC Project MARINA.* Fisheries Research Data Report No 20. Directorate of Fisheries Research, Ministry of Agriculture, Fisheries and Food, Lowestoft, England.

Crowther, James Arnold. (n.d.) *Ions, electrons, and ionizing radiations.* University Press, Cambridge, England.

D'Agata, John. (2010). *About a mountain.* W. W. Norton & Company, NY.

Dotto, Lydia. (1986). *Planet earth in jeopardy: Environmental consequences of nuclear war,* John Wiley and Sons.

Fackler, Martin. (September 4, 2013). Errors cast doubt on Japan's Nuclear Cleanup. *The New York Times.* pg. A1, A8. http://www.nytimes.com/2013/09/04/world/asia/errors-cast-doubt-on-japans-cleanup-of-nuclear-accident-site.html?pagewanted=all&_r=0.

Farber, S. A. and Hodgdon, A. D. (July 25, 1991). *Cesium-137 in wood ash: Results of nationwide survey.* PP 91-015. Presented at the Annual Meeting of the Health Physics Society. Yankee Atomic Electric Company, Boston, MA.

Ford, Daniel F. (1982). *The cult of the atom: The secret papers of the Atomic Energy Commission.* Simon and Schuster, NY.

Ford, Daniel F. and Kendall, Henry W. (1974). *An assessment of the emergency core cooling systems rulemaking hearings.* Union of Concerned Scientists, Cambridge, MA.

Gedikoglu, A. and Sipahi, B. L. (January 1989). Chernobyl radioactivity in Turkish tea. *Health Physics.* 56(1). pg. 97-101.

BB-Gibson, Pamela Reed. (2000). *Multiple chemical sensitivity: A survival guide.* New Harbinger Publications, Oakland, CA.

Glasstone, Samuel and Jordan, Walter H. (1980). *Nuclear power and its environmental effects.* American Nuclear Society, La Grange Park, IL.

Hardy, E. P., Jr. (July 1, 1980). *Environmental Measurements Laboratory: Environmental quarterly.* EML-374. Department of Energy, NY.

Hardy, E. P., Jr. (October 1, 1980). *Environmental Measurements Laboratory: Environmental quarterly.* EML-381. Department of Energy, NY.

Hardy, E. P., Krey, P. W. and Volchok, H. L. (February 1973). Global inventory and distribution of fallout plutonium. *Nature*. 241. pg. 444-445.

Helus, Frank, Ed. (1983). *Radionuclides production: Volume I.* CRC Press, Inc., Boca Raton, FL.

Hoopes, Roy. (1962). *A report on fallout in your food with tables and illustrations.* The New American Library, NY.

Hunt, G. J. (1987). *Aquatic environment monitoring report number 18: Radioactivity in surface and coastal waters of the British Isles, 1986.* Ministry of Agriculture, Fisheries and Food, Directorate of Fisheries Research, Lowestoft, England.

Ikaheimonen, T. K., Ilus, E. I. and Saxen, R. (1988). *Finnish studies on radioactivity in the Baltic Sea in 1987: Supplement 8 to Annual Report 1987 No. STUK-A74.* Report No. STUK-A82. Finnish Centre for Radiation and Nuclear Safety, Helsinki, Finland.

Ilus, E., Klemola, S., Sjoblom, K. L. and Ikaheimonen, T. K. (1988). *Radioactivity of Fucus vesiculosus along the Finnish coast in 1987: Supplement 9 to Annual Report 1987 (STUK-A74).* Report No. STUK-A83. Finnish Centre for Radiation and Nuclear Safety, Helsinki, Finland.

International Physicians for the Prevention of Nuclear War and the Institute for Energy and Environmental Research. (1991). *Radioactive heaven and earth: The health and environmental effects of nuclear weapons testing in, on, and above the earth.* The Apex Press, NY.

Lash, Terry R., Bryson, John E. and Cotton, Richard. (November 1975). *Citizens' guide: The national debate on the handling of radioactive wastes from nuclear power plants.* National Resources Defense Council, Inc., Palo Alto, CA.

Linsley, G. S., Simmonds, J. R. and Kelly, G. N. (December 1978). *An evaluation of the food chain pathway for transuranium elements dispersed in soils.* NRPB-R81. National Radiological Protection Board, Harwell, Didcot, Oxfordshire, England.

Long, Michael E. (July 2002). Half-life: The lethal legacy of America's nuclear waste. *National Geographic*. pg. 2-33.

MacKenzie, A. B. (April 2000). Environmental radioactivity: Experience from the 20th century – trends and issues for the 21st century. *The Science of the Total Environment*. 249(1-3). pg. 313-29.

Maine Yankee Atomic Power Company. (October, 1997). *Site characterization management plan.* Prepared by GTS Duratek, Inc., for the Maine Yankee Atomic Power Plant.

Maine Yankee Atomic Power Company. (April 16, 1998). *Appendix A: Spent fuel and other radioactive material stored in the Maine Yankee spent fuel pool.* MYPS-101, Rev. 0. Maine Yankee Atomic Power Company, Wiscasset, ME.

Maine Yankee Atomic Power Company. (April 1998). *GTS Duratek characterization survey report for the Maine Yankee Atomic Power Plant, revision 1.* 9 vols. Prepared by GTS Duratek, Inc. for the Maine Yankee Atomic Power Plant, Wiscasset, ME.

Markey, Edward J. and Waller, Douglas. (1982). *Nuclear peril: The politics of proliferation.* Ballinger Publishing Company, Cambridge, MA.

Ministry of Agriculture, Fisheries and Food and Scottish Environment Protection Agency. (September 1998). *Radioactivity in food and the environment, 1997.* RIFE-1. MAFF, London.

Moulder, John E. Frequently asked questions about static electromagnetic fields and cancer. *Stason.org.* http://stason.org/TULARC/health/static-fields-cancer/. Accessed March 2010.

National Council of Churches of Christ in the U.S.A. (September 1975). *The plutonium economy.* National Council of Churches of Christ in the U.S.A., NY, NY.

Nelkin, Dorothy. (1971). *Nuclear power and its critics: The Cayuga Lake controversy.* Cornell University Press, Ithaca, NY.

Ng, Kwan-Hoong. (October 20-22, 2003). Non-ionizing radiations–sources, biological effects, emissions and exposures. *Proceedings of the International Conference on Non-Ionizing Radiation at UNITEN (ICNIR2003) Electromagnetic Fields and Our Health.* http://www.who.int/peh-emf/meetings/archive/en/keynote3ng.pdf.

Nuclear Energy Agency. (April 1980). *Review of the continued suitability of the dumping site for radioactive waste in the north-east Atlantic.* NEA, Organisation for Economic Co-operation and Development, Paris, France.

Nuclear Energy Agency. (September 1981). *The environmental and biological behaviour of plutonium and some other transuranium elements.* NEA, Organisation for Economic Co-operation and Development, Paris, France.

Oak Ridge National Laboratory. (December 1997). *Integrated Data Base Report, 1996: U.S. Spent Nuclear Fuel and Radioactive Waste Inventories, projections, and characteristics.* Report No. DOE/RW-0006, Rev. 13. Oak Ridge National Laboratory, Oak Ridge, TN.

Puhakainen, M., Rahola, T. and Suomela, M. (1987). *Radioactivity of sludge after the Chernobyl accident in 1986: Supplement 13 to Annual Report STUK-A55.* Report No. STUK-A68. Finnish Centre for Radiation and Nuclear Safety, Helsinki, Finland.

Shannon, Sara. (1987). *Diet for the atomic age: How to protect yourself from low-level radiation.* Avery Publishing Group Inc., Wayne, NJ.

Shapiero, Fred C. (1981). *Radwaste: A reporter's investigation of a growing nuclear menace.* First Edition. Random House, NY.

Shapiro, Jacob. (1972). *Radiation protection: A guide for scientists and physicians.* Harvard University Press, Cambridge, MA.

Sternglass, Ernest. (1972). *Secret fallout: Low-level radiation from Hiroshima to Three-Mile Island.* Ballantine Books. Reprinted in 1981 by McGraw-Hill Book Company, NY.

Toombs, George L., Martin, Sylvia L., Culter, Peter B. and Dibblee, Martha G. (n.d.). *Environmental radiological surveillance report on Oregon surface waters 1961 - 1983: Volume I.* Radiation Control Section, Health Division, Oregon Department of Human Resources, Portland, OR.

United Nations. (1972). *Ionizing radiation: Levels and effects: A report of the United Nations Scientific Committee on the effects of atomic radiation to the General Assembly, with annexes: Volume I: Levels.* UN, NY.

United Nations. (1996). *Sources and effects of ionizing radiation: United Nations Scientific Committee on the Effects of Atomic Radiation UNSCEAR 1996 report to the General Assembly, with scientific annex.* UN, NY.

United Nations Environment Programme, the International Labour Organization, and the World Health Organization. (1983). *Selected radionuclides: Tritium, carbon-14, krypton-85, strontium-90, iodine, caesium-137, radon, plutonium*. Environmental Health Criteria 25. World Health Organization, Geneva.

Wasserman, Harvey and Soloman, Norman. (1982). *Killing our own: The disaster of America's experience with atomic radiation*. Dell Publishing, NY.

Handbooks and Guides

Allaby, Michael, Ed. (1994). *The concise Oxford dictionary of ecology*. Oxford University Press, Oxford, UK.

Centre for Environment, Fisheries & Aquaculture Science. (1997). *Monitoring and surveillance of non-radioactive contaminants in the aquatic environment and activities regulating the disposal of wastes at sea, 1994*. Science Series. Aquatic Environment Monitoring Report No. 47. CEFAS, Ministry of Agriculture, Fisheries and Food, Great Britain.

Counihan, Martin. (1981). *A dictionary of energy*. Routledge & Kegan Paul, Boston, MA.

Environment Information Center. (1980). *Toxic substances sourcebook: Series: 2*. EIC Co., NY, NY.

Helus, Frank, Ed. (1983). *Radionuclides production: Volume I*. CRC Press, Inc., Boca Raton, FL.

Helus, Frank, Ed. (1983). *Radionuclides production: Volume II*. CRC Press, Inc., Boca Raton, FL.

International Atomic Energy Agency. (1965). *Methods of surveying and monitoring marine radioactivity*. Safety Series No. 11, IAEA, Vienna.

Kase, Kenneth R. and Nelson, Walter R. (1978). *Concepts of radiation dosimetry*. Pergamon Press, NY.

McGraw-Hill. (2002). *McGraw-Hill encyclopedia of science & technology: 13: Par-Plan*. 9th Edition. McGraw-Hill, NY.

National Council on Radiation Protection and Measurements. (1971). *Basic radiation protection criteria*. NCRP Report No. 39., NCRP, Bethesda, MD.

New York Public Library. (1993). *The New York Public Library desk reference*. 2nd edition. Prentice Hall General Reference, NY.

Nuclear Energy Agency. (1989). *Nuclear accidents: Intervention levels for protection of the public*. NEA, Organization for Economic Co-operation and Development, Paris, France.

Organization of Economic Co-operation and Development. (2003). *Voluntary approaches for environmental policy*. Paris. www.oecd.org/document/58/0,2340,fr_2649_34375_238437_1-1-1-1,00.html

Shleien, Bernard, Slaback, Lester A., Jr. and Birky, Brian Kent, Eds. (1998). *Handbook of health physics and radiological health*. 3rd Edition. Williams & Wilkins, Baltimore, MD.

Stich, Hans. F., Ed. (1982). *Carcinogens and mutagens in the environment: Volume I: Food products*. CRC Press, Boca Raton, FL.

United Nations. (1972). *Ionizing radiation: Levels and effects: A report of the United Nations Scientific Committee on the effects of atomic radiation to the General Assembly, with annexes: Volume I: Levels.* UN, NY.

The original and definitive source of information on the biological effects of ionizing radiation.

United Nations. (1996). *Sources and effects of ionizing radiation: United Nations Scientific Committee on the Effects of Atomic Radiation UNSCEAR 1996 report to the General Assembly, with scientific annex.* UN, NY.

United Nations Economic Commission for Europe. (2007). *Globally harmonized system of classification and labeling of chemicals (GHS).* www.unece.org/traans/danger/publi/ghs_welcome_e.html.

US Department of Commerce. (June 5, 1959). *Maximum permissible body burdens and maximum permissible concentrations of radionuclides in air and in water for occupational exposure: Recommendations of the National Committee on Radiation Protection.* National Bureau of Standards Handbook 69, US Department of Commerce, Washington, DC.

US Department of Health and Human Services. (1999). *Toxicological Profile for Mercury.* Sec. 1.6. US Department of Health and Human Services, Agency for Toxic Substances and Disease Registry.

US Department of Health and Human Services. (March 1998). *Action levels for poisonous or deleterious substances in human food and animal feed.* US Department of Health and Human Services., Food and Drug Administration, Washington, DC.

US Environmental Protection Agency. (November 2000). *Guidance for assessing chemical contaminant data for use in fish advisories: Volume 2: Risk assessment and fish consumption limits.* Third Edition. Office of Science and Technology, Office of Water, U.S. Environmental Protection Agency, Washington, DC. www.epa.gov/ost/fishadvice/volume2/index.html.

US Occupational Safety and Health Administration. (2006). *A guide to the globally harmonized system of classification and labeling of chemicals (GHS).* www.osha.gov/dsg/hazcom/ghs.html.

West, Robert C., Ed. (1973-4). *CRC handbook of chemistry and physics.* 54th edition. CRC Press.

World Health Organization. (March 1986). *Environmental health: Dioxins and furans from municipal incinerators.* World Health Organization, Naples.

www.ingramcontent.com/pod-product-compliance
Lightning Source LLC
Chambersburg PA
CBHW051205200326
41519CB00025B/7008